Introduction to Homeland Security

Terrorism Prevention, Public Safety and Emergency Management

Third Edition

DAVID A. MCENTIRE, PHD
Utah Valley University

WILEY

Hardback ISBN: 9781394234325

Library of Congress Cataloging-in-Publication Data
Names: McEntire, David A., author.
Title: Introduction to homeland security : terrorism
  prevention, public safety and emergency management /
  David A. McEntire, PHD, Utah Valley University.
Description: Third edition. | Hoboken, NJ : John Wiley & Sons, 2024. |
  Includes bibliographical references and index. | Summary: "United States
  policy regarding international affairs, terrorism, and disasters has
  witnessed an ongoing tension between the security and emergency
  management points of view. Since the late 1940s, there has been
  recurring disagreement about the priority given to conflict events
  versus other types of hazards. Several events elevated the stakes in
  this debate and created additional urgency to find some sort of
  consensus about future priorities"— Provided by publisher.
Identifiers: LCCN 2024004838 (print) | LCCN 2024004839 (ebook) | ISBN
  9781394234325 (hardback) | ISBN 9781394234332 (epub) | ISBN
  9781394234349 (adobe pdf)
Subjects: LCSH: Terrorism—United States—Prevention. |
  Terrorism—Government policy—United States. | Emergency
  management—United States. | Computer security—Government
  policy—United States. | United States—Military policy.
Classification: LCC HV6432.M3854 2024 (print) | LCC HV6432 (ebook) | DDC
  363.340973—dc23/eng/20240131

LC record available at https://lccn.loc.gov/2024004838

LC ebook record available at https://lccn.loc.gov/2024004839

SKY10074560_050824

Introduction to Homeland Security

Third Edition

Introduction to
Homeland Security

Third Edition

For Mason, Madison, Kailey, and Ashley
and the future of children and grandchildren everywhere

Contents

About the Author

David A. McEntire, PhD, is a professor at Utah Valley University and the former dean of the College of Health and Public Service. Prior to these appointments, he was a professor in the Emergency Administration and Planning (EADP) Program in the Department of Public Administration at the University of North Texas.

Dr. McEntire has taught emergency management, homeland security, and national security courses in both undergraduate and graduate programs. His academic interests include emergency management theory, international disasters, community preparedness, response coordination, terrorism, and vulnerability reduction.

Dr. McEntire has received several grants funded by the Natural Hazards Center at the University of Colorado and the National Science Foundation, among others, which allowed him to conduct research on disasters and emergency management in Australia, Costa Rica, the Dominican Republic, Haiti, New Zealand, Peru, California, New York, Texas, and Utah.

Dr. McEntire is the author of *Disaster Response and Recovery* (Wiley) and *Comparative Emergency Management* (Federal Emergency Management Agency (FEMA)) and the editor of Disciplines, *Disasters and Emergency Management* (C.C. Thomas) and *The Distributed Functions of Emergency Management and Homeland Security* (CRC Press). His research has also been published in *Public Administration Review,* the *Australian Journal of Emergency Management, Disasters,* the *International Journal of Mass Emergencies and Disasters, the Journal of Emergency Management, International Journal of the Environment and Sustainable Development, Sustainable Communities Review, the International Journal of Emergency Management, the Towson University Journal of International Affairs, the Journal of the American Society of Professional Emergency Planners*, and the *Journal of International and Public Affairs,* among others. His articles in *Disaster Prevention and Management* received Highly Commended and Outstanding Paper awards.

Dr. McEntire completed an instructor guide for FEMA and is a contributing author to the *Handbook of Disaster Research* and the *Handbook of Disaster Management.* He also has a chapter in *Emergency Management,* a book published by the International City/County Management Association.

Dr. McEntire provided terrorism response training for FEMA in Arkansas and Oklahoma. He has been a contributing author for a study of Texas homeland security preparedness for the Century Foundation and two IQ reports for the International City/County Management Association. In addition, McEntire has presented papers in Hungary, Japan, Mexico, and Norway, and at the National Science Foundation, the National Academy of Sciences, and the Higher Education Conference at FEMA's Emergency Management Institute in Emmitsburg, Maryland. He is a former member of Congressman Michael C. Burgess's Homeland Security Advisory Council and of the Fire Protection Publications Advisory Board. He has reviewed books for Delmar Learning and is on the editorial staff for the *Journal of Emergency Management.*

In a prior position at the University of North Texas, Dr. McEntire served as an undergraduate coordinator, Ph.D. coordinator, assistant chair, associate dean, and director of summer sessions. Prior to his first academic appointment, Dr. McEntire attended the Graduate School of International Studies at the University of Denver. While pursuing his degree, he worked for the International and Emergency Services Departments at the American Red Cross.

Foreword

3rd Edition, *Introduction to Homeland Security*

Homeland security—both in professional practice and as an academic discipline—continues to mature from its relatively recent roots in the U.S. government restructuring following the attacks in New York, Washington, DC, and Pennsylvania on 9/11. The early focus was unambiguous, with terrorism as the field's unquestionable raison d'etre. This period is what some have termed "Homeland Security 1.0."

This priority and attention given to terrorism was blurred with the devastation wrought by Hurricane Katrina in 2005 and the national angst over the impacts of the *Global War on Terror*. "Homeland Security 2.0" ushered in a nominally more balanced, all-hazards approach to meeting the diverse security and emergency management challenges of the first decade of the century. As fear of another 9/11 magnitude attack waned, the emphasis shifted to high-likelihood, low-consequence events impacting U.S. communities. Planning, preparation, and exercising for diverse natural hazards (e.g., tropical storms, flooding, and earthquakes as well as a growing recognition of cyber threats) took priority as jurisdictions focused on the most probable needs of their communities. Conventional and unconventional military operations abroad also dispersed the threat of Al Qaeda and the Islamic State, and regional terror groups—Boko Haram, Al-Shabab, and Lashkar-e-Tayyiba, among others—appeared to pose limited challenges to the security of the U.S. homeland. Thus, emphasizing terrorism as a principal focus for the U.S. federal government, as well as state, local, tribal, and territorial jurisdictions and among partner nations, seemed less intuitive.

Yet *terrorism* has not gone away as borne out by recent events. Rather, the threat has diversified, metastasized, and matured, presenting an enduring and complex challenge for security policymakers and practitioners. As an example, Al Qaeda was indeed dispersed but reconstituted as franchise operations throughout North Africa, the Middle East, and Southwest Asia in loosely affiliated forces to be reckoned with, either independently or in loose collaboration with other regional terror groups.

At the time of this writing, homeland security faces perhaps the most demanding set of *wicked* problems ever. This includes brutal, industrial-age nation-state warfare in Europe with the possibility of global consequences; the potential for Great Power conflict in East Asia; proxy warfare in Yemen, Iraq, and Syria, with intermittent but continuing attacks on U.S. and allied personnel and facilities; mass migration and resultant border immigration crises; slow-onset potential mega-disasters related to climate change impacts; political conflict and uncertainty as the United States approaches another national election; the rapidly expanding use of information technologies that both shape public opinion and enable malign activities by sophisticated groups and nations; and, of course, the evolution of terrorism that takes advantage of the latter.

The most graphic reminder of the ongoing challenges and dynamic nature of terrorism was the horrific events in Israel on October 7, 2023, and the aftermath that will play out over the succeeding months and perhaps beyond. Is the homeland security enterprise up to the task of dealing with each of these vexing issues simultaneously?

The answer to the myriad challenges (as if one were ready at hand!) would seem to rest in our critical analysis of the threat and hazard landscape and the ability to identify, build, and deliver the multifaceted approaches to meet the diverse homeland security requirements in 2024 and 2025 and into future generations. The "how" of homeland security (although the precise method is both uncertain and subject to changes due to the threats we likely will face) must necessarily have its foundation in *understanding*; understanding that is based on the art and science of scholarly research applied to practice.

We should therefore congratulate and thank Dr. Dave McEntire for approaching "Homeland Security 3.0" with an unparalleled expertise in his clear-eyed exposition of the fundamental issues of homeland security in this third edition. Here, Dr. McEntire has refreshed our understanding of the professional field and academic discipline that may have diverged from the critical—indeed, life-threatening—impact that homeland security must address.

The third edition reintroduces students to essential elements of terrorism and the deadly intent of this form of violence in its modern form. Through an integrated approach to understanding homeland security, the author correctly encourages us to see the threats and hazards described earlier in the context of the larger system of national and global resilience and human security. The third edition is consequently comprehensive and complete and is a timely, single-source volume that provides the necessary vaccination to prevent a future failure of imagination.

Steve Recca
11.20.2023
Naval Postgraduate School
Center for Homeland Defense & Security
Monterey, California

Preface

U.S. policy regarding international affairs, terrorism, and disasters has witnessed ongoing tension between the security and emergency management points of view. Since the late 1940s, there has been recurring disagreement about the priority given to conflict disasters versus other types of hazards. Several events elevated the stakes in this debate and created additional urgency to find some sort of consensus about future priorities.

First, the 9/11 attacks and the recent Hamas violence in Israel underscore the fact that the threat of terrorism needs to be taken seriously by disaster scholars and emergency managers. No one should pretend that the world is the way it once used to be.

Second, Hurricane Katrina and a plethora of other disasters remind homeland security officials that they must not disregard human vulnerability to natural hazards. The frequency of natural, technological, and anthropogenic disasters is simply too great to ignore. In addition, the consequences of compound interactions are getting worse over time.

With these observations in mind, it is the opinion of this author that both homeland security and emergency management functions need to be addressed concomitantly in the future. Terrorist attacks are increasing in frequency, and their negative outcomes are widespread. While the possibility of terrorism involving weapons of mass destruction or cyberterrorism is uncertain, the impact of such attacks would indeed be overwhelming. More resources will be needed to address every type of threat.

However, government attention and resource distribution should also take into account the broad aspects of homeland security. Terrorism has been given the lion's share of public support in recent years, but this attention should not be allowed to overshadow the essential functions and contributions of emergency management.

Unfortunately, policy makers unintentionally created a substantial divide between the homeland security and disaster communities. Politicians may have overreacted to 9/11, and their decisions initially diminished the existing emergency management system of the 1990s. This has caused some ill feelings among emergency managers toward the military and the law enforcement communities, which is not a good situation when one considers the fact that terrorists have vowed to kill Americans everywhere (including at home).

Added to this ongoing discussion is the fact that additional risks have become more prevalent in recent years. This includes a rise in various types of crime, the occurrence of riots, the global COVID-19 pandemic, and concerns about climate change. Thus, those involved in homeland security and emergency management must also work collaboratively with other public safety, public health, and environmental agencies and officials to address the challenges of our day.

This book, *Introduction to Homeland Security: Terrorism Prevention, Public Safety and Emergency Management*, Third Edition, aims to provide a foundation that could assist in spanning the chasm between the disaster and terrorism communities. Its focus on terrorism may help to educate those who do not yet understand the need to prepare for this expanding threat. Its concentration on emergency management will remind homeland security officials that reinventing of the wheel is not only unnecessary but problematic. In addition, the book educates homeland security and emergency management professionals about other important considerations pertaining to unabated crime, social discord, physical well-being, and our precious natural resources.

Of course, taking this approach could result in increased antagonism between the different parties. It is also possible that the author has not adequately portrayed the specific details pertinent to all of the actors involved in the broad and interdisciplinary array of homeland security, emergency management, public safety, public health, and environmental protection activities. Nevertheless, it is the author's hope that this work will educate those working in each area and help promote a synergy of effort.

Chapter 1, **"Understanding a Global Priority: Diverse Threats, Terrorism, Homeland Security, and Emergency Management,"** examines the significant impacts of 9/11, the Hamas attacks in Israel, and other events in world history. It also defines homeland security and supplements homeland security with an emergency management perspective, thereby offering a broader view of how to deal with terrorist attacks.

Chapter 2, **"Identifying Terrorism: Ideologically Motivated Acts of Violence and Their Relation to Disasters,"** identifies the numerous definitions and perspectives of terrorism, comparing how these are both alike and dissimilar. This chapter also looks at the connections between terrorism and other types of disasters.

Chapter 3, **"Recognizing the Causes of Terrorism: Differing Perspectives and the Role of Ideology,"** explores what motivates people to participate in terrorism, paying special attention to how historical conflicts, mistakes in foreign policy, and extreme levels of poverty may impel some to engage in terrorist attacks.

Chapter 4, **"Comprehending Terrorists and Their Behavior: Who They Are and What They Do,"** assesses the nature of individual terrorists and those associated with groups and states and identifies how they finance operations, communicate with secret codes, and carry out attacks.

Chapter 5, **"Uncovering the Dynamic Nature of Terrorism: History of Violence and Change over Time,"** explores why terrorism initially emerged, how it evolved in other nations, and the ways it has manifested in the United States.

Chapter 6, **"Evaluating a Major Dilemma: Terrorism, the Media, and Censorship,"** looks at the difficult relationship between terrorism and the media, how to predict how reporters view terrorism, and the drawbacks and limitations of censorship.

In Chapter 7, **"Contemplating a Quandary: Terrorism, Security, and Liberty,"** you learn why, as a participant in homeland security, it is imperative that you assess the tradeoffs between security and rights and why terrorism exploits the tension between them.

Chapter 8, **"Preventing Terrorist Attacks: Root Causes, Law, Intelligence, and Counterterrorism,"** addresses the root causes of terrorism and explores primary ways of preventing attacks, like promoting laws that prohibit terrorism and punish those who support it, protecting all points of entry into the United States, and relying on human and other sources of intelligence to apprehend terrorists before they strike.

In Chapter 9, **"Securing the Nation: Border Control and Sector Safety,"** the permeability of the U.S. border is mentioned along with measures to prevent the infiltration of terrorists onto American soil. It also discusses the vulnerability of various economic sectors and describes ways to secure railways, air transportation, sea ports, and chemical facilities.

Chapter 10, **"Protecting Against Potential Attacks: Threat Assessment, Mitigation, and Other Measures,"** looks at the benefits of mitigation practices, such as working with others to assess threats posed to critical infrastructure, key assets, and soft targets, as well as differentiating between structural and nonstructural mitigation methods.

In Chapter 11, **"Preparing for the Unthinkable: Efforts for Readiness,"** it is revealed that preparing for terrorism is one of the central responsibilities of homeland security. To help your community prepare for possible terrorist attacks, you will need to be familiar with the executive orders and legislation issued by the president and congress and set the foundation for preparedness by creating an advisory council, passing ordinances, acquiring monetary resources, and establishing an EOC.

Chapter 12, **"Responding to Attacks: Important Functions and Coordination Mechanisms,"** examines effective ways to react to terrorist attacks, including the numerous functions involved, such as investigation, the protection of first responders, and the treatment of the victims of terrorist attacks.

Chapter 13, **"Recovering from Impacts: Short- and Long-term Measures,"** addresses the variety of recovery measures that need to be performed after a terrorist attack takes place, including declaring a disaster or state of emergency, addressing mass fatality issues, disposing debris, and providing emotional support for those who have been emotionally impacted by the event.

Chapter 14, **"Assessing Significant Threats: WMD and Cyberterrorism,"** assesses the probability that terrorists will launch more unique and devastating attacks. It identifies the threat of radiological, nuclear, biological, and chemical weapons along with numerous recommendations to counter such assaults. The chapter also describes the risk of cyberterrorism and mentions the measures being taken to increase preparedness in this area.

Chapter 15, **"Evaluating Other Pressing Problems: Criminal Activity, Social Disturbances, Pandemics, and Climate Change,"** explores additional risks that must be taken into consideration today by both homeland security and emergency management professionals. These risks include a variety of problems relating to organized crime, illegal drugs, human trafficking, protests, riots, COVID-19, and global warming.

In Chapter 16, **"Looking Toward the Future: Challenges and Opportunities,"** the need for accountability in homeland security is identified. A discussion about policy occurs, and recommendations are provided for both researchers and practitioners.

Pre-Reading Learning Aids

Each chapter of *Introduction to Homeland Security: Terrorism Prevention, Public Safety and Emergency Management*, Third Edition features a number of learning and study aids, described in the following sections, to activate students' prior knowledge of the topics and orient them to the material.

Do You Already Know?

This bulleted list focuses on the *subject matter* that will be taught. It tells students what they will be learning in this chapter and why it is significant for their careers. It also helps students understand why the chapter is important and how it relates to other chapters in the text.

The online assessment tool in introduces chapter material, but also helps students anticipate the chapter's learning outcomes. By focusing students' attention on what they do not know, the self-check test provides students with a benchmark against which they can measure their own progress. The pre-test is available online at www. wiley.com/go/mcentire/homelandsecurity3e.

What You Will Learn and What You Will Be Able to Do

This bulleted list emphasizes *capabilities and skills* that students will learn as a result of reading the chapter and notes the sections in which they will be found. It prepares students to synthesize and evaluate the chapter material and relate it to the real world.

Within-Text Learning Aids

The following learning aids are designed to encourage analysis and synthesis of the material, support the learning process, and ensure success during the evaluation phase.

Introduction

This section orients the student by introducing the chapter and explaining its practical value and relevance to the book as a whole. Short summaries of chapter sections preview the topics to follow.

In the Real World

These boxes tie section content to real-world organizations, scenarios, and applications. Engage stories of professionals and institutions—challenges they faced, successes they had, and their ultimate outcome.

Summary

Each chapter concludes with a summary paragraph that reviews the major concepts in the chapter and links back to the "Do You Already Know" list.

Key Terms and Glossary

To help students develop a professional vocabulary, key terms are bolded when they first appear in the chapter and are also shown in the margin of the page with their definitions. A complete list of key terms with brief definitions appears at the end of each chapter and again in a glossary at the end of the book. Knowledge of key terms is assessed by all assessment tools (see next).

Evaluation and Assessment Tools

The evaluation phase consists of a variety of within-chapter and end-of-chapter assessment tools that test how well students have learned the material and their ability to apply it in the real world. These tools also encourage students to extend their learning into different scenarios and higher levels of understanding and thinking. The following assessment tools appear in every chapter.

Self-Check

Related to the "Do You Already Know" bullets and found at the end of each section, this battery of short-answer questions emphasizes student understanding of concepts and mastery of section content. Though the questions may be either discussed in class or studied by students outside of class, students should not go on before they can answer all questions correctly.

Understand: What Have You Learned?

This online post-test should be taken after students have completed the chapter. It includes all of the questions in the pre-test so that students can see how their learning has progressed and improved. The post-test is available online at www.wiley.com/go/mcentire/homelandsecurity3e.

Applying This Chapter

These questions drive home key ideas by asking students to synthesize and apply chapter concepts to new, real-life situations and scenarios.

Be a Homeland Security Professional

Found at the end of each chapter, "Be a ..." questions are designed to extend students' thinking and are thus ideal for discussion or writing assignments. Using an open-ended format and sometimes based on web sources, they encourage students to draw conclusions using chapter materials applied to real-world situations, which foster both mastery and independent learning.

Instructor and Student Package

Introduction to Homeland Security: Terrorism Prevention, Public Safety, and Emergency Management, Third Edition, is available with the following teaching and learning supplements. All supplements are available online at the text's Book Companion Website, located at www.wiley.com/go/mcentire/homelandsecurity3e.

Instructor's Resource Guide

The Instructor's Resource Guide provides the following aids and supplements for teaching a Homeland Security course:

- **Text summary aids:** For each chapter, these include a chapter summary, learning objectives, definitions of key terms, and answers to in-text question sets.
- **Teaching suggestions:** For each chapter, these include at least three suggestions for learning activities (such as ideas for speakers to invite, videos to show, and other projects) and suggestions for additional resources.
- **PowerPoints:** Key information is summarized in 10–15 PowerPoints per chapter. Instructors may use these in class or choose to share them with students for class presentations or to provide additional study support.
- **Test Bank:** The test bank features one test per chapter, as well as a midterm and two finals–one cumulative and one noncumulative. Each includes true/false, multiple-choice, and open-ended questions. Answers and page references are provided for the true/false and multiple-choice questions, while page references are given for the open-ended questions. Tests are available in Microsoft Word and computerized formats.

Acknowledgments

I express appreciation to those individuals who have made substantial contributions to *Introduction to Homeland Security*. I am indebted first and foremost to Laura Town, Brian Baker, Summer Scholl, and Judy Howarth (editors at Wiley) for their critical assessment of the text and their useful recommendations for improvement. I am also grateful to the additional members of the Wiley staff—Stefanie Volk, Katrina Maceda, J. Hari Priya, and Rajeev Kumar—who helped format the text and find pictures for the entire document. Others including Kailey Birchall, Ashley Layton, Tres Layton, and Madison Rojas provided valuable assistance in the preparation of this manuscript.

I am likewise appreciative of several reviewers for their beneficial suggestions on this version and earlier drafts of the manuscript. This includes:

- Robin Ebemyer, Utah Valley University
- Vincent J. Doherty, Naval Postgraduate School
- John J. Kiefer, The University of New Orleans
- Scott D. Lassa, Milwaukee Area Technical College
- David W. Lewis, University of Maryland
- Steve Recca, Naval Postgraduate School
- Ronald P. Timmons, University of North Texas
- Brandon Amacher, Utah Valley University
- Ryan Vogel, Utah Valley University

While I am solely responsible for the content of this book, I am thankful for all those who have shared valuable insights and unique perspectives. Their areas of expertise and years of experience have undoubtedly assisted me during the publication process.

Finally, I would be remiss if I did not recognize the many scholars and practitioners whom I have come across during my involvement in this field. Your knowledge and professionalism have not only helped to educate me about terrorism and disasters; you have also underscored the significant need for homeland security and emergency management functions in our society. More importantly, I am cognizant that your persistent efforts will enable our nation to reduce vulnerability and increase our ability to react effectively when exigency exists. For this I am truly grateful.

David A. McEntire, PhD

About the Companion Website

This book is accompanied by a companion website: **www.wiley.com/go/mcentire/homelandsecurity3e**

The instructor website includes:

- Instructor's resource material
- Powerpoint slides
- Pre-test
- Post-test
- Test bank
- Image gallery

The student website includes:

- Pre-test
- Post-test
- Image gallery

CHAPTER 1

Understanding a Global Priority

Diverse Threats, Terrorism, Homeland Security, and Emergency Management

DO YOU ALREADY KNOW?

- Why we should be concerned about threats to our security
- If terrorist attacks are becoming more frequent
- Why 9/11 changed the world
- How to define homeland security
- Why many professions, including emergency management, can help us deal more effectively with terrorist attacks

For additional questions to assess your current knowledge of terrorism and homeland security, go to **www.wiley.com/go/mcentire/homelandsecurity3e**

WHAT YOU WILL LEARN	WHAT YOU WILL BE ABLE TO DO
1.1 Threats facing the United States	• List terrorist activity in recent years
1.2 The probability of additional attacks in the future	• Evaluate the consequences of terrorism
1.3 The far-reaching effects of 9/11	• Convey how the world changed after the 9/11 hijackings
1.4 Definitions of homeland security	
1.5 The breadth of organizations involved in homeland security	• Describe the mission of the homeland security profession
	• Assess how various disciplines can help practitioners deal with terrorism

Introduction

Welcome to the important discipline and profession of homeland security! If you are interested in this important field of study and desire to work in this area of practice, it is imperative that you comprehend the variety of threats facing the United States and what can be done to deal with them in an effective manner. This book has the purpose of helping you achieve these goals by exploring the fundamental concepts and principles of the homeland security profession. While reading this introductory chapter, you will gain a basic understanding of numerous security concerns in general and the perplexing problem of terrorism in particular. You will learn how the terrorist attacks on September 11, 2001, changed the world and initiated a new era in global history. You will acquire knowledge about the mission and scope of homeland security along with the challenges it currently faces. The importance of approaching homeland security from a holistic perspective is then mentioned, enabling you to recognize why many professions—including emergency management—must form an integral part of efforts to deal with terrorist attacks and other threats.

1.1 Diverse Threats and Terrorism are the New Normal

Anytime you watch the news, search the Internet, or scroll through social media, you will likely see several stories depicting threats to the security, safety, and well-being of people, organizations, and nations around the world. For instance:

- Mass shootings are all too frequently witnessed at school, in the workplace, or any location where many people congregate. Individuals and families are worried about their safety in the mall, at church, or while attending a sporting event. This concern appears to be warranted due to the fact that there were 689 mass shootings in 2021, 645 in 2022, and over 500 at the time of this writing in 2023 (Nazzaro 2023).

- The commission of crimes—including physical assaults and property theft—rose dramatically from 2020 to 2022 (Grawert and Kim 2023). People feel unsafe at home or in the streets, and businesses are worried about their profitability and future viability under these disturbing conditions.

- The manufacturing and distribution of illegal drugs in the United States have risen to epic proportions. Government reports reveal that more than 100,000 persons die annually from illicit substances like fentanyl (National Institute on Drug Abuse 2023).

- Human trafficking is enslaving thousands of people in the United States and elsewhere. Exploited sex workers and other forced laborers suffer as a result of this, the most significant human rights violation of our time. For instance, since the inception of the National Human Trafficking Hotline in 2007, more than 82,000 cases have been reported in the United States, which have involved over 164,000 victims (see **https://humantraffickinghotline.org/en/statistics**).

- Social demonstrations have become increasingly common in recent years, and this has caused Verisk Maplecroft (a strategic consulting firm) to place the United States within the high-risk category for social unrest (Campbell and Hribernik 2020). Some of the recent protests over the death of George Floyd and the 2020 election results exploded into riot behavior, which resulted in the loss of life, hundreds of injuries, and the destruction of property valued at billions of dollars.

- Public health emergencies have been seen repeatedly over the past few decades. Our recent experience with the Covid-19 pandemic illustrates the significant impact contagious diseases can have on lives and livelihoods. There is evidence that our public health institutions are inadequately prepared for deadly outbreaks that may occur in the future (Baker and Ivory 2021).

- Global warming and climate change are frequently discussed by politicians and environmental activists. Concerns about the possibility of further environmental disasters appear to be warranted and suggest a future of additional and aggravated risks (see **https://www.usgs.gov/faqs/how-can-climate-change-affect.natural-disasters**).

As can be seen, each of these threats may result in injuries, death, and/or physical, economic, emotional, and other types of harm. While these problems are typically addressed by law enforcement personnel, public health organizations, and environmental agencies, they have not always been the focus of a holistic, concerted, or systematic effort. This is changing, however, because local, state, and federal governments recognize that more must be done to prevent such occurrences and react to them in a more effective manner.

Source: TapTheForwardAssist/Wikimedia Commons

In The Real World

Disasters on the Rise?

According to the National Oceanic and Atmospheric Administration, the United States experienced 18 weather and climate-related disasters in 2022, which caused more than $1 billion in damage each (Smith 2023). This included one winter storm, one wildfire, one heatwave/drought, one flood, two tornadoes, three tropical cyclones, and nine severe weather/hail events. Taken together, these disasters killed 474 people and totaled over $165 billion in losses in a single year. Evidence seems to suggest that the risk of disasters is increasing at alarming rates. Disasters are just one of many different threats facing the United States today and into the future.

Although there are grave concerns about criminal activity, social disturbances, public health emergencies, and climate-related disasters, another threat has garnered significant attention from policy makers than those mentioned earlier. This issue has transformed the nature of government in the United States over the past two decades. That concern is known as terrorism.

In simple terms, **terrorism** is the use or threat of violence to support ideological purposes. Individuals, groups, and even nation-states participate in terrorism to achieve social, political, economic, and other objectives. Recent events might cause you to think that these terrorist attacks are more frequent and deadly than in prior decades. Your instinct is certainly justified according to the London Institute for Economics and Peace (Cassidy 2015). Although there are a whole host of problems that must be addressed by politicians and public safety personnel, terrorism appears to be ever present and its impacts cannot be denied. Attacks are not only more common than in the past—they are also more consequential. Five cases illustrate the "new normal" we are facing today.

1.1.1 Boston Marathon Bombing

On April 15, 2013, Tamerlan and Dzhokhar Tsarnaev—Chechen brothers who were residing in the United States—detonated two homemade pressure cooker bombs at the Boston Marathon in Massachusetts. The bombs exploded in the late afternoon about 200 yards apart near the finish line of the race on Boylston Street. Three people died from the blasts, and over 260 others were injured. The marathon was suspended while athletes and bystanders were directed to safety and treated by law enforcement and emergency medical personnel. A massive manhunt was soon underway to find those involved in the attack. One of the brothers was killed during a confrontation with police. The other was apprehended and is now incarcerated in a maximum-security prison in Florence, Colorado.

1.1.2 San Bernardino Regional Center Shooting

On December 2, 2015, a mass shooting took place in San Bernardino, California, at the Inland Regional Center. The perpetrators were Syed Rizwan Farook and his wife, Tashfeen Malik. This married couple killed 14 people and injured 22 when they opened fire at a holiday party being held by the county's Department of Public Health.

While this was clearly devastating, the incident could have been more consequential. Three pipe bombs found at the scene failed to explode. The terrorists were killed in a shoot-out with police a short time later.

1.1.3 Orlando Nightclub Shooting

On June 12, 2016, a security guard named Omar Mateen instigated one of the most despicable acts of violence in U.S. history (see Figure 1-1). With the use of a pistol and a semiautomatic rifle, he killed 49 people and wounded 53 others at the Pulse—a gay nightclub in Orlando, Florida. After a 3-h standoff, the terrorist was shot and killed by the officers from the Orlando Police Department. The carnage was the result of one of the worst mass shootings in the United States to date.

1.1.4 Bombings in New York and New Jersey

From September 17 to September 19, 2016, several bombs were detonated and found unexploded in New York and New Jersey. One pipe bomb was placed in a trash can and exploded during a 5K race in the community of Seaside Park. No one was injured in the incident, but additional bombs were discovered in the area and disposed of. That same day, another bomb exploded in a dumpster in New York City near a construction site. Thirty-one people were injured as a result of this pressure cooker bomb that was filled with metal bearings. Another bomb was identified nearby and reported to the authorities. The following day, additional bombs were discovered in Elizabeth, New Jersey. Ahmed Khan Rahimi, an Islamic extremist, was captured after fingerprints, a mobile phone, DNA evidence, and purchase receipts revealed who the culprit was.

FIGURE 1-1 Mass shootings like the Orlando nightclub shooting reveal the significant impact a lone terrorist can have on innocent citizens. *Source: Orlando Police Department/Wikimedia Commans/Public Domain.*

1.1.5 Hudson River Bike Path Attack

On October 31, 2017, a man named Sayfullo Habibullaevic Saipov used a rented pickup truck as a weapon on a bike path in a park near the Hudson River. Over a 1-mile stretch, Saipov ran into and over cyclists and joggers, killing 8 people and injuring 11 others. He eventually crashed the vehicle into a bus and then engaged in a shoot-out with police. Saipov was shot and arrested. During the investigation, it was determined that Saipov launched the attack in support of the Islamic State of Iraq and the Levant (ISIL). This terrorist was convicted of murder and is incarcerated for life.

1.1.6 Other Notable Attacks

The aforementioned attacks are not isolated. The list of such events has increased over the past 25 years. For instance, Hesham Mohamed Hadayet opened fire at the El Al ticket counter at the Los Angeles Airport. His attack killed two people and injured four others on July 4, 2002. On March 3, 2006, Mohammed Reza Taheri-azar intentionally drove a vehicle into a crowd at the University of North Carolina at Chapel Hill. He injured nine people. In November 2009, Nidal Malik Hasan murdered 14 people at the Fort Hood military base in Texas.

On February 18, 2010, a disgruntled man named Joseph Stack flew his private plane into an Internal Revenue Service building in Austin, Texas. In the small town of Moore, Oklahoma, Alton Nolen beheaded a woman at a Vaughan Foods plant on September 24, 2014. Robert Lewis Dear killed three people at a Planned Parenthood clinic in Colorado Springs, Colorado, on November 27, 2015. Micah Xavier Johnson killed five police officers on July 7, 2016. This disturbing sniper attack took place in Dallas, Texas, at a Black Lives Matter protest. It wounded nine others, including two civilians.

A Short List of Attacks

The list of terrorist attacks seems to be never ending. Fifteen incidents illustrate the prevalence of attacks along with a diversity of targets and divergent methods:

1. Bruce Edwards Ivins mailed several letters containing anthrax spores to news outlets and two Democratic senators. Five people were killed, and seventeen others were injured from September 18 to October 9, 2001.

2. John Allen Williams and Lee Boyd Malvo (the "Beltway Snipers") murdered 10 people and injured 3 others in Washington, D.C., Maryland, and Virginia over a three-week period in October 2002.

3. John Patrick Bedell injured two police officers at the entrance of the Pentagon on March 4, 2010.

4. Wade Michael Page gunned down six people at a Sikh Temple in Wisconsin on August 5, 2012.

5. Ali Muhammad Brown killed three civilians in Seattle, Washington, during a period from April 27, 2014, to June 1, 2014.

6. Zale Thompson attacked two police officers with a hatchet in Queens, New York, on October 23, 2014.

7. Ismaaiyl Brinsley killed two police officers in an ambush in Brooklyn, New York, on December 20, 2014.

(continued)

(Continued)

8. Elton Simpson and Nadir Soofi opened fire at a conference that was hosting an exhibition of a cartoon of Prophet Muhammad in Garland, Texas, on May 3, 2015.

9. Muhammad Youssef Abdulazeez shot and killed four marines and a sailor at a military base in Chattanooga, Tennessee, on July 16, 2015.

10. Faisal Mohammad attacked students with a knife at a university in Merced, California, on November 4, 2015.

11. Samuel Woodward, a member of a far-right/neo-Nazi group named the Atomwaffen Division, stabbed a gay Jewish University of Pennsylvania student on January 2, 2018.

12. John T. Earnest, a 19-year-old young man who was sympathetic to alt-right and anti-Semitic thought, set the Dar-ul-Arqam Mosque on fire in Escondido, California, on March 24, 2019.

13. Mohammed Saeed Alshamrani, a Saudi Air Force Second Lieutenant who was being trained in the United States, shot and killed three U.S. Navy sailors and injured eight others in a classroom at the Naval Air Station in Pensacola, Florida, on December 6, 2019.

14. Anderson Lee Aldrich, a 22-year-old man with anti-LGBT sentiment, killed five people and injured twenty-five others in a shooting at the Club Q bar in Colorado Springs, Colorado, on November 19, 2022.

15. Audrey Hale, a 28-year old transgender man, used a firearm to kill three adults and three children at the Covenant Christian School in Nashville, Tennessee, on March 27, 2023.

The terrorist attacks mentioned earlier were not the first to occur in the United States or around the world. Nor will they be the last ones to take place in this country or elsewhere. There have been many unsuccessful attacks in New York as well as in Arkansas, Dallas, Florida, Illinois, Michigan, Missouri, New Jersey, Washington, D.C., and so on. This is to say nothing about terrorist attacks initiated in other nations, which are even more prevalent and deadly.

One of the most consequential bouts of terrorism occurred on November 13, 2015, when terrorists carried out a number of coordinated attacks in France and Belgium. Six locations were targeted in the assaults, ranging from the Stade de France stadium to popular bars in and around Paris. The bloodshed began when suicide bombers wearing explosive devices detonated them near a major soccer match being played between France and Germany. A few minutes later, gunmen began unleashing heavy gunfire at several restaurants in Paris. The most fatal of the attacks occurred in the Bataclan theater. Three perpetrators entered the concert hall and fired assault rifles into the audience. Some members of the crowd were able to escape through exits, but 89 people lost their lives and many more were wounded. By the time each of the individual but coordinated attacks concluded, 130 people were killed and 368 were injured.

Some of the gunmen, who were affiliated with ISIL, were neutralized in the firefight with police, and others (including Abdelhamid Abaaoud) were arrested. Unfortunately, the same cell responsible for the attacks in November also launched additional suicide bombings in Belgium on March 22, 2016. On this particular day, these terrorists killed 32 innocent civilians and wounded over 300 others at the Brussels Airport and a Brussels metro station. A few of the perpetrators died in the incident, and law enforcement personnel were able to apprehend some of the other participants. This series of events was one of the worst attacks in Europe. More attacks—whether successful or unsuccessful—will certainly be undertaken in the United States and elsewhere.

In The Real World

Failed Attacks

A number of attacks have been thwarted since the start of the new millennium. Seven are particularly noteworthy:

- On December 22, 2001, Richard Colvin Reid (also known as the "shoe bomber") attempted to detonate explosives hidden in his shoes on an American Airlines Flight from Paris, France, to Miami, Florida. Fortunately, Reid was subdued by other passengers and flight attendants before he could successfully light the fuse.

- A terrorist plot involving homemade liquid explosives (disguised as sports drinks) was thwarted before it could be carried out on several commercial airline flights in 2006. Over 20 suspects were arrested after British police uncovered the scheme.

- On May 1, 2010, Faisal Shahzad (also known as the "Times Square Bomber") attempted to detonate a car bomb in New York City. Fortunately, the explosives failed to detonate, and security was notified when people noticed smoke coming from a car.

- Robert Lorenzo Hester, Jr., aka Mohammed Junaid Al Amreeki, was charged for his attempt to provide material support to a foreign terrorist organization on February 17, 2017. Hester believed he was helping Islamic State of Iraq and Syria to launch an attack, but, in reality, he was communicating with undercover FBI agents. This federal law enforcement agency became aware of Hester's intentions and investigated the matter further after he posted several statements regarding his desire to attack the United States.

- In 2019, Christopher Paul Hasson, a lieutenant in the U.S. Coast Guard, was arrested for illegal possession of firearms. Hasson was a white nationalist who had stockpiled 15 weapons and more than 1,000 rounds of ammunition. He was arrested by the FBI before he was able to shoot left-leaning politicians, celebrities, and journalists.

- On April 29, 2019, Mark Domingo was captured before he could launch an attack against a white nationalist group in Southern California. Domingo was a U.S. Army Veteran who supported radical Islamic ideologies. He wanted to seek revenge for the mosque shootings in Christchurch, New Zealand.

- The Federal Bureau of Investigation arrested 13 members of the Wolverine Watchmen militia on October 8, 2020. These right-wing extremists planned to kidnap a politician named Gretchen Whitmer. The plot was foiled before these individuals were able to snatch the Governor of Michigan and hold her hostage.

Self-Check

1. There are a variety of threats facing the United States. True or false?
2. Terrorism is not considered the "new normal" in the United States or elsewhere. True or false?

3. Terrorist attacks have increased over the past few years. True or false?
4. Terrorist attacks have occurred at what locations in the past?
    a. Bars and restaurants
    b. Military bases
    c. Government buildings
    d. Sports stadiums
    e. All of the above
5. What is an example of a recent terrorist attack?

1.2   A Growing Risk

If you browse the Internet for news, you will probably find several articles discussing the rising menace of terrorism (see Figure 1-2). Headlines frequently highlight possible threats and recent attacks:

- Terrorists Infiltrate the United States
- Man Attempts to Detonate Shoe on Plane
- Aviation Security Still Weak
- Oregon Professor Charged with Terrorism
- Sea-born Cargo a Likely Target
- Eco-terrorism Occurs in California
- Officials Detain Man after Filming Chicago Bridge
- Explosives Missing in Georgia
- Agro-terrorism a Real Possibility
- Industrial Security Still Lacking
- Pipelines Targeted in Possible Attack
- Cruise Ship Receives Threatening Letter
- Bombs Obliterate Spanish Resort
- Australia Weary about Potential Terrorists
- Plot Busted in Pakistan
- Bus Ripped Apart by Blast in London
- Children Taken Hostage in Russia
- Cartoon of Mohamed Inflames Terrorists in Europe
- Iran Seeks Nuclear Weapons
- Terrorists Set Sights on Olympics

In addition, the media has provided numerous reports about the **Islamic State of Iraq and Syria (ISIS)**. This particular group sought and continues to establish an Islamic government in the Middle East and has been one of the most-feared and recognized terrorist organizations in the world (Cockburn 2016; Weiss and Hassan 2016). Their actions show no mercy toward victims, and their methods involve the most brutal forms of violence imaginable. These terrorists have illustrated their

FIGURE 1-2   The news is dominated by stories about terrorism and terrorist attacks. *Source: Neville Elder/Shutterstock.*

willingness to kill hundreds and thousands of enemies through mass executions with diverse means: using weapons, with power saws, via drowning, and by dousing people with gasoline and lighting them on fire. ISIS members have thrown homosexuals off rooftops and placed the severed heads of their victims on railings or posts. On January 4, 2024, ISIS detonated two bombs at the burial site of Iranian military commander Qasem Soleimani. The coordinated attacks killed nearly 100 people and injured 300 more. The actions of ISIS are regarded to be atrocious, intentional efforts to induce migration, conduct genocide, and force policy change.

Case Study

Hamas Attack on Israel

On October 7, 2023, the Palestinian terrorist group Hamas launched a sneak attack from the Gaza Strip against Israel. This quickly became the worst terrorist attack in Israeli history and the deadliest day in Jewish history since the Holocaust. And, unfortunately, the series of violent actions came without any advanced warning and took Israel by surprise.

To understand this case, it is important to know the context of the attack. Hamas was formed as an Islamic resistance movement in 1987 during the first intifada—a political uprising of the Palestinian people against Israeli occupation of the West Bank and the Gaza Strip. But this conflict in the Middle East started long before the late 1980s. For instance, Arabs and Jews have each claimed ownership of the territory in and around Jerusalem for centuries. This long-standing disagreement was aggravated in 1947 when the United Nations adopted Resolution 181, which divided British-controlled lands into a Jewish state and other Arab countries. Political leaders around the world felt sympathy

(continued)

Case Study   *(Continued)*

for Jews after the Holocaust and wanted to find a location where this religious group could live in peace. Unfortunately, Arabs opposed the action of the United Nations, and groups like Hamas were created to eliminate the state of Israel. In fact, the charter of Hamas declares that Palestine is an Islamic land, and it mandates jihad (a holy war) to liberate this territory. Peaceful solutions are rejected in the charter, and Muslims are called upon to fight and kill Jews until liberation is achieved. Several skirmishes have therefore occurred between Hamas and Israel over time, and Iran has supported this terrorist organization financially and with small arms and other military-grade weapons.

After an unknown period of intensive planning, thousands of Hamas fighters launched a coordinated and multifaceted attack on Israel during their observance of Simchat Torah (a holiday to celebrate their reading of the Torah). Several members of this violent group entered Israel with motorized paragliders or by boat. They began to attack Israeli military personnel on bases near the border. Meanwhile, heavy equipment was used to breach the fence separating the Gaza Strip and Israel. Terrorists then entered Israel by ground with several vehicles and numerous motorcycles.

These Hamas fighters then began to attack armed guards and shot civilians wherever they were found—in cars, in the streets, while attending parties, and in designated safe rooms at home. During the military response and the subsequent investigation, ample forensic evidence of torture and rape was found, and there were allegations of beheadings (although there has been some debate about the accuracy of this latter assertion).

Nevertheless, the mass slaughter of men and women as well as the elderly and children was bloody and gruesome and turned out to be even more disturbing since Hamas fighters portrayed and celebrated their violent activities live on social media platforms. In addition, over 240 people were abducted and taken hostage by Hamas, including at least 10 Americans. While numerous hostages have been released at the time of this writing, the future well-being of the others is uncertain.

If the attack and abductions were not bad enough, Hamas also fired off anywhere over 8,500 Iranian rockets, homemade missiles, and mortar shells against southern and central Israel. The Jewish people were notified of this air raid via public warning alarms and sought safety in designated bomb shelters throughout the country. Although many of these missiles were shot down by Israel's "Iron Dome," the series of batteries overwhelmed the mobile air defense system at times and many Hamas bombs struck buildings and caused unimaginable damage. Emergency measures were undertaken to care for those impacted by the attack.

When all was said and done, over 1,100 Israelis were killed and another 4,500 were injured. This caused Prime Minister Benjamin Netanyahu to confirm that a murderous surprise attack had occurred against the state of Israel and its citizens. He denounced Hamas and vowed to fight back against and destroy this terrorist organization so this type of attack would not occur again in the future.

After warning Palestinians to evacuate Gaza, Israel began to launch retaliatory missile strikes against Hamas leaders, their offices, and related strongholds. At the time of this writing, it is estimated that over 30,000 people have been killed in Gaza and another 69,000 have been injured. As many as 1.9 million have been displaced thus far due to the barrage of missiles being fired from Israel.

Israel also amassed troops for a ground assault, which encountered an elaborate tunnel system (with estimates of 1,300 tunnels spanning 500 kilometers in total). During the altercation with Hamas, Hezbollah—another anti-Israeli terrorist group—launched attacks from the north from their bases in Lebanon. In addition, Iran has threatened to become involved in the conflict, which could transform the terrorist attack and counter-terrorism operation into an all-out war in the Middle East.

(continued)

Case Study *(Continued)*

While most countries have expressed sympathy to Israel for the attacks, there have been demonstrations by Arabs and Muslims throughout the world to protest the retaliation against Hamas. Jewish embassies and synagogues were put on high alert. Some of the verbal animosity turned violent and antisemitic attacks have surged since October 7, 2023. As an example, a Jewish leader was found stabbed to death in Detroit, Michigan. Many other Jews have been assaulted and their property has been vandalized. This situation encouraged many people to denounce anyone who supported the actions of Hamas. University students and administrators have been harshly criticized for their verbal support of violence against the Jewish people on the part of these radical Palestinians. In fact, donors started to cut back funding of various academic institutions and some university presidents have been removed from their positions.

In response to the escalating problem, President Biden traveled to Israel to express support for Israel. He warned others not to intervene in the problem, and he also promised to send relief to the Palestinian people. Over the next several months, President Biden started to pressure Israel to leave the Gaza strip and halt hostilities. Republicans questioned Biden's loyalty to Israel and encouraged the president to reverse the aid he committed to Iran. Political opponents of the president even recommended that Biden reinstate Trump's economic sanctions against the Iranian regime for its support of Hamas.

This case brings up many questions that the Israeli government and others will need to consider as nations reflect on what has transpired. For instance:

- Was the Hamas attack a violent protest, terrorism, or war?
- What factors led to this attack, and can they be minimized?
- Who is Hamas, and how did they plan and carry out this attack?
- Why did Hamas use the media to portray the attack in real time?
- What can Israel do to protect the security of its people?
- Why was there no advanced intelligence about this planned attack?
- How can Israel protect itself from the violence perpetrated by Hamas?
- What can Israel do to prepare for future attacks?
- How could response and recovery operations be improved going forward?
- Should Israel anticipate other attacks from Hamas that involve other types of weapons of mass destruction?
- What additional challenges and opportunities does Israel and other nations need to consider for the future?

As you read this book, you should consider these types of questions and how they might relate to other terrorists and the potential threats they pose.

Recognizing issued threats and actual terrorist activity, many conclude that politically motivated acts of violence will be more common in the future. Several years ago, Senator Richard Lugar, R-Ind, stated, "The bottom line is this: For the foreseeable future, the United States and other nations will face an existential threat from ... terrorism." In 2016, FBI Director James Comey reiterated this warning by stating that terrorists will infiltrate Western Europe and the United States and that future attacks will be on "an order magnitude greater" than those of the past. Today, and in light of the attack on the U.S. Capitol on January 6, 2021, there is rising concern about home-grown radicals—particularly right-wing terrorist groups—who engage in violence (Byman 2022).

1.2.1 Reasons to Anticipate More Attacks

There are numerous reasons why we may witness additional and more consequential attacks in the future. For instance, the promise of Western forms of economic development has not materialized in many nations, and poverty may be associated with increased terrorist activities. Put differently, the poor nations 50 years ago are predominantly the poor countries today, and they are breeding grounds for terrorist organizations. In addition, the end of the Cold War and current political realities have resulted in the resurgence of deep-seated ethnic or political rivalries. Chechnya desires autonomy and independence from Russia, while China is worried about extremist activities from the Uyghur and other Turkic Muslims. Furthermore, U.S. military power and involvement in the Middle East has angered many Arabs. Many Palestinians and others view the American presence in their region as a form of colonialism. Also, there is fear that countries like Iran and North Korea will develop, use, or share nuclear weapons and radioactive materials with terrorist organizations. Furthermore, some religious and social movements have become more extreme over time. Fundamentalist Muslims and right-wing interest groups want change now and are willing to promote it through violent behavior. What is more, the divisions separating republicans and democrats in the United States have expanded recently, and conflict is increasingly seen as the legitimate means to achieving political objectives. Furthermore, protecting each of the vulnerable locations that terrorists could attack is virtually impossible. Government buildings, ports, shopping malls, and schools are all likely targets. Furthermore, training and preparedness for terrorism response could be inadequate. As an example, we do not know enough about how to deal with biological weapons and hazardous materials used by terrorists.

Five other factors may ultimately lead terrorists to enact their deadly craft in the years to come:

1. Prior military conflicts among nation-states persist, and patience to resolve them is growing thin (e.g., the creation of the state of Israel several decades ago has resulted in ongoing tensions in the Middle East).
2. Citizens are frustrated with the harsh conditions of dictatorship in many countries or the unresponsiveness of certain democratic governments around the world (e.g., they desire political change and think that their needs are not being met in an expeditious manner).
3. It is extremely difficult for intelligence analysts to know who the "enemy" is (e.g., how can one pinpoint a terrorist when they often blend into the crowd?).
4. Technology and education will allow terrorists to develop and use more sophisticated weapons (e.g., even typical household chemicals can be combined in such a way as to make bombs).
5. The ideology of terrorists has become so radical that their acceptance of brutality knows no boundaries.

Should more and worse terrorist attacks occur as predicted, the United States can expect an increased loss of life, financial hardship, social disruption, political change, and other negative consequences. As an example, it is not out of the possibility to have casualties in the thousands, hundreds of thousands, or even higher due to terrorists who seek modern weapons that employ today's advanced knowledge and technology. The economy will surely suffer after major attacks, and financial losses could total millions or billions. Travel and shopping may be severely hampered as well when people worry about their safety, and impending attacks could be geared toward altering people's daily lives.

Terrorism could likewise result in a massive transformation of the government and the introduction of new laws pertaining to security, travel, and immigration. Further consequences and changes will certainly be undertaken when terrorists strike again in the United States and elsewhere. All of this is to say that terrorism is now recognized as a consistent feature of our time, and it cannot be discounted or ignored. In short, "terrorism has become the plague of the twenty first century" (Franks 2006, p. 1).

In The Real World

Bin Laden's War Against the United States

Terrorists, such as the now-deceased Osama bin Laden, often declare war against Western nations. In the case of Al-Qaeda, bin Laden and his followers disapproved of the foreign policy of the United States in the Middle East, and they stated it is the responsibility of all Muslims to attack the "infidels." Reports from intelligence analysts indicate that terrorist groups like ISIS have worked hard to launch attacks in the United States and elsewhere around the world. Most experts believe that the efforts of terrorists will be successful unless significant counterterrorism measures are undertaken in the future.

Self-Check

1. There are very few reports of terrorist threats in newspapers. True or false?
2. Terrorist attacks create several negative consequences ranging from death and injuries to social and economic disruption. True or false?
3. Reasons to be concerned about the future of terrorism include:
   a. Resurgence of ethnic rivalries
   b. Poverty in many nations around the world
   c. More extreme religious attitudes
   d. Availability of weapons
   e. All of the above
4. Will we have more attacks in the future? If so, why?

1.3 9/11: A Wake-Up Call

The most consequential attack up to the time of this publication occurred on September 11, 2001. **9/11**, as it is known, will forever be remembered as the terrorist attacks involving four hijacked planes against the United States. It ushered in a new era in world history and illustrates why terrorism must be taken seriously.

After years of planning, 19 hijackers affiliated with Osama bin Laden and **Al-Qaeda** (an extreme Islamic fundamentalist organization) boarded four commercial planes to initiate a massive campaign of terror against the United States (see Figure 1-3). American Airlines Flight 11, departing from Boston to Los Angeles, was overtaken by men with box cutters or other sharp instruments. It was then deliberately flown into the North Tower of the World Trade Center in New York City. United Airlines Flight 175, also departing from Boston to Los Angeles, was diverted and used as a missile to kill people working in the South Tower of the World Trade Center. Within minutes, American

FIGURE 1-3 Terrorists used passenger jets to attack the United States on September 11, 2001. *Source:* © FEMA.

Airlines Flight 77, departing from Dulles to Los Angeles, struck the Pentagon in Arlington, Virginia. Another plane, United Airlines Flight 93, departing from Newark to San Francisco, was also hijacked. However, by this time, passengers on board became aware of other incidents and attempted to take back the aircraft. Unfortunately, the plane crashed into the ground a short time later in an empty field southeast of Pittsburgh.

The brave efforts of passengers on Flight 93 amounted to a symbolic victory for the United States. Nevertheless, the hijackers succeeded in their goal of bringing attention to their hatred of Western culture and disapproval of American foreign policy. At least 266 passengers and crew were killed in the orchestrated attacks. Over 2,500 other people died in the subsequent collapses of the World Trade Center towers in New York, in the fire at the Pentagon in Virginia, and at the crash site near Shanksville, Pennsylvania. In addition to the loss of life, America experienced fear near or on par with that witnessed during the attack on Pearl Harbor. Economic disruption occurred on an unprecedented scale, and damages alone totaled more than $40 billion. Terrorism had certainly captured the attention of the United States.

When informed of the situation, President Bush requested the grounding of all flights to prevent further loss of life and damage. He also had to make the difficult decision to issue an order to the Air Force to shoot down any other hijacked planes, although none were fortunately encountered. Almost immediately, firefighters, police officers, paramedics, hospital personnel, and government officials began to evacuate the World Trade Center and address the needs of the victims of the attacks. When flights resumed a few days later, new measures were taken at U.S. airports to minimize the probability of similar events in the future. Volunteers, businesses, and numerous local, state, and federal agencies also arrived in New York, Virginia, and Pennsylvania to consider how they would address long-term rebuilding activities and memorialize the deceased.

In the meantime, and after determining who was responsible for these attacks, U.S. troops were sent into Afghanistan to topple the Taliban. The **Taliban** is the name of the government that provided safe haven for Al-Qaeda. Intelligence efforts were also augmented, and a successful manhunt was undertaken to find Osama bin Laden,

CAREER OPPORTUNITY | Central Intelligence Agency Analyst

The information that pointed to the culpability of Al-Qaeda and led to the demise of Osama bin Laden was collected by employees of the Central Intelligence Agency (CIA). The CIA does not fall under the Department of Homeland Security. However, this agency fulfills vital homeland security functions relating to intelligence and counterterrorism operations.

There are a variety of employment opportunities with the CIA, and one of the most essential positions is given the title of analyst. CIA analysts gather information from individual interviews, group communications, government documents, media reports, and satellite imagery among other international sources. Analysts then determine if the information is accurate, cross-check it with other knowledge, make sense of the implications, write up policy memos, and share the findings with supervisors and other decision makers in oral briefings.

Not all analysts are the same, however. There are analysts that specialize in information pertaining to regional history, political leaders, terrorism threat analysis, military operations, financial tracking, weapons proliferation, and many other topics.

Analysts can make around $60,000 to start off, and salaries may reach over $150,000 and beyond. Because of the rise of diverse threats, growth in this profession is anticipated to expand slowly but steadily in the coming decade.

Requirements to become a CIA analyst are similar to other jobs in homeland security: age (at least 18 years old), American citizenship, completion of application, passing of polygraph tests, background investigations, psychological examinations, and physical and medical evaluations. Desired qualifications for applicants include critical thinking, solid communication skills, graduate degrees, foreign language proficiency, and a willingness to live in Washington, D.C., or abroad.

For further information, see Adapted from **https://www.cia.gov/careers/cia-requirements**.

the leader of the Al-Qaeda terrorist network. In time, Congress passed numerous laws to repel terrorist activity by improving border control, increasing public security, and promoting readiness for future terrorist plots. Elected officials, public servants, law enforcement agencies, corporations, and many others are now working together to prevent further terrorist attacks or react effectively should they occur.

The narrative describing 9/11 and its aftermath brings up three central questions that will be addressed in the remainder of this book:

- What is terrorism?
- Why and how does terrorism occur?
- What can and should be done to deal with it in an effective manner?

Self-Check

1. Terrorism may be described as the pursuit of ideological purposes through violent means or the threat of violence. True or false?
2. 9/11 is the name given to the terrorist attacks that occurred in New York, Washington, D.C., and Pennsylvania. True or false?

3. The attacks on September 11, 2001, involved:
   a. Explosives
   b. Guns
   c. Hijacked airplanes
   d. Hand grenades
   e. None of the above
4. Why did 9/11 change the world?

1.4   The Nature of Homeland Security

The foregoing discussion indicates the need for what is now known as "homeland security." Interestingly, discussions about this field and emerging profession did not begin after 9/11. For instance, concerns about terrorism have been around for decades. President Clinton acknowledged the threat of terrorism after a number of attacks were initiated in the 1990s. Later on, President Bush created an office in the White House to assess the growing threat of terrorism after his election. However, homeland security did not move to the forefront of the policy agenda until after 9/11. The events of this day revealed the reality of what was heretofore unthinkable and the need to address terrorism in a systematic fashion. Today, homeland security is a "primary public policy area just like education, healthcare, environment, nation defense, and others" (Jones 2008, p. 95).

1.4.1   Definitions of Homeland Security

When **homeland security** was initially conceived by national leaders, it was defined as "a concerted national effort to prevent terrorist attacks within the United States, reduce America's vulnerability to terrorism, and recover from and minimize the damage of attacks that do occur" (Office of Homeland Security 2002, p. 2). While this definition captures the essence of current efforts to deal with terrorism, consensus on the term is not universal. For instance:

- *Citizens believe homeland security refers to the federal agency in charge of preventing terrorist attacks in the United States.* The **Department of Homeland Security (DHS)** was created after 9/11, and this organization is now composed of over 260,000 employees from 22 federal agencies. Its mission is to prevent terrorist attacks and react effectively to those that may occur.
- *Elected officials view homeland security as a policy framework.* Its purpose is to organize "the activities of government and all sectors of society to detect, deter, protect against, and if necessary, respond to domestic attacks such as 9/11" (Kamien 2006, p. xli).
- *Scholars see homeland security as an area of study and an emerging or even established academic discipline.* It is considered a multi- or interdisciplinary research endeavor that involves academic fields such as international relations, criminal justice, public administration, and even medicine.
- *Practitioners regard homeland security to be a function or collection of functions performed in response to the terrorist threat.* In this sense, homeland security

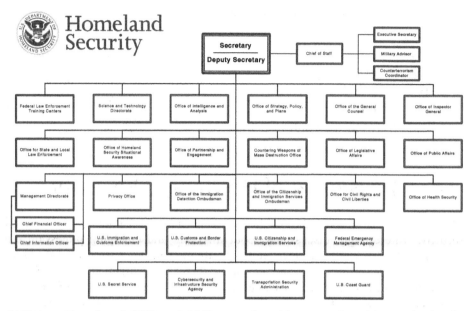

CAPTION: Even though DHS was initially created to address terrorism, this organizational chart illustrates the many complex responsibilities that fall under this large government department. DHS covers functions ranging from immigration and maritime rescues to the Secret Service and disaster recovery. *Source: DHS.*

deals with intelligence gathering, border control, airport security, fire suppression, public health, and emergency medical care.

- *The military asserts that homeland security is the new priority in the post-Cold War era.* Because the Cold War ended in the late 1980s, attention given to national security has shifted in many ways toward individual terrorists, terrorist organizations, and the states that support terrorism. Of course, it is important to recognize that there are now growing anxieties about the belligerence of Russia and China in the Ukraine, Taiwan, and elsewhere around the world.

1.4.2   Agreement About Homeland Security

Even though homeland security means different things to different people, there are several points of agreement. First, homeland security was initially created to counter the threat of terrorism in the United States, and it is consequently a unique blend of national security and emergency management. According to the initial National Strategy for Homeland Security, there were six essential missions of homeland security when it was established. These include:

- **Mission Area 1: Intelligence and Warning.** One goal of homeland security is to identify possible terrorist attacks before they occur. This eliminates surprises and permits the implementation of protective measures if potential targets can be identified.

- **Mission Area 2: Border and Transportation Security.** Another purpose of homeland security is to prevent the infiltration of terrorists into the United

States. Protecting our land, water, and air transportation systems from attack is also a major objective of homeland security.

- **Mission Area 3: Domestic Counterterrorism.** This aim focuses on interdicting terrorist activity and prosecuting those who fund or engage in terrorism. The goal here is to thwart terrorist plans and apprehend those involved in attacks against America.

- **Mission Area 4: Protecting Critical Infrastructures and Key Assets.** This strategy desires to defend vital buildings, roadways, utilities, technology, and so on. Steps must also be taken to prevent attacks against important monuments, valued industries, and national symbols (e.g., the Statue of Liberty).

- **Mission Area 5: Defending Against Catastrophic Threat.** The intention of this objective is to prevent the proliferation of dangerous weapons. Homeland security also wants to quickly detect and deal with the impact of major attacks.

- **Mission Area 6: Emergency Preparedness and Response.** The final priority of homeland security is to plan, train, and equip police, fire, and paramedics to react successfully to terrorism. There is also a need to promote recovery with the assistance of disaster specialists.

In the 2010 Quadrennial Homeland Security Review, the missions of homeland security were revised slightly and reflected a more specific focus on the DHS (rather than the broad functions pertinent to the goals of homeland security). The mission at this time included: (a) preventing terrorism and enhancing security, (b) securing and managing our borders, (c) enforcing and administering our immigration laws, (d) safeguarding and securing cyberspace, (e) ensuring resilience to disasters, and (f) maturing and strengthening the homeland security enterprise.

Today, the DHS continues to give attention to these priorities with some additional variation. It seeks to counter terrorism and other threats that may impact the safety and security of Americans. The DHS also aims to protect U.S. borders, secure cyberspace and critical infrastructure, preserve and uphold the nation's prosperity and economic performance, strengthen preparedness and resilience, and champion its workforce and strengthen the department.

A second and widely held view espoused by the DHS is that this expanding endeavor requires integrated efforts on the part of many people. Homeland security is therefore a major undertaking and is more of a function and enterprise than a single government department. Put differently, homeland security undoubtedly requires a comprehensive approach (Martin 2017).

According to Richard Falkenrath, an expert on international conflict:

Men and women from dozens of different disciplines—regional experts, terrorism analysts, law enforcement officials, intelligence officers, privacy specialists, diplomats, military officers, immigration specialists, customs inspectors, specific industry experts, regulatory lawyers, doctors and epidemiologists, research scientists, chemists, nuclear physicists, information technologists, emergency managers, firefighters, communications specialists, and politicians, to name a few—are currently involved in homeland security. (In Kamien 2006, p. xxvi)

In other words, there are a variety of participants in homeland security. Some may represent the government at local, state, and national levels. Many cities, counties, and states now have homeland security agencies like the DHS. Tribal governments are also involved in homeland security efforts. Others from the business and non-profit communities will assist in fulfilling homeland security responsibilities. Corporations play a huge role in transportation and shipping, while organizations like the American Red Cross help to educate the public about terrorism preparedness. Even citizens may fulfill homeland security functions by notifying officials of potential terrorist activity (e.g., "see something, say something" public education campaign). Although much of the activity in homeland security occurs within the domestic arena among individuals, businesses, and cities or states, the assistance of national and international organizations and partners is also required. National intelligence agencies share information about terrorists operating abroad, and the United Nations has passed resolutions on how the international community should confront terrorism.

In The Real World

National Plans to Deal with Disasters and Terrorism

After 9/11, the federal government developed a new strategy for dealing with terrorist attacks. Rather than building upon or altering the prior Federal Response Plan, a new strategy was created. The National Response Plan (NRP) added layers of bureaucracy to federal response operations and obfuscated responsibility for numerous disaster functions. The plan was criticized while it was being created and especially after it failed to be implemented successfully in the aftermath of Hurricane Katrina. Part of the problem was a result of placing too much attention on terrorism and downplaying other types of hazards. The director of FEMA also lost direct ties to the president after 9/11, which hindered communication. The challenges that resulted from the creation of the NRP indicate why the Federal Emergency Management Agency (FEMA) should be involved in the formation of homeland security policies. This is because FEMA plays a key role in preparing for and coordinating post-disaster response and recovery operations. Fortunately, efforts have been made to clarify the relationship between DHS and FEMA in all types of disasters. The National Response Framework is a new planning document that corrects many of the weaknesses of the NRP.

A third area of agreement is that tensions have reemerged or resulted at times from homeland security initiatives (Canton 2016). The most visible examples concern the problems homeland security unintentionally created for those responsible for dealing with disasters. For example, the **Federal Emergency Management Agency (FEMA)**—the national entity in charge of disaster management—lost much of its budget and autonomy when it was integrated into the newly created DHS. A significant portion of the operating funds from FEMA's small budget (at least by federal standards) were poured into the DHS to cover start-up costs, and the ability of this disaster organization to influence the direction of policy was severely hampered. FEMA, which had cabinet-level status in a prior administration, saw its direct ties to the president severed when its director was placed under DHS. Furthermore, FEMA's interest in all types of hazards, disaster mitigation programs, and even certain preparedness functions was overlooked. Under the DHS, terrorism seemed to take precedence over all other concerns, and efforts to address other types of disasters were neglected. The heavy military and law enforcement approach to homeland security also had an impact on interagency collaboration. Information sensitivities as well as command and control/top-down communication structures hindered coordination across organizations horizontally and among governments vertically.

FIGURE 1-4   This shield illustrates that the Federal Emergency Management Agency is an organization within the Department of Homeland Security. *Source: © FEMA.*

Morale at FEMA started to deteriorate under these conditions, and many knowledgeable disaster professionals retired or switched careers as a result. Such problems were in part responsible for the slow and disjointed response to Hurricane Katrina in fall 2005. Neither FEMA nor DHS officials could effectively coordinate important post-disaster functions like warning, evacuation, sheltering, and mass care. After several congressional investigations into these failures, efforts have been made to correct them. In particular, there is growing recognition that homeland security cannot focus on the threat of terrorism alone or without the help of organizations like FEMA. In fact, the mission of DHS was adapted in 2007 to include a greater emphasis on all types of disasters along with a recognition of the importance of preparedness.

Finally, it is important to point out that individuals like Representative Jeff Duncan (R-SC) have argued that homeland security has had mixed results during its short existence. On the one hand, the United States has been successful in preventing major terrorist attacks like 9/11 against the homeland. Achievements in this area are to be commended because several terrorist plots against Americans have been foiled. This, probably more than any other factor, is a major accomplishment in the war on terrorism. On the other hand, DHS has gone through several growing pains because of its hasty creation and the enormous challenges it has faced as the organization has tried to implement its broad mission. For instance, there have been numerous allegations that the start-up funding devoted to homeland security lacked careful controls to prevent fraud, waste, and abuse. At least some of the money designated for homeland security may have been spent on questionable purchases. Stories abound of communities using homeland security money to buy dump trucks, polo shirts, and other items that seem at first glance to be unrelated to terrorism. Other problems, like border control, are yet to be resolved due to the politics relating to illegal immigration. For these reasons, the impact of homeland security is somewhat inconclusive. Of course, it is necessary to recognize that it is not easy to assess what success means in the context of homeland security. This is because government leaders cannot always publicize threats or evaluate responses to attacks that have been thwarted. Regardless of these concerns, homeland security remains a vital function, and it is essential for national interests.

FEMA's Strategic Plan

In 2022, the Federal Emergency Management Agency created a new strategic plan to guide emergency management into the future. While these types of strategic plans change periodically and with new presidential administrations, the current focus is on three key priorities. Deanne Criswell, the FEMA administrator, shared these goals in a letter to all stakeholders who have an interest in reducing the impact of disasters. They include (a) instilling equity as a foundation of emergency management, (b) leading the whole community in climate resilience, and (c) promoting and sustaining a ready FEMA and a prepared nation. In other words, FEMA desires to better assist those who are most vulnerable, promote policies and actions to address global warming and ensure we have the capacity and capability to deal effectively with future disasters. Such measures are essential since the number of disasters FEMA is managing is now 311 annually as compared to 108 in the past. These disconcerting statistics underscore the need for strategic planning and forward thinking. For further information on FEMA's strategic plan, see Adapted from **https://www.fema.gov/about/strategic-plan**.

Self-Check

1. Homeland security was initially defined as efforts to prevent terrorist activity, reduce vulnerability, and recover from attacks. True or false?
2. Everyone views homeland security in the same way. True or false?
3. The goals of homeland security are to:
   a. Gather intelligence
   b. Protect borders and infrastructure
   c. Prepare for major catastrophes
   d. Answers a and b only
   e. Answers a, b, and c
4. Why is it important to take a broad view of homeland security?
5. Has homeland security been effective thus far? Why or why not?

1.5 Disciplines Involved in Homeland Security and the Emergency Management Profession

If you are to work in the important field of homeland security, you must be aware of its academic underpinnings and the disciplines that contribute to its knowledge base. It is true that some have questioned if homeland security is an academic discipline (Falkow 2013). The argument is that homeland security is still emerging and does not yet have an agreed-upon set of concepts and theories. However, others assert that homeland security is a "meta" discipline. This suggests that homeland security is a combination of many areas of study including international relations, criminal justice, public administration, and public health. These fields and many others offer important insights into terrorism and homeland security:

- **International relations** focuses on the conflicts among nation-states and nonstate actors. It identifies why terrorism occurs and what governments are doing about it.
- **Criminal justice** is interested in intelligence gathering, terrorist investigation, and prosecution. It also has a relation to border control and other security measures.
- **Public administration** directs attention to the formation and implementation of policy. It also helps to identify the best form of organization to deal with difficult societal problems.
- **Public health** concentrates on understanding diseases and how to treat them. It plays an important role in preparing for terrorists' use of nuclear, biological, chemical, or radiological weapons.

Other academic disciplines are also important to the study of terrorism and homeland security. National security and military studies explore intelligence gathering and lessons from counterterrorism operations. Anthropology enables an understanding of the culture of terrorism. Sociology facilitates comprehension of human behavior in crisis situations. Political science and law address policy making and human rights issues, which are vital as democratic governments fight terrorism. Journalism permits comprehension of terrorists' use of the media for increased publicity. Engineering provides valuable advice on protecting buildings and critical infrastructure from possible attacks. The physical sciences permit the discussion of nuclear material, chemical reactions, and biological contagion. Computer scientists explore the threats posed by cyberattacks and the means to counter them. Environmental scientists study the degradation of our natural resources and its impact on the occurrence of disaster.

Because these and other fields are vital to homeland security, this book will approach the subject of terrorism from a holistic perspective. However, the book focuses to a great extent on the discipline and profession of emergency management. As will be seen, emergency management plays an especially important role in homeland security.

In The Real World

Resources to Help You Understand Terrorism

Learning about terrorism is essential if you are working in homeland security, and it should be a lifelong process. With this in mind, you may want to be aware of and tap into several resources that are available to you.

1. The Department of Homeland Security has funded several Centers of Excellence that are university-led research networks. Current Centers of Excellence include:
   - Center for Accelerating Operational Efficiency (Arizona State University)
   - Cross-Border Threat Screening and Supply Chain Defense (Texas A&M University)
   - Criminal Investigations and Network Analysis (George Mason University)

(continued)

In The Real World *(Continued)*

- o Critical Infrastructure Resilience Institute (University of Illinois at Urbana-Champaign)
- o Coastal Resilience Center (University of North Carolina at Chapel Hill)
- o Master of Business Administration—Security Technology Transition (George Washington University)
- o National Counterterrorism Innovation, Technology and Education Center (University of Nebraska at Omaha)
- o Soft-Target Engineering to Neutralize the Threat Reality (Northeastern University)

2. There are also many emeritus Centers of Excellence, and they are listed on the Department of Homeland Security Website (see **https://www.dhs.gov/science-and-technology/centers-excellence**). One of the most helpful Centers of Excellence is the National Consortium for the Study of Terrorism and Responses to Terrorism. The website for this consortium (see **https://www.start.umd.edu**) provides information and publications on a variety of topics as well as a number of data sets and resources for education and training.

3. Several academic and professional journals are available for you to read. A partial list of important examples is provided here:

Critical Studies on Terrorism

Disasters

Disaster Prevention and Management

Government Security

Homeland Defense Journal

Homeland Security Affairs

Homeland Security Professional

Homeland Security Today

Intelligence and National Security

Journal of Applied Security Research

Journal of Business Continuity & Emergency Planning

Journal of Counterterrorism & Homeland Security International

Journal of Emergency Management

Journal of Homeland Security

Journal of Homeland Security and Emergency Management

Journal of Homeland Security Education

Journal of Hazards, Crisis and Public Policy

Journal of Security, Intelligence, and Resilience Education

Journal of Terrorism Research

Natural Hazards Review

Security Journal

Studies in Conflict and Terrorism

Terrorism and Political Violence

1.5.1 The Role of Emergency Management

Emergency management is a profession that specifies how to prevent or react successfully to various types of disasters (McEntire 2022). It includes four functional phases described as the life cycle of disaster: mitigation, preparedness, response, and recovery. Each of these concepts is important for the study of terrorism, and they have unique relationships to the homeland security profession.

Mitigation is an activity that attempts to avoid disasters or minimize negative consequences. Mitigation is also closely associated with two terms that are frequently discussed in homeland security:

- **Prevention** includes counterterrorism operations (such as intelligence gathering and covert military activities) or other functions like border control.
- **Protection** incorporates antiterrorism operations such as infrastructure protection and increased security surveillance at locations like airports and sporting events.

There are three other important phases in emergency management. **Preparedness** includes readiness measures in anticipation of a disaster. Planning, training, and exercises are examples of preparedness initiatives. **Response** is the immediate reaction to an emergency situation like a terrorist attack or a disaster. In homeland security, response refers most often to evidence collection and emergency medical care functions. **Recovery** refers to long-term activities to rebound after disasters or terrorist attacks. It includes emotional recovery and rebuilding with future hazards and threats in mind.

Since its inception, homeland security has focused most of its attention on preventing, protecting against, and prosecuting terrorist activities. These measures were initially labeled as **crisis management.** However, this concept did not necessarily encapsulate functions related to preparedness, response, and recovery operations. These undertakings were therefore called **consequence management**. While it is imperative to perform both crisis and consequence management functions, government leaders began to recognize that these two priorities should not be treated as isolated actions. Doing this only led to coordination difficulties. In addition, while it was crucial to stress prevention, protection, and prosecution measures, the assumption that this would be possible 100% of the time was questioned. For these reasons, emergency management is seen as an increasingly vital component of homeland security.

Emergency management has a long history of dealing with a plethora of natural, technological, and anthropogenic disasters. This discipline and profession have generated important recommendations for dealing with conflict and collective stress situations (Drabek 1986). What is more, scholars such as Bill Waugh (2001) and McEntire et al. (2001) indicate the close relation between terrorism and emergency management. Others also see unique ties between emergency management and homeland security (Bullock et al. 2005).

In spite of this close relation, homeland security policies did initially or sufficiently draw from the research and practice of emergency management. The failure to adequately integrate these efforts has created many challenges pertaining to terrorism (see Figure 1-5 or see option on p. 27). Some of today's problems are reminiscent of those that existed during the civil defense era (Alexander 2002). **Civil defense** is the name given to the government's initiative to prepare communities and citizens to react effectively to nuclear war against the Soviet Union. The primary focus of the Cold War was on responding to nuclear weapons with a top-down, military, command, and control approach. Throughout this period, natural and technological hazards as well as collaboration with nonmilitary organizations were not always given serious consideration.

Emergency Management

Definition, Vision, Mission, Principles

Definition

Emergency management is the managerial function charged with creating the framework within which communities reduce vulnerability to hazards and cope with disasters.

Vision

Emergency management seeks to promote safer, less vulnerable communities with the capacity to cope with hazards and disasters.

Mission

Emergency management protects communities by coordinating and integrating all activities necessary to build, sustain, and improve the capability to mitigate against, prepare for, respond to, and recover from threatened or actual natural disasters, acts of terrorism, or other man-made disasters.

Principles

Emergency management must be:

1. **Comprehensive**—Emergency managers consider and take into account all hazards, all phases, all stakeholders, and all impacts relevant to disasters.
2. **Progressive**—Emergency managers anticipate future disasters and take preventive and preparatory measures to build disaster-resistant and disaster-resilient communities.
3. **Risk driven**—Emergency managers use sound risk management principles (hazard identification, risk analysis, and impact analysis) in assigning priorities and resources.
4. **Integrated**—Emergency managers ensure unity of effort among all levels of government and all elements of a community.
5. **Collaborative**—Emergency managers create and sustain broad and sincere relationships among individuals and organizations to encourage trust, advocate a team atmosphere, build consensus, and facilitate communication.
6. **Coordinated**—Emergency managers synchronize the activities of all relevant stakeholders to achieve a common purpose.
7. **Flexible**—Emergency managers use creative and innovative approaches in solving disaster challenges.
8. **Professional**—Emergency managers value a science and knowledge-based approach based on education, training, experience, ethical practice, public stewardship, and continuous improvement.

Homeland security officials have made similar mistakes in recent years. Those in charge of policy focused initially and almost exclusively on terrorism and favored a heavy law enforcement or paramilitary approach. Leaders failed to recognize that the United States is prone to many different types of hazards (Mileti 1999). Homeland security also ignored to its own peril the research that suggests that coordination with others is of paramount importance if responses to disasters are to be successful (Auf der Heide 1987).

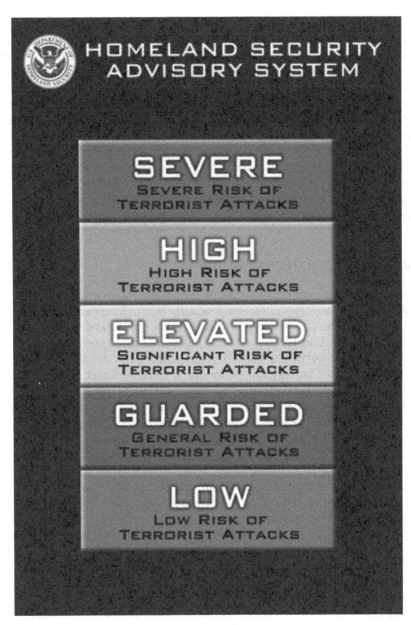

FIGURE 1-5 The Homeland Security Advisory System did not warn citizens about what to do if a terrorist attack should occur. *Source: US States Department of Homeland Security/Public Domain.*

The most vivid example of these mistakes was the creation of the **Homeland Security Advisory System (HSAS)** (see Figure 1-5). The HSAS was the nation's method for warning the population of potential and actual terrorist attacks. It illustrated the problems that result when the emergency management community was not consulted and lessons from prior disasters were not incorporated into new policies. For instance, research on disasters and emergency management has provided useful advice to improve warning messages for nearly 50 years (McEntire 2014). Evidence from decades of research illustrates that warnings should be clear, consistent, and credible.

Crisis communication messages must also help communities and citizens understand exactly what they are supposed to do when disasters and terrorist attacks occur.

In contrast to these recommendations, many argue that HSAS lacked clarity as well as specific and useful information for citizens (Aguirre 2004; Knight 2005). As a case in point, what was implied when the threat level was raised from yellow to orange? Did it mean an attack had occurred? How should citizens react? Why would a change in color status help promote successful responses? Since the HSAS was not based on existing emergency management research, it had difficulty in successfully providing answers to these questions. The HSAS therefore became the focus of many jokes on late-night television and has since been replaced with a different warning system. This situation and others reveal that emergency management is an important discipline for homeland security. There is definitely a need to integrate the knowledge of emergency management into homeland security policy (Kiltz 2012).

1.5.2 Important Terminology

While emergency management knowledge can help in many areas, one of the greatest potential contributions to homeland security is in reference to this profession's views about hazards, vulnerability, and risk. The term **hazard** was introduced by geographers, and it almost always refers to physical or other agents that may trigger or initiate disaster events and processes (Alexander 2002, p. 29). While hazards such as an earthquake, industrial plant explosion, or terrorist attack are real and menacing, focusing on them excessively can create many problems for those involved in homeland security and emergency management. For instance, giving priority to certain hazards often leads to dramatic and detrimental shifts in policies as we have recently seen with the almost exclusive emphasis on terrorism (Waugh 2004). Placing ultimate priority on hazards likewise downplays the human role and responsibility in all types of disasters (McEntire 2004). What is more, since we cannot eliminate or control all extreme events, there is growing recognition that vulnerability is a stronger determinant of disaster than the hazards themselves (Alexander 2006, p. 2; Cutter 2005, p. 39). For these and other reasons, many recommend moving from an "agent-centered approach" to one that gives greater attention to a broad conceptualization of vulnerability (Perry 2006, p. 9; Weichselgartner 2001).

As defined in the research literature, **vulnerability** implies a high degree of disaster proneness and/or limited disaster management capabilities. One school of thought on this matter suggests that vulnerability is the likelihood of a disaster occurring and that individuals or the community as a whole will experience negative impacts from hazards (e.g., injuries, death, property damage, financial losses, social disruption). There are several scholars who accept this viewpoint (Anderson 1995, p. 41; Bolin and Stanford, 1998, p. 9; Boulle et al., 1997, p. 179; Maskrey, 1989, p. 1; Mitchell, 1999, p. 296; Salter, 1997/98, p. 28; Wisner et al., 2004, p. 11).

Another perspective on vulnerability relates to capacity or capability. This line of thought centers on the ability or inability of people, organizations, and social systems to anticipate, prevent, prepare for, cope with, respond to, or recover from the impact of a hazard. It is also supported by many researchers (Schroeder 1987, p. 33; Vasta 2004, pp. 10–11; Warmington 1995, p. 1; Wisner et al. 2004, p. 11). Vulnerability is thus regarded as a multifaceted concept that the literature almost always conveys in terms of proneness and capabilities (Chambers 1989; Comfort et al. 1999; Pelling and Uitto 2001; Watts and Bohle 1993).

Hazards and vulnerability are closely associated with the concept of risk. Some scholars assert that hazards and vulnerability are determinants of risk or the likelihood of occurrence (Mileti 1999). Others assert that risk deals with exposure to disaster agents or possible losses (Alexander 2002). The truth of the matter is that risk is determined by both types of variables. **Risk** is therefore a measure of probability and consequences alike. The concept of risk permits an understanding of what can happen and how bad it could be. Although it is difficult to know how much weight to give to probability versus consequences, the notion of risk is valuable to those working in emergency management.

Interestingly, this same framework of risk can also be applied to terrorism and homeland security. For instance, limited intelligence, porous borders, and weak security are factors that must be corrected if the probability of attacks is to be minimized. Furthermore, inadequate prevention and preparedness abilities will likely increase the consequences of attacks during response and recovery operations. Probabilities and consequences thus seem to be extremely important concepts for both the homeland security and emergency management professions. Consequently, this book will approach terrorism and other threats by discussing many themes that relate to and determine both probability and consequences.

In The Real World

Risk and 9/11

The attacks on 9/11 clearly illustrated the risk of terrorism against the United States. Terrorists managed to enter the country in virtually an unnoticed manner. They trained for their attacks in American flight schools and were able to smuggle box cutters onto planes. Once the hijackings were underway, a system was not fully in place to interdict the hijacked aircraft. After the planes were flown into the World Trade Center, firefighters and police had a difficult time communicating with each other. Many died in part because information could not be shared among intelligence agencies as well as police and fire departments. After the buildings collapsed, it took some time before different sources of intelligence could be combined to determine who was responsible for the attacks and how they were funded. Efforts were undertaken to improve incident command at the scene of terrorist attacks. Recovery also took time and taxed many government agencies and the businesses that were impacted. The 9/11 attacks showed that a variety of efforts are needed to minimize the probability of attacks and successfully deal with their consequences.

1.5.3 A Preview of Subsequent Chapters

To meet the goals of this book that pertain to both homeland security and emergency management, the remaining chapters will proceed as follows.

Chapter 2, "Identifying Terrorism: Ideologically Motivated Acts of Violence and Their Relation to Disasters," presents numerous definitions of terrorism and compares how these perspectives are both alike and dissimilar. This chapter also looks at the connections between terrorism and other types of disasters.

Chapter 3, "Recognizing the Causes of Terrorism: Differing Perspectives and the Role of Ideology," explores what motivates people to participate in terrorism. It pays special attention to how historical conflict, mistakes in foreign policy, and extreme levels of poverty may impel some people to engage in this type of violence.

Chapter 4, "Comprehending Terrorists and Their Behavior: Who They Are and What They Do," assesses the nature of individual terrorists and those associated with supporting groups and states. It identifies how these people and organizations finance operations, communicate with secret codes, and carry out attacks.

Chapter 5, "Uncovering the Dynamic Nature of Terrorism: The History of Violence and Change Over Time," explores why terrorism initially emerged, how it evolved in other nations, and the ways it has been manifested in the United States.

Chapter 6, "Evaluating a Major Dilemma: Terrorism, the Media, and Censorship," looks at the symbiotic relationship between terrorism and the media. It predicts how reporters view terrorism and mentions the drawbacks and limitations of censorship.

In Chapter 7, "Contemplating a Quandary: Terrorism, Security, and Liberty," you will learn why, as a participant in homeland security, it is imperative that you assess the tradeoffs between security and rights and why terrorism exploits the tension between these two important values.

Chapter 8, "Preventing Terrorist Attacks: Root Causes, Law, Intelligence, and Counterterrorism," tackles the reasons why terrorism occurs and explains ways to prevent attacks. It mentions laws that prohibit terrorism and punish those who support it. It discusses the need to rely on human and other sources of intelligence to neutralize or apprehend terrorists before they strike.

In Chapter 9, "Securing the Nation: Border Control and Sector Safety," the permeability of the U.S. border is mentioned along with measures to prevent the infiltration of terrorists onto American soil. It also discusses the vulnerability of various economic sectors and describes ways to secure railways, air transportation, seaports, and chemical facilities.

Chapter 10, "Protecting Against Potential Attacks: Threat Assessment, Mitigation, and Other Measures," looks at the benefits of proactive practices, such as working with others to evaluate threats posed to critical infrastructure, key assets, and soft targets. The chapter also differentiates between structural and nonstructural mitigation activities.

In Chapter 11, "Preparing for the Unthinkable: Efforts for Readiness," you will learn that preparing for terrorism is one of the central responsibilities of homeland security. This chapter specifies how your community, state, and nation may prepare for possible terrorist attacks by becoming familiar with the executive orders and legislation issued by the president and Congress. This chapter also puts forward the foundation of preparedness through the creation of an advisory council, the passing of ordinances, the acquisition of monetary resources, and the establishment of an Emergency Operations Center.

Chapter 12, "Responding to Attacks: Important Functions and Coordination Mechanisms," examines effective ways to react to terrorist attacks. This includes numerous functions such as investigation, the protection of first responders, and the care for victims of terrorist attacks.

Chapter 13, "Recovering from Impacts: Short-Term and Long-Term Actions to Promote Resilience," addresses the variety of recovery activities that need to be performed after a terrorist attack takes place. Such measures include declaring a disaster or state of emergency, addressing mass fatality issues, disposing of debris, and providing emotional support to those who have been impacted by the event.

Chapter 14, "Assessing Significant Threats: WMD and Cyberterrorism," evaluates the probability that terrorists will launch more unique and devastating attacks. It covers the risk of radiological, nuclear, biological, and chemical weapons with numerous recommendations to counter such assaults. The chapter also describes the increasing prevalence of cyberterrorism and mentions the actions being taken to increase preparedness in this area.

In Chapter 15, "Evaluating Other Pressing Problems: Criminal Activity, Social Disturbances, Pandemics, and Climate Change," you will gain a better understanding of various concerns including mass shootings, illegal drugs, human trafficking, protests, riots, contagious diseases, and global warming. Solutions to these problems are mentioned as well.

Finally, in Chapter 16, "Looking Toward the Future: Challenges and Opportunities," the need for accountability in homeland security is mentioned. A discussion about policy strengths and weaknesses occurs, and recommendations are provided for both researchers and practitioners.

As you proceed through the remainder of this book, keep in mind the broad nature of the threats we face and the complexity of the homeland security enterprise. Consider what you can do now and, in the future, to help this vital profession be successful.

Self-Check

1. International relations and criminal justice are related to homeland security. True or false?
2. Homeland security and emergency management have no relationship whatsoever. True or false?
3. Vulnerability implies:
   a. An ability to deal with terrorism effectively
   b. A high degree of proneness and limited capabilities
   c. A low degree of proneness and enhanced capabilities
   d. That terrorism will not occur
   e. That we can respond successfully
4. What is meant by the terms "liability reduction" and "capacity building?"

Summary

In this chapter, you have learned about the threats that may jeopardize our security. In addition, facts about terrorist attacks that have recently plagued the United States have been provided. You have been exposed to evidence and arguments which suggest that further attacks will take place in the future. The enormous impact of the terrorist attacks on 9/11 in relation to world history was discussed. The chapter defined homeland security as a concerted national effort to prevent terrorist attacks within the United States, reduce America's vulnerability to terrorism, and recover from and minimize the damage of attacks that do occur. It also discussed the broader mission of homeland security and the need for a holistic framework. Various disciplines, including emergency management, were argued to be important components of homeland security. By supplementing homeland security with an emergency management perspective, you will be better able to deal with the threats and impacts of terrorist attacks.

Assess Your Understanding

Understand: What Have You Learned

Go to **www.wiley.com/go/mcentire/homelandsecurity3e** to assess your knowledge of terrorism and homeland security.

Summary Questions

1. The diversity of threats facing the United States includes mass shootings and rising crime but not global warming and disasters. True or false?

2. The United States has not had a number of terrorist attacks in the past 10 years. True or false?

3. Terrorism is not an important topic in today's world. True or false?

4. Osama bin Laden and Al-Qaeda were responsible for the 9/11 terrorist attacks. True or false?

5. The 9/11 hijackers flew two of the four hijacked planes into the World Trade Center buildings. True or false?

6. Scholars, elected officials, military personnel, practitioners, and citizens have a common view of homeland security. True or false?

7. The Department of Homeland Security performs all functions relating to the prevention of terrorism. True or false?

8. Corporations and nonprofits do not play an important role in homeland security. True or false?

9. Emergency management addresses the prevention of and reaction to different types of disasters. True or false?

10. The Homeland Security Advisory System did a good job of informing citizens on how to take action in a crisis period. True or false?

11. There is or should be an important relationship between emergency management and homeland security. True or false?

12. A community that has a low degree of disaster proneness and sufficient access to resources has a high degree of vulnerability. True or false?

13. Which of the following threats are of concern to law enforcement, public health, and environmental officials?

    a. Illegal drugs

    b. Disease outbreaks

    c. Climate change

    d. All of the above

14. What government organization was responsible for supporting the 9/11 attacks?

    a. Hamas

    b. The Taliban

    c. Hezbollah

    d. China

15. Included in the National Strategy for Homeland Security's six missions are:
    a. Border and Transportation Security
    b. Extensive Academic Research on Terrorism
    c. Defending against Catastrophic Threat
    d. a and c
16. Which of the following increase(s) vulnerability to terrorism?
    a. Secure borders
    b. Limited intelligence
    c. Preparedness
    d. Both b and c
17. Since the formation of the Department of Homeland Security:
    a. There have been no terrorist attacks whatsoever.
    b. There have been allegations that money devoted to homeland security lacks controls to prevent fraud, waste, and abuse.
    c. The Department of Homeland Security has positively impacted FEMA's ability to perform effectively.
    d. All of the above.
18. Emergency management deals with:
    a. Natural disasters
    b. Technological disasters only
    c. All types of disasters
    d. Terrorist attacks only
19. Homeland security involves which of the following disciplines?
    a. International relations
    b. Criminal justice
    c. Public administration
    d. All of the above

Applying This Chapter

1. What threats are confronting the United States, and how does this impact the safety and security of those residing in this country?
2. Why were the terrorist attacks on 9/11 so significant? Explain how this event has had an impact on the American way of life.
3. In this chapter, there is a list of reasons why we may witness a greater number of more violent attacks in the future. Pick a terrorist attack found in recent events and explain it in terms of one or more of the reasons listed earlier.
4. In what ways is homeland security a specific government organization? In what ways is it a function that is performed by many government agencies and even others in the private and nonprofit sectors?
5. Discuss what the response to a terrorist attack could look like with successful collaboration between citizens, cities, states, nonprofits, businesses, and the federal government. How would it be different if there was no collaboration?
6. State the importance of "liability reduction" and "capacity building" in homeland security. How do these ideas tie back into the concept of vulnerability?

Be a Homeland Security Professional

Explaining Homeland Security

You work for the Department of Homeland Security as a public information officer. During an interview, you noticed that the press is struggling to understand what
homeland security is. How would you define it for them? What is the mission of
homeland security? How could you describe it as a function or agency? What else
could you say to help them understand this concept?

Educating the State Legislators

As the lead member of the New York Division of Homeland Security and Emergency
Management, you have been assigned to speak in front of the legislators to defend
your budget. You must clearly state why it is important to have lots of resources at
your disposal. Make a case as to why terrorism is a significant threat and why the
state needs to take it seriously.

Tensions in Homeland Security

Homeland security illustrates some tensions between a law enforcement and emergency management perspective. Explain why both viewpoints are needed and how
their goals may complement one another.

Key Terms

Al-Qaeda   An extreme Islamic fundamentalist terrorist organization

Civil defense   The government's initiative to prepare communities and citizens to
react effectively to a nuclear exchange during the Cold War

Consequence management   An emergency management function that stresses
planning, emergency medical response and public health, disaster relief, and
restoration of communities

Criminal justice   A discipline and profession interested in intelligence gathering,
terrorist investigation, prosecution, border control, and other security measures

Crisis management   A law enforcement function that concentrates on identifying,
anticipating, preventing, and prosecuting those involved in terrorism

Department of Homeland Security (DHS)   A government organization created
to prevent terrorist attacks or react effectively

Emergency management   A discipline and profession that addresses how to
prevent or react successfully to various types of disasters

Federal Emergency Management Agency (FEMA)   The national entity in
charge of disaster management

Hazard(s)   The physical or other agent(s) that may trigger or initiate disaster
events and processes

Homeland security   A concerted national effort to prevent terrorist attacks within
the United States, reduce America's vulnerability to terrorism, and recover from
and minimize the damage of attacks that do occur

Homeland Security Advisory System (HSAS)   The nation's method for warning the population of potential and actual terrorist attacks

International relations   A discipline and profession that deals with the conflicts among nation-states and nonstate actors (e.g., why terrorism occurs and what governments are doing about it)

Islamic State of Iraq and Syria (ISIS)   A group that seeks to establish an Islamic government and is now the most-feared and well-known terrorist organization in the world

Mitigation   Activity that attempts to avoid disasters or minimize negative consequences

9/11   The terrorist attacks involving hijacked planes against the United States

Preparedness   Readiness measures in anticipation of a disaster

Prevention   Counterterrorism operations such as intelligence gathering and preventive strike activity

Protection   Antiterrorism operations such as border control and infrastructure protection

Public administration   A discipline and profession that directs attention to the formation policy and the best organization to deal with difficult societal problems

Public health   A discipline and profession that concentrates on understanding diseases and how to treat them (e.g., identifying how to react from a medical standpoint to the use of nuclear, biological, chemical, or radiological weapons)

Recovery   Long-term activities to rebound after disasters or terrorist attacks

Response   The immediate reaction to an emergency situation, like a terrorist attack

Risk   A measure of probability and consequences

Taliban   The name of the government that provided a safe haven for Al-Qaeda

Terrorism   The use or threat of violence to support ideological purposes

Vulnerability   A high degree of disaster proneness and/or limited disaster management capabilities

References

Aguirre, B.E. (2004). Homeland security warnings: lessons learned and unlearned. *International Journal of Mass Emergencies and Disasters* 22: 103–115.

Alexander, D. (2002). *Confronting Catastrophe*. New York: Oxford University Press.

Alexander, D. (2006). Globalization of disaster: trends, problems and dilemmas. *Journal of International Affairs* 59 (2): 1–24.

Anderson, M.B. (1995). Vulnerability to disaster and sustainable development: a general framework. In: *Disaster Prevention for Sustainable Development: Economic and Policy Issues* (ed. M. Munasinghe and C. Clarke), 41–60. Washington, DC: International Decade for Natural Disaster Reduction/World Bank.

Auf der Heide, E. (1987). *Disaster Response: Principles of Preparation and Coordination*. St. Louis, MO: Mosby.

Baker, M. and D. Ivory. (2021). Why public health faces a crisis across the U.S. *The New York Times*. October 21, 2021. https://www.nytimes.com/2021/10/18/us/coronavirus-public-health.html. (Accessed September 19, 2023).

Bolin, R. and Stanford, L. (1998). *The Northridge Earthquake: Vulnerability and Disaster*. New York: Routledge.

Boulle, P., Vrolijks, L., and Palm, E. (1997). Vulnerability reduction for sustainable urban development. *Journal of Contingencies and Crisis Management* 5: 179.

Bullock, J.A., Haddow, G.D., Coppola, D. et al. (2005). *Introduction to Homeland Security*. New York: Butterworth-Heinemann.

Byman, D. (2022). Assessing the right-wing terror threat in the United States a year after the January 6, insurrection. Brookings Institute. January 5. https://www.brookings.edu/articles/assessing-the-right-wing-terror-threat-in-the-united-states-a-year-after-the-january-6-insurrection/. (Accessed September 21, 2023).

Campbell, T. and M. Hribernik. (2020). A dangerous new era of civil unrest if dawning in the United States and around the world. Verisk Maplecroft. December 10, 2020. https://www.maplecroft.com/insights/analysis/a-dangerous-new-era-of-civil-unrest-is-dawning-in-the-united-states-and-around-the-world/. (Accessed September 21, 2023).

Canton, L.G. (2016). Emergency management vs. homeland security: can we work together? *Emergency Management* (September 30).

Cassidy, J. (2015). The facts about terrorism. *The New Yorker* (November 24). http://www.newyorker.com/news/john-cassidy/the-facts-about-terrorism (accessed October 10, 2017).

Chambers, R. (1989). Editorial introduction: vulnerability, coping, and policy. *IDS Bulletin* 2 (2): 1–7.

Cockburn, P. (2016). *The Rise of the Islamic State: ISIS and the New Sunni Revolution*. New York: Verso.

Comfort, L., Wisner, B., Cutter, S. et al. (1999). Reframing disaster policy: the global evolution of vulnerability communities. *Environmental Hazards* 1: 39–44.

Cutter, S. (2005). Are we asking the right question? In: *What Is a Disaster? New Answers to Old Questions* (ed. R.W. Perry and E.L. Quarantelli), 39–48. Philadelphia, PA: Xlibris.

Drabek, T.E. (1986). *Human System Responses to Disaster: An Inventory of Sociological Findings*. New York: Springer-Verlag.

Falkow, M.D. (2013). Does homeland security constitute an emerging academic discipline? Thesis. Monterey, CA: Naval Post Graduate School.

Franks, J. (2006). *Rethinking the Roots of Terrorism*. New York: Palgrave Macmillan.

Grawert, A. and N. Kim. (2023). Myths and realities: Understanding recent trends in violent crime. Brenan Center for Justice. May 9. https://www.brennancenter.org/our-work/research-reports/myths-and-realities-understanding-recent-trends-violent-crime. (Accessed September 19, 2023).

Jones, D. (2008). Homeland security: emerging discipline challenges, and research. In: *Homeland Security Handbook* (ed. J.Pinkowski), 95–127. Boca Raton, FL: CRC Press.

Kamien, D. (ed.) (2006). *The McGraw-Hill Homeland Security Handbook*. New York: McGraw Hill.

Kiltz, L. (2012). The benefits and challenges of integrating emergency management and homeland security into a new program. *Journal of Home-land Security Education* 1 (2): 6–28.

Knight, A.J. (2005). Alert status red: awareness, knowledge and reaction to the threat advisory system. *Journal of Homeland Security and Emergency Management* 2 (1): Article 9.

Martin, G. (2017). *Understanding Homeland Security*. Thousand Oaks, CA: Sage Publications, Inc.

Maskrey, A. (1989). *Disaster Mitigation: A Community Based Approach*, Development Guidelines, vol. 3. Oxford: Oxfam.

McEntire, D.A. (2004). Tenets of vulnerability: an assessment of a fundamental concept. *Journal of Emergency Management* 2 (2): 23–29.

McEntire, D.A. (2005). Revisiting the definition of "hazard" and the importance of reducing vulnerability. *Journal of Emergency Management* 3 (4): 9–11.

McEntire, D.A. (2022). *Disaster Response and Recovery: Strategies and Tactics for Resilience*. New York: Wiley.

McEntire, D.A., Robinson, R., and Weber, R.T. (2001). Managing the threat of terrorism. *IQ Report* 33: 1–19.

Mileti, D.S. (1999). *Disasters by Design: A Reassessment of Natural Hazards in the United States*. Washington, DC: Joseph Henry Press.

Mitchell, J.K. (ed.) (1999). *Crucibles of Hazard: Megacities and Disasters in Transition*. Tokyo: United Nations University Press.

National Institute on Drug Abuse. (2023). Drug overdose rates. National Institute on Drug Abuse. June 30. https://nida.nih.gov/research-topics/trends-statistics/overdose-death-rates. (Accessed September 19, 2023).

Nazzaro, Miranda. (2023). US surpasses 500 mass shootings in 2023: Gun Violence Archive. *The Hill*. September 17. https://thehill.com/policy/national-security/4209371-us-surpasses-500-mass-shootings-in-2023-gun-violence-archive/. (Accessed September 19, 2023).

Office of Homeland Security (2002). *National Strategy for Homeland Security*. Washington, DC: Office of Homeland Security.

Pelling, M. and Uitto, J.I. (2001). Small island developing states: natural disaster vulnerability and global change. *Environmental Hazards* 3: 49–62.

Perry, R.W. (2006). What is a disaster? In: *Handbook of Disaster Research* (ed. H.Rodriguez, E.L.Quarantelli and R.R. Dynes), 1–15. New York: Springer.

Salter, J. (1997/1998). Risk management in the emergency management context. *Australian Journal of Emergency Management* 12 (4): 22–28.

Schroeder, R.A. (1987). *Gender Vulnerability to Drought: A Case Study of the Hausa Social Environment*. Madison, WI: University of Wisconsin.

Smith, A. (2023). 2022 U.S. billion-dollar weather and climate disasters in historical context. Climate.gov. January 10, 2023. https://www.climate.gov/news-features/blogs/beyond-data/2022-us-billion-dollar-weather-and-climate-disasters-historical. September 19, 2023.

Vasta, K.S. (2004). Risk, vulnerability, and asset-based approach to disaster risk management. *International Journal of Sociology and Social Policy* 24 (10/11): 1–48.

Warmington, V. (1995). *Disaster Reduction: A Review of Disaster Prevention, Mitigation and Preparedness*. Ottawa: Reconstruction and Rehabilitation Fund of the Canadian Council for International Cooperation.

Watts, M.J. and Bohle, H.G. (1993). The space of vulnerability: the causal structure of hunger and famine. *Progress in Human Geography* 17: 43–67.

Waugh, W.L. (2001). Managing terrorism as an environmental hazard. In: *Handbook of Crisis and Emergency Management* (ed. A.Farazmand), 659–676. New York: Marcel Dekker.

Waugh, W.L. (2004). The "all-hazards" approach must be continued. *Journal of Emergency Management* 2 (1): 11–12.

Weichselgartner, J. (2001). Disaster mitigation: the concept of vulnerability revisited. *Disaster Prevention and Management* 10 (2): 85–94.

Weiss, M. and Hassan, H. (2016). *ISIS: Inside the Army of Terror*. New York: Regan Arts.

Wisner, B., Blaikie, P., Cannon, T., and Davis, I. (2004). *At Risk: Natural Hazards, People's Vulnerability and Disasters*. New York: Routledge.

CHAPTER 2

Identifying Terrorism
Ideologically Motivated Acts of Violence and Their Relation to Disasters

DO YOU ALREADY KNOW?

- How to define terrorism
- The nature of terrorism
- Types of terrorism
- The relation of terrorism to other types of disasters

For additional questions to assess your current knowledge of how to identify terrorism, go to **www.wiley.com/go/mcentire/homelandsecurity3e**

WHAT YOU WILL LEARN

2.1 What terrorism is

2.2 The common characteristics of terrorism

2.3 Examples of terrorism

2.4 The relation of terrorism to other disasters

WHAT YOU WILL BE ABLE TO DO

- Explain the different perspectives on terrorism
- Predict the typical features of terrorism
- Catalog the distinct manifestations of terrorism
- Compare how terrorism is similar to and different from other disasters

Introduction

If you are to work in and contribute to homeland security, the first and most important step for you to reduce the probability and consequences of terrorism is to understand exactly what the phenomenon is. The chapter provides several definitions of terrorism and compares the similarities and differences among the divergent viewpoints. The common characteristics of terrorism are then mentioned, thereby helping you determine if a certain activity can be considered such a phenomenon. The types of terrorism are then discussed to help you discover its various manifestations. Finally, this chapter explores the relationship between terrorism and other types of disasters.

2.1   Defining Terrorism

There are literally hundreds of definitions of terrorism but little agreement on a single concept that captures the essence of what the term actually means (Jenkins 1980). This is a major and ongoing concern since it is hard to resolve a problem if it cannot be clearly defined in the first place. In fact, Boaz Ganor, the director of the International Policy Institute for Counter-Terrorism, asserts that an objective definition of terrorism is "indispensable to any serious attempt to combat" it (in Schmid 2004, p. 375).

Much of the difficulty in defining terrorism is a result of the fact that the term is emotionally charged, pejorative, and laden with negative value judgments (Capron and Mizrahi 2016, p. 5). That is to say, no one wants to be labeled a "terrorist." Those engaging in terrorism might argue that they are simply responding to an intolerable injustice that has yet to be resolved through peaceful means (Laquer 1999, p. 9). Thus, it is a common adage that "one person's terrorist is another person's freedom fighter" (Martin 2003, p. 22). A case in point is the U.S. patriots seeking independence from Britain in the late 1700s. Did the activities of these citizens during the Revolutionary War constitute terrorism? Those in England and in the New World may have had different perspectives on this question. The British most likely viewed the Boston Tea Party and the unconventional combat tactics as being similar to the terrorist activities prevalent today. In contrast, the Americans regarded their behavior as a legitimate form of protest or an effective method to level the playing field against the better-trained and equipped redcoats.

The problem of defining terrorism continues today among practitioners and scholars alike, and it remains an issue that has not been resolved domestically or internationally (McElreath et al. 2021, p. 305). For example, the U.S. government has developed several definitions of terrorism:

- The **Department of State (DOS)** is the federal agency in charge of diplomatic relationships among nations. It asserts that terrorism is "premeditated, politically motivated violence perpetrated against noncombatant targets by sub-national groups or clandestine agents, usually intended to influence an audience."

- The **Department of Defense (DOD)** is the public entity responsible for the military and its operations. It declares that terrorism is "the calculated use of violence or the threat of violence to inculcate fear, intended to coerce or to intimidate governments or societies in the pursuit of goals that are generally political, religious or ideological."

- The **Federal Bureau of Investigation (FBI)** is a government organization that enforces U.S. federal laws. It states that terrorism is "the unlawful use of force against persons or property to intimidate or coerce a government, the civilian population, or any segment thereof, in the furtherance of political or social objectives."

- The **Department of Homeland Security (DHS)** is the organization charged with preventing terrorist attacks and reacting effectively to their adverse consequences. In one of its planning documents, DHS describes terrorism as "any activity that (1) involves an act that (a) is dangerous to human life or potentially destructive of critical infrastructure or key resources; and (b) is a violation of the criminal laws of the United States or of any State or other subdivision of the United States; and (2) appears to be intended (a) to intimidate or coerce a civilian population; (b) to influence the policy of a government by intimidation or coercion; or (c) to affect the conduct of a government by mass destruction, assassination, or kidnapping" (Department of Homeland Security, 2004, p. 73).

Each of these definitions reflects the unique mission and perspective of the sponsoring agency. The DOS takes an international view and concentrates attention on non-state actors around the world. The DOD recognizes that terrorism impacts national security and has the objective of influencing foreign policy. The FBI reveals that terrorism is illegal since their violent behavior is not permitted under the law. Finally, the DHS focuses heavily on the destruction of critical infrastructure. While each definition provides an important view of terrorism, it is important to recognize the benefit of a holistic perspective. A reliance on any single definition from a government agency may limit our understanding of the phenomena and therefore hinder efforts to deal with the threat of terrorism.

Other countries also have distinct definitions of terrorism, and the United Nations has likewise struggled to find consensus on the matter at times (Jenkins 1980, p. 377). For instance, France declares that terrorism is violence that creates a climate of insecurity and has the purpose of overthrowing the government (McElreath et al. 2021, p. 305). The United Kingdom asserts that terrorism is an event that is designed to "interfere seriously with or substantially disrupt" infrastructure (McElreath et al. 2021, p. 305). The European Union states that terrorism is a violent act that destabilizes or destroys the "fundamental political, constitutional, economic, or social structures of a country or an international organization" (McElreath et al. 2021, p. 305). The United Nations affirmed that terrorism is any action "that is intended to cause death or seriously bodily harm to civilians or non-combatants, when the purpose of such an act, by its nature of context, is to intimidate a population or to compel a government or an international organization to do or to abstain from doing any act."

Perhaps because of these numerous definitions, scholars have also attempted to clarify what is meant by the term terrorism. But their descriptions of the phenomena likewise diverge in unique ways. Brian Jenkins "calls terrorism the use or threatened use of force designed to bring about a political change" (in White 2002, p. 8). Walter Laqueur says, "[T]errorism constitutes the illegitimate use of force to achieve a political objective by targeting innocent people" (in White 2002, p. 8). And Cindy Combs

FIGURE 2-1   Terrorists often launch attacks in visible locations to achieve maximum visibility and publicity. *Source: UPI/Alamy Stock Photo.*

believes that terrorism "is the synthesis of war and theater, a dramatization of the most proscribed kind of violence—that which is perpetrated on innocent victims—played out before an audience in the hope of creating a mood of fear, for political purposes" (Combs, 2000, p. 8) (see Figure 2-1).

The differences among these definitions are noteworthy. As a case in point, Jenkins accepts the threat of force as terrorism, whereas Laqueur and Combs do not. The definitions by Laqueur and Combs focus on noncombatants as victims. Their inclusion of innocent people reiterates that governments are not the only ones targeted by terrorists. Furthermore, Combs implies that the media is utilized to broadcast messages to further the aims and intentions of terrorists. Fear is therefore mentioned or implied in some of the definitions and not in others. Only one of the definitions stresses the influence of ideology on terrorism, even though people's attitudes and values seem to have a significant role in this type of violent activity (issues that will be taken up in Chapter 3).

Although there are distinct and divergent ways of looking at terrorism, the definitions provided earlier by governments and scholars share remarkable similarities. Each of these descriptions acknowledges the use or threat of force and the goal of obtaining specific purposes or objectives. Thus, terrorism is clearly related to violent activity or the threat of violence, which has the goal of influencing the behavior of others. At the same time, none of the definitions recognize the disruption society experiences because of terrorism. This is ironic in that terrorism may impact government policies (e.g., due to the creation of new laws), business activities (e.g., as a result of economic losses), and citizen behavior (e.g., with people's reluctance to shop or travel when attacks have occurred). Disruption is therefore another major, but unrecognized, aspect of terrorism.

Self-Check

1. There are many different definitions of terrorism. True or false?
2. Definitions of terrorism do not reflect the values or missions of different people and organizations. True or false?
3. In Combs' definition of terrorism, the term "theater" implies:
   a. The location of the terrorist attack
   b. The widespread publicity of an attack by the media
   c. The importance of national security
   d. The disruption caused by terrorism
4. Compare and contrast the different definitions of terrorism.

2.2 Common Characteristics of Terrorism

Taking the previous definitions and considerations into account, we may conclude that there are at least five crucial components of terrorism. Terrorism is often:

- **Violence or the threat of violence that severely disrupts society.** It is not like a protest, a sit-in, or a strike with a picket line. Instead, it involves illegitimate violent activities that are not sanctioned by law. Terrorism creates many social, political, and economic problems for targeted populations, societies, and the international community. Some of the violent means of terrorism may include the use of guns or detonation of bombs.
- **Performed by an individual, group, or state that espouses an ideology.** Terrorism results from unique ways of looking at the world that attempt to rationalize illegal behavior. For instance, the ideology of terrorism often neglects or subverts negotiation, diplomacy, and democratic processes.
- **Conducted against governments or citizens as targets and shown to an audience.** Certain people may be injured, killed, or otherwise affected by terrorism. But terrorism also intends to spread adverse impact far beyond those who were immediately targeted in attacks. Victims may be produced directly through terrorism or indirectly through publicity of the attacks.
- **Accompanied by fear and coercion.** Terrorism relies on shock, outrage, horror, discouragement, disruption, and intimidation to impel some sort of activity on the part of its victims. This includes both those targeted and the audience observing the attacks.
- **Directed toward the attainment of goals and objectives.** Terrorism may have various aims like promoting political independence, social justice and human rights, environmental protection, religious freedoms or doctrinal dominance, or other objectives (e.g., free reign of drug cartels or the halting of abortions).

Consequently, when actions and events incorporate many or all of these traits, you may have increased confidence that the activity is terrorism. Conversely, if these features are not present, the incident in question cannot be considered in this light. Activities without these characteristics are more likely related to social movements or other types of violence like riots or crime.

CAREER OPPORTUNITY | FBI profiler

Identifying who is likely to commit or who has committed a terrorist attack based on ideological leanings is the responsibility of FBI profilers.

FBI profilers are law enforcement specialists who seek to understand the minds of terrorists and other criminals. These employees are able to evaluate existing facts to form theories about the age, gender, race, and other details about possible perpetrators, thereby helping to apprehend suspects before or after they engage in violent or illegal behavior. FBI profilers are similar to CIA agents in some ways, although FBI profilers generally assess information acquired from within the United States while CIA analysts gather information from abroad.

FBI profilers start off making over $50,000 annually, and salaries expand significantly over time based on experience. The job is therefore coveted. However, it can be somewhat difficult to become an FBI profiler. The position requires a 4-year degree, 10 years of experience in law enforcement, and the passing of many background, physical, and emotional tests.

The good news is that there is also a plethora of similar positions available in municipal, county, and state law enforcement agencies nationwide. In fact, job growth is solid and is anticipated to expand by around 7% over the next decade.

For further information, see **https://www.fbijobs.gov**.

Self-Check

1. Terrorists have a goal that they promote through violent activity. True or false?
2. Only groups are involved in terrorism. True or false?
3. Which of the following is not a common characteristic of terrorism?
    a. An act of disruption or violence
    b. Performed by someone who espouses an ideology
    c. Accompanied with fear
    d. A strike at the company headquarters
4. Restate why terrorists target innocent citizens.

2.3 Types of Terrorism

In spite of the common features described earlier, terrorism may take on unique forms at any given time. Feliks Gross, an expert on terrorism in Russia and Eastern Europe, affirms that terrorism manifests itself in five different ways (Gross, 1990, p. 8). These include mass terror, dynastic assassination, random terror, focused terror, and tactical terror. Each one will be discussed in turn.

- **Mass terror** is terrorism by the government in power against its own citizens. This is a situation in which the ruling regime uses violence or the threat of violence to suppress the opposition and maintain control. The efforts of Saddam Hussein and his Republican Guard in Iraq fall into this category. This leader and his armed forces used poison gas on the Kurds in the northern part of the country and killed or imprisoned anyone who dared to speak out against the dictatorship and military in this nation.

- **Dynastic assassination** is the murder of the head official in the government. Some may assert that the shooting of Abraham Lincoln falls into this category. After shooting the president, Booth shouted "Sic semper tyrannis," a Latin phrase that means "Thus always to tyrants."

- **Random terror** is an attack on large numbers of people wherever they gather (see Figure 2-2). The coordinated attacks on Spanish trains on March 11, 2004, are an example of random terror. These blasts killed nearly 200 people and injured over 2000 others.

- **Focused terror** is terrorism directed toward a specific group of people deemed as the enemy. Some may assert that the Jewish Underground, an organization that opposed German occupation during World War II, practiced focused terror. On December 22, 1942, the group detonated explosives at a café where Nazi officers were dining (Combs 2000, p. 9).

- **Tactical terror** is the use of attacks against the government for revolutionary or other purposes. The 1995 bombing of the Murrah Federal Building in Oklahoma City by Timothy McVeigh is an example of tactical terror due to his concerns about the second amendment.

Besides the five types of terrorism described by Gross, it may be necessary to mention and describe other useful categories. You are probably aware of the frequent terrorist attacks on American troops in Iraq or Afghanistan. These events typically involved a sudden and surprise strike with small arms or bombs and a quick retreat. How would you describe this type of terrorism? Many suggest that this is guerilla warfare. **Guerilla** is a Spanish term for little war, which is an armed protest of occupying forces (Simonsen and Spindlove 2000, p. 35). Guerilla warfare is similar to **asymmetrical warfare,** which implies terrorist attacks on the part of the militarily weak against those who are powerful.

FIGURE 2-2 Car bombings, such as this one in 2017 in the Ukraine, are examples of random terror. *Source: A_Lesik/Shutterstock.*

Typologies of Terrorism

The study of terrorism is replete with various categories of the phenomena. According to Alex Schmid (2023, pp. 5–6), terrorism often falls into one of four types:

1. **Perpetrator-centered typologies:** Regime repressive state terrorism, nonstate terrorism, anarchist terrorism, social-revolutionary left-wing terrorism, racist/xenophobic right-wing terrorism, ethno-nationalist separatist or irredentist terrorism, vigilante revenge terrorism, lone wolf/actor terrorism, organized crime related narco-terrorism, and state-sponsored foreign terrorism.
2. **Methods and tactics-centered terrorism:** insurgent civil war terrorism, warfare interstate war terrorism, nuclear terrorism, suicide terrorism, sexual terrorism, and cyber-terrorism.
3. **Motive-centered typologies:** political terrorism, revolutionary-centered terrorism, religious fundamentalist terrorism, theoterrorism, eco-terrorism, single-issue terrorism, and idiosyncratic (e.g., mental illness–related terrorism).
4. **Location-centered typologies:** domestic (national) terrorism, urban terrorism, transnational terrorism, and international terrorism.

Terrorism has also been categorized in other ways. **Domestic terrorism** is defined as terrorism that occurs within a single country. The release of nerve gas on March 19, 1995, in a subway by the Japanese religious group Aum Shinrikyo is an example of this type of terrorism. **International terrorism**, on the other hand, is terrorism that spans two or more nations. This type of terrorism is initiated by individuals, groups, or the government within one country and targets the people or the leaders of another. The bombing of the USS Cole in Yemen on October 12, 2000, can be considered international terrorism.

Although classifying terrorism as being domestic or international in origin or scope is helpful, it is not always so clear-cut. Sometimes it is difficult to determine where an attack was instigated and the extent of the incident. The bombings of two Russian planes on August 24, 2004, by a Chechen field commander and the Bali bombings in Indonesia on October 12, 2002, by Jemaah Islamiyah are examples of this dilemma. Although these attacks took place within the aforementioned countries, the events included terrorists and victims from many nations around the world.

In The Real World

Classifying the 1993 WTC Bombing

Suppose you were working as a CIA analyst when the World Trade Center was bombed on February 26, 1993. Ramzi Yousef, along with nine other Arab Islamic terrorists, devised a plan to park a van full of urea nitrate (a fertilizer explosive) in the underground garage of the North Tower. The goal of these terrorists was to destroy the foundation of Tower I so that this building would then collapse onto Tower II. Although their intentions ultimately failed, the bomb did kill 6 people and injured more than 1,000. If you had to write an after-action report on this attack, how would you classify it? Is there more than one way to look at this terrorist event?

Thus, there are also numerous complexities relating to each of the aforementioned categories of terrorism. For instance, the bombing of the Federal Murrah Building by Timothy McVeigh was tactical terror, but it could also be described as asymmetrical warfare and domestic terrorism. In contrast, the violence that has been witnessed in Iraq was focused and international in scope and not just a guerilla variant. Therefore, caution should be used when classifying terrorism into a single and distinct alternative.

Self-Check

1. All terrorist attacks manifest themselves in the same fashion. True or false?
2. The shooting of the president is an example of tactical terrorism. True or false?
3. Which type of terrorism is implemented by the government to control citizens?
   a. Focused terror
   b. Mass terror
   c. International terrorism
   d. Random terror
4. List and differentiate the five types of terrorism proposed by Gross.

2.4 Relation of Terrorism to Other Disasters

Terrorism has generally been studied by political scientists, those involved in criminal justice, and scholars of international relations. However, terrorism can also be viewed from the perspective of emergency management. Some scholars, such as Quarantelli (1993), assert that terrorism is a "conflict disaster" and that it is qualitatively different from "consensus disasters" (see Figure 2-3). A **conflict disaster** is an event that involves a riot, violence, or some type of warfare. It often illustrates division within or across societies. A **consensus disaster**, instead, is an event like an earthquake or tornado that brings the community together. After natural disasters, for example, people join efforts to respond to victims and help them recover. Quarantelli asserts that a major difference between these two classifications is that conflict disasters include a person or group that is intentionally trying to inflict harm or distress on others. Alternatively, a consensus disaster is characterized by individuals and groups that join efforts to solve mutual challenges. According to Quarantelli, the first type of disaster divides society, while the other unites it.

Quarantelli's study raises an interesting question about how to classify terrorism. Is terrorism a conflict disaster or a consensus disaster? A case can be made for both arguments. Terrorism is obviously based on social conflict, but it may also lead people to work together to react to the devastation. Regardless of the answer to this question, Gary Webb (2002) provides evidence that social behavior in terrorism is very similar to consensus disasters. As an example, he declares that there is very little panic in most consensus disasters and that people also act rationally even within major events

FIGURE 2-3   9/11 has features of both conflict and consensus disasters. It was initiated by
international terrorists but unified American sentiment. *Source:* © *Getty Images. Reproduced with
permission of Getty Images.*

such as 9/11. Webb relays the fact that police and fire personnel are remarkably reliable in virtually all disasters and that no one abandoned their post after the World
Trade Center was attacked. Professor Webb also states that response operations are
characterized by improvisation in most disasters, and this was certainly the case as
organizations adapted to difficult circumstances in New York City. While his research
recognizes the need for additional studies on terrorism, Webb asserts that the defining features of 9/11 are remarkably similar to consensus disasters.

Research by Lori Peek and Jeannette Sutton (2003) likewise seems to illustrate
that terrorism does not always fall easily into either the conflict or consensus category. These scholars discuss six hypotheses to make their argument:

- **Proposition 1:** Both pro- and antisocial behavior occurred after 9/11. In most
  disasters, people forget prior divisions in society, and they work jointly to reach
  important goals. This behavior was evident on 9/11 since people helped one
  another evacuate the World Trade Center towers and donated blood to help
  potential victims. However, it is true that there was also some intolerance and
  hostility toward Muslims and others of Arab descent after the attacks in 2001.

- **Proposition 2:** People experienced negative consequences owing to 9/11. In
  consensus disasters, many people will not suffer lengthy psychological disturbances or certain health repercussions. Preliminary evidence suggests that this
  was not the case with the attacks on New York City. Many people experienced
  post-traumatic stress and depression, and emergency personnel developed
  respiratory problems due to the dust debris they encountered from the collapsed
  buildings.

- **Proposition 3:** The response to 9/11 was made more difficult due to mounting
  security concerns. During consensus disasters, organizations often struggle to
  accomplish the monumental tasks that they are required to meet. Things were
  even more complicated in New York City. After the attacks on the World Trade

Center, air, water, and road transportation systems were shut down. In addition, control over the disaster area was strictly guarded in case further attacks were to occur. These measures created even more challenges while responding to the disaster.

- **Proposition 4:** Political change was dramatic after 9/11. Most consensus disasters do not often result in massive transformations in society. This was not the case after the 2001 terrorist attacks. New legislation was quickly devised and passed, and the government underwent a massive transformation to deal with homeland security issues. These changes were impressive and even overwhelming in terms of speed and scope.

- **Proposition 5:** Preparing for terrorism involves measures beyond consensus-type disasters. Natural and technological disasters require planning, training, and exercises. Terrorist attacks such as 9/11 introduce a new complication in that it is very difficult to anticipate and predict where the next attack will occur. In this sense, it could be more challenging to prepare for terrorism than consensus-type disasters.

- **Proposition 6:** Long-term change will be a defining feature of the 9/11 attacks. Most disasters do not impact or influence societies in any significant way. The events in New York resulted in substantial financial losses, produced notable levels of unemployment, created disagreement about rebuilding plans, and led to the implementation of new security measures around the nation.

Based on these six propositions, Peek and Sutton therefore believe that terrorism is a complex phenomenon that exhibits features of both conflict and consensus-type disasters.

In The Real World

Responding to the Boston Bombing

After the 2012 Boston Marathon bombing, people from all walks of life worked together in a harmonious fashion to treat the injured and find the perpetrators. Police officers, healthcare professionals, and citizens rushed to the scene and risked their own safety to assist those in need. The injured were assessed based on the severity of their injuries. Wounds were quickly dressed with anything available, including the shirts off people's backs. Hospitals were notified of the event within two minutes so they could prepare for the distribution and treatment of victims. The 264 patients were transported from the scene by 64 ambulances within 22 minutes, and they received expert care from emergency medical technicians, paramedics, nurses, and doctors. In fact, everyone who made it to a hospital survived. Furthermore, within 30 minutes after the bombing, scores and scores of individuals and businesses had shared footage of suspects with the FBI. Local, state, and federal law enforcement initiated a major manhunt, and citizens in the community shared information regarding the whereabouts of the terrorists. The 129-page after-action report by the Massachusetts Emergency Management Agency called the coordinated response a "great success." In spite of a few minor areas for improvement, the Boston Marathon bombing is an excellent example of a consensus-type disaster.

McEntire and his colleagues accept the findings from Webb as well as Peek and Sutton. They state that "managing the threat of terrorism is both similar to and different from the management of other types of disasters" (McEntire et al. 2002, p. 1). On the one hand, those involved in planning for or reacting to terrorism must be able to perform functions that are routine to almost any disaster. This would include communicating with the population, providing emergency medical care, evacuating the affected area, and sheltering impacted residents as needed.

On the other hand, terrorism poses new challenges that have not always been addressed by emergency management personnel in the past. For instance:

- Intelligence gathering is a relatively new and important task for terrorism prevention and has now filtered down to local police departments.
- Planning in homeland security must place a greater amount of emphasis on the roles of the military, the FBI, and public health.
- Response operations incorporate law functions (e.g., investigation) to a greater extent than in the past.
- Recovery may be impossible or delayed if an area has been contaminated by deadly radiation or harmful chemicals.

Thus, the nature of terrorism complicates traditional emergency management operations and makes working harmoniously in homeland security even more imperative. Even though you might generally expect people to behave as they normally do in consensus disasters, we probably do not have sufficient evidence to predict with full certainty how they will react to terrorist events that are unfamiliar (e.g., a large-scale biological attack). More research will be required to improve understanding of terrorist disasters in the future.

Self-Check

1. Terrorism is always exactly like any other type of disaster. True or false?
2. A conflict disaster is often characterized by a riot or some type of violent behavior. True or false?
3. Which of the following is true about the 9/11 terrorist attacks?
    a. There was no indication of antisocial behavior afterward.
    b. It resulted in new laws and significant changes in the government.
    c. Security measures did not impact response operations.
    d. Muslims were not the target of hatred or discrimination.
4. Quarantelli asserts that terrorism is a conflict disaster. Support or critique this proposition.

Summary

In this chapter, you learned that one of the first steps to addressing the probability and consequences of terrorism is to comprehend the nature of this phenomenon. After reviewing numerous definitions of the complicated concept of terrorism, this chapter compared how the various perspectives are both alike and dissimilar. The prevalent features of terrorism were revealed, which will enable you to appraise if terrorism has occurred or is about to take place. Although there are recurring features of terrorism, this should not imply that every situation is similar. This would be a mistake since many different types of terrorist attacks could occur. The chapter also examined the connections between terrorism and other types of disasters. Based on the evidence provided, it appears that terrorism has features of both conflict and consensus-type disasters.

Assess Your Understanding

Understand: What Have You Learned

Go to **www.wiley.com/go/mcentire/homelandsecurity3e** to assess your knowledge of how to identify terrorism.

Summary Questions

1. Definitions of terrorism are all objective and free of value judgments. True or false?
2. Terrorists want to affect only those directly targeted by the attack. True or false?
3. Feliks Gross says that terrorism manifests itself in five different ways: mass terror, tactical terror, focused terror, dynastic assassination, and random terror. True or false?
4. Guerilla warfare is the opposite of asymmetrical warfare. True or false?
5. It can sometimes be difficult to determine whether a terrorist attack is categorized as domestic or international. True or false?
6. Consensus disasters involve some kind of conflict or warfare. True or false?
7. According to Proposition 4 in Peek and Sutton's research, it is not common for extreme political and societal changes to occur after a disaster. True or false?
8. Each of the following agencies has established definitions for terrorism except:
    a. The Department of Defense
    b. The Department of Homeland Security
    c. The Department of Housing and Urban Development
    d. The Department of State
9. The definition of terrorism provided by the Department of Homeland Security emphasizes:
    a. The importance of infrastructure
    b. The illegal nature of terrorism
    c. Terrorism's effects on foreign policy
    d. The role of nonstate actors in the terrorist event
10. Which of the following BEST describes "random terror?"
    a. Terrorism by a government against its citizens
    b. Terrorism against a group of people who are considered the enemy
    c. Terrorism used for revolutionary purposes
    d. A terrorist attack on a large group of people in any gathering place
11. Each of the following is a crucial component of the definition of terrorism except:
    a. An act of disruption, violence, or the threat of violence
    b. Carried out on a large scale by many terrorists
    c. Accompanied by fear and coercion
    d. Directed toward the attainment of goals and objectives

12. Which of the following definitions BEST describes guerilla warfare?
    a. A large military attack
    b. A small, planned military attack on those who are less powerful
    c. An armed protest against occupying forces
    d. An anticipated attack from terrorists
13. Which of the following is not an aspect of a consensus disaster?
    a. It brings the community together.
    b. It often takes the form of a natural disaster.
    c. It is socially disruptive.
    d. It involves a person or group intentionally causing harm to others.

Applying This Chapter

1. What is meant by the adage "one person's terrorist is another person's freedom fighter?" How does this complicate the process of defining or dealing with terrorism?
2. It is suggested that terrorism is directed toward both victims and an audience. How do the victims and target audience, in the view of terrorists, impact the ability of terrorists to reach their goals and objectives?
3. There are many different classifications for terrorist activity. Why is it so difficult to put a single terrorist attack into one specific category?
4. Briefly discuss terrorism in terms of conflict disasters and consensus disasters. What are some examples of how terrorist attacks take on characteristics of both kinds of disasters? Are all terrorist acts the same?

Be a Homeland Security Professional

Defining Terrorism in a Public Document

You have been assigned to write the strategic plan for the police department. Your boss wants you to discuss terrorism. How would you define terrorism? Is it similar to, or different than, other types of illegal behavior? How so?

Writing a Report on Terrorism

As an analyst for the FBI, you have been asked to write a report that explains the common characteristics of terrorism and the different manifestations it has had over time. The report will be given to members of Congress who oversee national security issues. What would you include in your statement?

Assessing Terrorism and Disasters

While attending a conference on homeland security, you observe disagreement among scholars and practitioners regarding terrorism. Some suggest that the consequences are similar to natural and technological disasters, while others state that it has the defining features of conflict events. Could you make a comment that would satisfy both sides of the argument? What would you say?

Key Terms

Asymmetrical warfare   Terrorist attacks on the part of the militarily weak against those who are powerful

Conflict disaster   A socially disruptive and divisive event that involves a riot or some type of warfare

Consensus disaster   A socially disruptive event like an earthquake or tornado that brings the community together

Department of Defense (DOD)   The government agency responsible for the military and its operations

Department of Homeland Security (DHS)   The organization charged with preventing terrorist attacks or reacting effectively to their adverse consequences

Department of State (DOS)   The government agency in charge of diplomatic relationships among nations

Domestic terrorism   Terrorism that occurs within a single country

Dynastic assassination   The murder of the head official in the government

Federal Bureau of Investigation (FBI)   The government agency that enforces U.S. federal law

Focused terror   Terrorism directed toward a specific group of people deemed as the enemy

Guerilla   Spanish term for little war, which is an armed protest of occupying forces

International terrorism   Terrorism that spans two or more nations

Mass terror   Terrorism by the government in power against its own citizens

Random terror   An attack on large numbers of people wherever they gather

Tactical terror   The use of attacks against the government for revolutionary or other purposes

References

Capron, T.A. and S.B. Mizrahi. (2016). *Terrorism and Homeland Security: A Text/Reader.* Thousand Oaks, CA: Sage.

Combs, C.C. (2000). *Terrorism in the Twenty-First Century.* Upper Saddle River, NJ: Prentice Hall.

Department of Homeland Security (2004). *National Response Plan.* DHS: Washington, DC.

Gross, F. (1990). *Political Violence and Terror in Nineteenth and Twentieth Century Russia and Eastern Europe.* New York: Cambridge University Press.

Jenkins, B. M. (1980). The Study of Terrorism: Definitional Problems. Rand Corporation. https://www.rand.org/content/dam/rand/pubs/papers/2006/P6563.pdf. (Accessed September 27, 2023).

Laquer, W. (1999). *The New Terrorism: Fanaticism and the Arms of Mass Destruction.* New York: Oxford University Press.

Martin, G. (2003). *Understanding Terrorism: Challenges, Perspectives and Issues.* Thousand Oaks, CA: Sage Publications.

McElreath, D.H., Doss, D.A., Russo, B., Etter, G., Van Slyke, J., Skinner, J., Corey, M., Jensen III, C.J., Wigginton, Jr., M., and R. Nations. (2021). *Introduction to Homeland Security.* Boca Raton, FL: CRC Press.

McEntire, D.A., Robinson, R.J., and Weber, R.T. (2002). Managing the threat of terrorism. *IQ Rep. 33* (12). Washington, DC: ICMA.

Peek, L.A. and Sutton, J.N. (2003). An exploratory comparison of disasters, riots and terrorist attacks. *Disasters* 27 (4): 319–335.

Quarantelli, E.L. (1993). Community crises: an exploratory comparison of the characteristics and consequences of disasters and riots. *Journal of Contingencies and Crisis Management* 1 (2): 67–78.

Schmid, A. (2023). *Defining Terrorism*. International Centre for Counter-Terrorism. https://www.icct.nl/publication/defining-terrorism (Accessed September 27, 2023).

Schmid, A. (2004). Terrorism—the definitional problem. *Case Western Reserve Journal of International Law* 36 (2): 375–419.

Simonsen, C.E. and Spindlove, J.R. (2000). *Terrorism Today: The Past, The Players, The Future*. Upper Saddle River, NJ: Prentice Hall.

Webb, G.R. (2002). Sociology, disasters, and terrorism: understanding threats of the new millennium. *Sociological Focus* 35 (1): 87–95.

White, J.R. (2002). *Terrorism: An Introduction*. Belmont, CA: Wadsworth.

CHAPTER 3

Recognizing the Causes of Terrorism

Differing Perspectives and the Role of Ideology

DO YOU ALREADY KNOW?

- Why history, foreign affairs, and economic conditions lead to terrorism
- How political variables are associated with terrorist attacks
- If religion and culture cause terrorism
- The role of ideology in terrorism

For additional questions to assess your current knowledge of the causes of terrorism, go to **www.wiley.com/go/mcentire/homelandsecurity3e**

WHAT YOU WILL LEARN

3.1 **The diverse causes of terrorism**

3.2 **How numerous political variables influence the likelihood of attacks**

3.3 **Why terrorism results from religious and cultural beliefs**

3.4 **How ideology impacts the occurrence of terrorism**

WHAT YOU WILL BE ABLE TO DO

- Evaluate explanations about the causes of terrorism
- Differentiate among the various political origins of terrorist attacks
- Predict how terrorism may be influenced by religious values and cultural norms
- Appraise the impact of ideology on the phenomena of terrorism

Introduction

If you are to reduce the likelihood and impact of terrorism, it is necessary to understand its causes so you may take steps to alleviate them. Ian Lesser, an expert on terrorism, asserts that this phenomenon "has systemic origins [or root causes] that can be ameliorated" (1999, p. 127). You should therefore be aware of how historical conflicts, foreign policy, and poverty may aggravate terrorism. Internal political factors, culture, and religion may also influence people to engage in terrorist attacks. While there are many possible causes of terrorism, you must pay special attention to the role of ideology. People's values and beliefs always have a significant bearing on terrorism. The following chapter will help you understand why terrorism occurs so you can do something to reverse or minimize such causes.

3.1 Frequently Mentioned Causes of Terrorism

You've probably wondered what drives people to engage in terrorist attacks—even to the point that they are willing to sacrifice their lives to kill others (Hoffman 2003). It is commonly said that there are perhaps as many causes of terrorism as there are terrorists. Some of the frequent explanations include historical grievances, foreign policy decisions, and poverty (Sultalan 2008). Each of these will be explained in turn.

3.1.1 Historical Grievances

One perspective of why terrorism occurs focuses on the wrongs people feel they have encountered over time (Sizoo, Strijbos, and Glas 2022). The assertion here is that small initial conflicts among different groups of people have been perpetuated and aggravated over time. There is ample evidence to support this claim, and the Middle East provides a strong case in point.

Continual conflicts have occurred between Arabs and the Jewish people in this part of the world over time (see Figure 3-1). The historical narrative in the Bible is full of such struggles, and it records several violent acts among different groups of people. Numbers 25:1, 6–8 reveals how Phineas killed an Israelite man and a Midianite woman who were involved in prohibited sexual relations. In Joshua 10:1–14, we learn that Joshua attacked the city of Hazor, killed all of the inhabitants, burned the buildings, and took all of the remaining livestock.

During the Middle Ages, there were frequent wars in the Middle East as Christians tried to spread their religion during the Crusades. The **Crusades** were wars endorsed by the pope to recapture the Holy Land of Jerusalem from the control of **Muslims**—people who worship Allah, follow The Prophet Muhammad and adhere to the religion of Islam. These wars, which had religious and other motivations, took place between 1100 and 1300 A.D. Because of the intolerance and cruelty exhibited in the Crusades, Christians were opposed by those espousing Islam. Nevertheless, the Christians pushed back against and overtook the Muslims, who lost their empire and leadership in science, art, and trade.

FIGURE 3-1 Many Arabs oppose the establishment of Israel in the Middle East. *Source: mikhail/Shutterstock.*

Throughout history, other events would add to the frustration of those residing in the Middle East. During World War II, the German leader, Adolf Hitler, denounced the Jews in Europe and desired to eradicate those professing this faith. The hatred against this religious group resulted in the **Holocaust**—the extermination of approximately six million Jews by the Nazi regime. Recognizing the plight of the Jewish people under the Nazis, the United Nations established the state of Israel in 1948. This country was created with the purpose of returning more Jews to this part of the world and serve as a safe haven for this persecuted group of people. Unfortunately, this move of compassion intensified a long-standing dispute over territory that is sacred to both Jews and Muslims. Conflict and terrorism have therefore been prevalent features of this part of the world for centuries. Both sides claim self-defense against the enemy aggressor. It seems as if terrorism has been omnipresent in the Middle East because of the events of the past.

3.1.2 U.S. Foreign Policy

Another explanation for terrorism focuses on the foreign policy decisions of the United States (Milton 2012). Under this view, American activities abroad are to blame for the terrorist attacks against us. Many Islamic terrorists have made this explicit in their justifications for their aggression against the United States.

For instance, Osama bin Laden, the leader of Al-Qaeda, said he and his followers were justified in attacking Americans for at least six reasons. He asserted that the United States:

1. **Does not understand the history and desires of Middle Eastern countries.**
   The United States does not recognize the preference of Muslims for **theocracy**, a government run by clerics in the name of God.

2. **Participates in colonialism in the area and is involved in exploitative policies.** The United States has military bases in Saudi Arabia, and bin Laden declared that American reliance on oil has caused meddling in the affairs of sovereign nations.

3. **Supports puppet governments that are repressive regimes.** Saddam Hussein was initially an ally of the United States during the Iran–Iraq war, and he was a brutal leader.

4. **Founded the state of Israel and continues to fund its military capability.** U.S. support of the Israeli apparatus has resulted in the loss of Arab land and is assumed to pose a threat to neighboring countries.

5. **Neglects human rights.** Palestinians assert they have fewer political rights than their Israeli counterparts.

6. **Sends American troops to the Middle East.** This desecrates sacred land and is associated with increased conflict and war.

Some of bin Laden's claims appear to have merit, but others might be considered debatable or even hypocritical. It is true that people in the United States do not fully comprehend the Middle East. It is a very complicated region socially, politically, culturally, and economically. Also, America has often relied heavily on oil from the Middle East. And the United States supported Saddam Hussein as a means to counter the military threat posed by Iran after this country experienced revolution and sanctioned the taking of hostages from the U.S. embassy. At the same time, the intentions of America and other nations to find a safe haven (i.e., the state of Israel) for persecuted Jews after World War II were understandable. The democratic form of the government in America provides more freedom and rights to its citizens than theocratic governments (i.e., women may not have the same political status as men in some Arab nations). In addition, bin Laden was allied with the Taliban, a Sunni government that restricted the rights of women and oppressed the Hazara (Shia) minority.

In spite of these observations, it is certainly possible that U.S. foreign policy may drive some of the terrorist activity around the world. Paul Pillar has examined this issue in detail. He states that many "of the issues underlying ... terrorism are to be found overseas" and that decisions on how to prevent or react to terrorism "must be formulated as an integral part of broader US foreign policy" (2001, p. 10).

3.1.3 Poverty

Another common causal explanation for terrorism is poverty (Krueger 2007). It is painfully evident that destitution is omnipresent around the world. But the lack of economic prosperity is not found equally in all societies. This brings up the notion of **relative poverty**, meaning that some people are less wealthy than their fellow citizens or peers in other countries. However, absolute poverty is even more concerning. **Absolute poverty** implies that people lack so many resources that they cannot even meet basic necessities such as food, clothing, and shelter. In this case, the deficiency and want are so significant that life is full of misery, illness, and even death. Unfortunately, there are millions of people around the world who fall into this latter category. For instance, over one billion people make less than $400 annually. It is possible that this number could grow in the future, particularly in Africa (Weatherby et al. 2000, p. 14).

The causal link between poverty and terrorism has been noted by many individuals including scholars and the leaders of various countries. While this argument has been questioned (Jager 2018), it is reported that "the head of the World Bank even proclaimed that terrorism will not end until poverty is eliminated" (Francis 2002, p. 1). The argument here is that people become so frustrated with their impoverished conditions that they express their aggravation through violent activity including terrorism.

There is evidence that seems to support this viewpoint in certain situations. Alberto Abadie notes that "much of the modern-day transnational terrorism seems to generate from grievances against rich countries" (Lozada 2005, p. 1). A study of the Basques in Spain revealed that the lower economic classes were more likely to engage in political violence than the wealthy (Crenshaw 2010). A 2011 study in Germany revealed that unemployment was related to right-wing extremist crimes (Falk et al. 2001). Many other examples could be given around the world. Poverty can therefore be a significant cause of terrorism. However, it is important to remember that not all poor people engage in terrorist activity. In addition, per capita income and other economic variables are not always associated with terrorism. Research reveals that "protest, violence, and even terrorism can follow either a rising or declining economic tide" (Krueger and Maleckova in Francis 2002, p. 1). Regardless, poverty needs to be addressed, and it is a logical explanation for the occurrence of terrorism.

Self-Check

1. Historical events may lead some people to engage in terrorism. True or false?
2. Bin Laden endorses terrorism because U.S. foreign policy:
    a. Disavows support of puppet governments
    b. Places U.S. troops in the Middle East
    c. Does not support Israel
    d. Benefits poor countries
3. Why would poor people be involved in terrorism?

3.2 Political Causes

Terrorism has an inherent relation to political disagreements (Schmid and Jongman 2005). It may also result from the nature of political systems as well as their functions and structures. Conflict over priorities, the inability of political systems to adapt, the performance of government, and the nature of the structural arrangements of the ruling regime may all impact the possibility of terrorist attacks.

3.2.1 Politics

Political disputes are one of the likely explanations for terrorism. **Politics** has been defined as the authoritative allocation of values and resources in society (Easton 1953). Politics therefore concerns the process of determining who gets what in a community or nation through the creation and enforcement of law. Since there are often conflicting views on values (such as what constitutes "good government"?) and because designing and running an effective government is inherently a political process, disagreements and even conflict may occur.

Early and modern political philosophy reveals numerous suggestions on how to best govern societies (Holmberg and Rothstein 2012). For some, the establishment of good government is based on the acquisition of knowledge, beneficent leaders, the promotion of justice, and the avoidance of corruption. According to the United Nations Economic and Social Commission for Asia and the Pacific (no date), good governance has eight characteristics. These features include participation, consensus orientation, accountability, transparency, responsiveness, effectiveness and efficiency, equity and inclusiveness, and adherence to the rule of law.

While most people would agree with these recommendations, there are other points of significant dispute. For instance, some assert that majority rule should be promoted, while others believe the rights of minorities must be guaranteed. Political equality is espoused by many as a way to ensure that all viewpoints are taken into consideration, yet others assume that some individuals are better suited to shape public policy because of their leadership experience and communication skills. Armed revolution against corrupt leaders is regarded as a plausible method for the foundation of good government by certain groups, but an opposing view is that order and stability are necessary to promote safety and security. Another disagreement centers on the contributions of leaders and citizens. One perspective is that wise elected officials will promote good government. Another school of thought is that it is citizen involvement that makes political decisions legitimate.

The major implication of this explanation is that values often determine what constitutes good government and these differing viewpoints are issues of considerable debate (Macridis and Hulliung 1996). This being the case, it is to be expected that terrorist attacks could possibly occur under any of these perspectives. In other words,

CAREER OPPORTUNITY | Social Science Professor

Understanding the cause of terrorism and other criminal behavior is essential if it is to be prevented or reduced in the future.

A variety of university professors—such as historians, political scientists, and psychologists as well as scholars of international relations, homeland security, emergency management, and criminal justice—study the reasons why individuals and groups engage in violent and illegal activities. These professors also conduct research, publish articles and books, teach courses, and speak at conferences. In addition, professors fulfill other administrative duties (e.g., revising curriculum, attending faculty meetings, advising students, providing consulting services).

University professors in the social sciences often make between $75,000 and $150,000 a year, and because of forthcoming retirements, the job outlook is generally positive (at about 8% over the coming decade).

However, becoming a professor is not easy. This career requires a master's degree at a minimum but more than likely demands a Ph.D. Such degrees are time consuming and expensive, but the education helps professors understand their discipline, improve critical thinking skills, and enhance written and oral communication abilities.

For further information about openings for professor positions, see Adapted from **https://www.jobs.chronicle.com**.

if individuals or groups do not feel that their values and priorities are given attention, then it follows that they could engage in violent activity to increase the chance that their preferences will be sufficiently recognized and implemented (Schmid and Jongman 2005). For instance, if a person is disturbed by environmental degradation and believes that society is not protecting natural resources, then there is a possibility that this individual may act violently against those who cause or profit from such activity. In fact, there have been several cases where extreme environmentalists have burned SUVs on dealers' lots as a way to protest the vehicle's gas mileage (Madigon 2003).

3.2.2 Political Systems

Another thought about what causes terrorism is based on the ability of political systems to meet needs, adapt, and change. A **political system** is a governing body that operates in a self-contained environment (e.g., a national territory). David Easton (1953), one of the most recognized political scientists in the United States, suggests that political systems are composed of inputs (i.e., taxes and demands for service), decision-making processes (i.e., debates about proposed pieces of legislation), and outputs (i.e., policy choices and government programs). Easton also notes that political systems are influenced by feedback—meaning that the inputs, decision-making processes, and outputs can change over time and adjust to new needs and preferences. In Easton's view, this feedback loop is essential for the successful performance of political systems. In other words, if the political system does not adapt to unfolding situations, it could fail see Figure 3-2.

Applying this model to terrorism, it may be argued that terrorism is more likely to occur when a political system ignores the preferences of citizens or when national leadership has lost the capability to govern and in essence has become a "failed" state (Pasagic 2020). An example of disagreements about preferences would be the desire of people to own a gun. If the system is viewed as not protecting citizen rights to possess weapons, then some may become disgruntled and engage in acts of terrorism as a result. The Viper Militia in Arizona is a group that attempted to engage in terrorism due to concerns over weapons rights. Its members plotted to blow up several federal offices in Arizona in 1996. Fortunately, this group was caught before the plans could be implemented (Hoffman 1998, p. 109). A case of a failed state is Syria. A whole

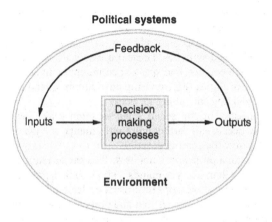

FIGURE 3-2 Without sufficient feedback and adaptation, a political system may be affected by terrorism.

host of social, political, and economic problems led to the collapse of this government. This situation has resulted in many terrorist attacks and a contentious civil war (Bakkour and Sahtout 2023).

3.2.3 Political Functions

A closely related view is that the nature of the performance of government functions may also be regarded as a cause of terrorism. The argument here is that terrorism is likely to take place in a "blocked" society where the government is resistant to adjustment, change, and innovation (Crenshaw 1981, p. 384). Scholarly work in this area has some relation to the research by Easton (1953) and possibly draws from others including Gabriel Almond and James S. Coleman. In their book *The Politics of the Developing Areas*, Almond and Coleman suggest that it is imperative to examine how the government works (1960). In particular, they note that the most important functions of the government pertain to system requirements, inputs, outputs, and maintenance.

- **System requirements** are functions that must be established to maintain the operation of a political system. They include routinization of the way things are done (e.g., elections and policy debates), integration of parts (e.g., political parties working together), and goal attainment (e.g., the satisfying of wants).
- **Input functions** are activities that influence priorities in the political system. These include interest articulation (e.g., agenda setting) and interest aggregation (e.g., formation of political parties).
- **Output functions** are activities emanating from the political system. They incorporate rule making (e.g., policy decisions), rule application (i.e., policy enforcement), and rule adjudication (e.g., court decisions about policy).
- **Maintenance** is the feedback function of the political system. It is composed of socialization (e.g., teaching rules of appropriate political behavior), political recruitment (e.g., the joining of interest groups), communication (e.g., calling your senator or State of the Union speeches), and symbolism (e.g., a parade to increase pride and trust).

Applying this theoretical perspective to terrorism, it is presumed that political systems that do not perform these functions to a certain standard and/or are unwilling or unable to adapt would be more prone to attack. For instance, if an individual does not believe that law adequately protects animals in medical testing and there is no indication of a change in the future, then this person may be more inclined to engage in violence against the government and others who are accused of harming dogs, monkeys, livestock, and so on. In essence, this perspective maintains that the functioning of the government determines if terrorism is likely to occur. A government that successfully performs functions will have fewer terrorist attacks, while a government that is regarded to be ineffective is more likely to be the target of terrorists.

3.2.4 Political Structure

A final political explanation for terrorism is related to the structure of government in society. The argument here is that violent conflict may result if certain groups that are vying for political power are unable to influence decisions because the government is closed to internal or external influence or because the political system is unwilling to resolve the disaffection that is felt by some of its members (Crenshaw 1981, p. 383). Thus, **structure** refers to the organizational relationships within the political system in reference to the ability to acquire and wield power. There are at least four different structures that can be identified:

1. The **group competition model** asserts that interest groups interact with or counteract one another in their attempt to sway government policy. Some might assert that the United States is an example of this particular perspective.

2. The **economic class model** suggests a division of society based on the amount of wealth one possesses (e.g., upper, middle, and lower classes). Depending on your viewpoint, the United States or Latin American countries might also be seen in this light.

3. The **political elite model** is a situation where the leaders are ruling over the masses. Iraqi citizens, under the control of Saddam Hussein and his Republican Guard, would fall under this category.

4. The **corporatist model** stresses the integration of various components of society into the government (e.g., close ties to businesses, churches, and clubs). Germany, during World War II, would be a logical example of this model.

The implication of the political structure argument is that certain forms of government will experience violence in different ways and for different reasons. For instance, if a special interest organization feels that its needs are not being met due to competition from other groups, then it might participate in terrorism. As an example, in 2006, some Muslims in France burned McDonald's restaurants due to a law that would allow the company to fine them for violating corporate policy. In another case, if one economic class is being exploited by another, then terrorism may result. For instance, the Tupamaros, a revolutionary organization in Uruguay, engaged in terrorism to nationalize economic resources and distribute them in such a way as

In The Real World

State Terrorism in Russia

Although the Russian Revolution of 1917 had the goal of promoting economic equality, it soon became apparent that the political change resulted in a dictatorship that centralized political power. Vladimir Lenin promoted the Bolshevik takeover, and his government began to repress those who voiced opposition. Many artists, intellectuals, and clergy soon became targets under the new regime. When Lenin died, Joseph Stalin acquired power as the absolute ruler in what would become the Soviet Union. He was the communist leader from 1929 to 1953. During his tenure, Stalin also ruled with an iron fist. He introduced new laws to crack down on those who dissented. His secret police executed thousands and sent millions to labor camps. It is estimated that he killed 800,000 people. In addition, approximately 1.5 million others died due to the extreme conditions experienced in labor camps. This experience seems to suggest that authoritarian or totalitarian political structures, where an elite group rules over the citizens, may be closely associated with mass terror.

to reduce poverty. Other political structures, including the political elite and corporatist models, may be accompanied by state-sponsored terrorism against its own citizens. Joseph Stalin used violence as a way to maintain his power over the masses in Russia, and Francisco Franco executed many opponents of his fascist regime after the Spanish Civil War. Alternatively, terrorist activity may result in this structure as people rise up against the government. The movie "Valkyrie" depicts how some military leaders in Germany attempted to topple the Nazi regime through bombings and other activities.

Self-Check

1. Because politics is about competing values, it is not associated with terrorism. True or false?
2. Which part of the political system is most likely to aggravate terrorism?
   a. Inputs
   b. Outputs
   c. Feedback loop
   d. Decision making
3. If the government always met the needs of individuals, would terrorism occur?
4. Is terrorism related to political structure? If so, how?

3.3 Cultural and Religious Causes

Other potential explanations for terrorism focus on culture and religion (Kluch and Vaux 2015; Badey 2002). **Culture** is the lifestyle of groups based on their shared history, language, religion, and moral system. **Religion** specifically deals with the beliefs and practices espoused by those sharing a common spiritual faith. Both culture and religion are assumed to be related to terrorist activity, and there is ample literature on these subjects that makes this case.

3.3.1 Cultural Dimensions of Terrorism

One of the most well-known discussions about the impact of culture comes from Samuel Huntington, an American political scientist. In 1993, he wrote an article in *Foreign Affairs* describing how the world had changed when the Cold War ended. He asserted that the conflicts of the past were a result of: (1) kings seeking power, safety, or wealth; (2) nations seeking territory or sovereignty; and (3) governments seeking to promote democracy or communism around the world. Huntington also examined how conflict may change over time. He believed violence of the future would revolve around culture.

Specifically, Huntington declared that there are numerous cultural civilizations around the world (e.g., Western, Confucian, Japanese, Islamic, African, Hindu,

Slavic, and Latin American). Although there are conflicts within these civilizations, he was more concerned about the possibility that some of the world's civilizations may oppose Western culture because the United States is so powerful militarily, economically, and politically. He also suggested that other cultures may counter American values because of their focus on individualism, constitutional government, human rights, and democracy. As a result, Huntington affirms that conflict could result from this "clash of civilizations."

Others, including Benjamin Barber (1996), have followed up on this theme. In his book *Jihad vs. McWorld*, Barber asserted that the tribal values of certain cultural groups (e.g., Islamic fundamentalists who support Jihad) are diametrically opposed to those who espouse the Western capitalist economic system (which is symbolized by the fast-food chain McDonald's). Such differences of opinion are reasons that conflict exists today. In particular, Barber insinuated that the people who oppose globalization "argue the virtues of ancient identities, sometimes in the language of bombs" (Barber 1992, p. 5). He declares that "the aim of many of these small-scale wars is to redraw boundaries, to implode states and re-secure parochial identities. ... War ... [is also] an emblem of identity, an expression of community, and an end in itself" (Barber 1992, p. 5).

Huntington and Barber bring up some intriguing points related to the cultural dimensions of terrorism. It is certainly clear that there are significant disagreements about culture and that acts of violence have emanated because of opposing traditions and viewpoints. As an example, the Basques in Spain created a terrorist organization known as the Euskadi ta Askatasuna. Their desire was political separatism to protect their language and other cultural attributes and preferences.

3.3.2   The Role of Religion in Terrorism

Religion has also been cited as a reason why conflict occurs, and there appears to be a unique relationship between faith and terrorism (Juergensmeyer 1993; Hoffman 1995). Terrorism has occurred by those professing Christian, Jewish, Muslim, and other faiths.

Terrorism in the name of Christianity has taken place many times in history. In Ireland, Catholics and Protestants used "religion to identify people with politics" (White 2002, p. 57). Terrorists on both sides relied on their faith as the means to promote Irish independence or continue national association with the United Kingdom. Terrorism in the name of Christianity has also occurred elsewhere. There have been situations where those professing Christianity have bombed abortion clinics and killed doctors involved in such practices.

Terrorists in the United States have at times intertwined Christianity with racist or anti-government overtones (Stanton 2021). *The Turner Diaries*, written by a white supremacist named William Pierce (1978), discusses the fictitious story of a man who joins a terrorist organization. The man—named Earl Turner—learns about God's alleged plan for a pure race and becomes violent toward others. Sadly, this made-up and disappointing story has had real-life implications. In fact, Timothy McVeigh was influenced by this book (see Figure 3-3). Although it is unclear to what extent he had racist sentiment, McVeigh and other people relied on such teachings to reinforce their hatred toward the government. It is alleged that McVeigh got the idea of how to bomb the Murrah Federal Building in Oklahoma City while reading this work.

FIGURE 3-3   Timothy McVeigh was influenced by *The Turner Diaries*, and this led him to bomb the Murrah Federal Building in Oklahoma City.

According to Jonathan White, groups like Aryan Nations, Posse Comitatus, and the American Institute of Theology have adopted a religious view that claims that whites are superior to other races. Jonathan White states that the fringe Christian Identity movement "is strongly anti-Semitic, claiming that humans originated from two seed lines. Whites are direct descendants from God, while Jews originated from an illicit sexual union between the devil and the first white woman. Under this disturbing viewpoint, nonwhite races are assumed to have evolved from animals and are categorized as sub-humans. Identity Christians believe that biblical covenants apply only to the white race and that Jesus of Nazareth was not a Jew, but the white Israelite son of God" (White 2002, p. 59). It is extremely appalling that people support such attitudes and that these beliefs translate into actions that are discriminatory and even violent.

In The Real World

Racist Attack

On June 17, 2015, a 21-year-old named Dylann Roof opened fire at the Emanuel African Methodist Episcopal Church in Charleston, South Carolina. He killed nine people and wounded one other. This hate crime was racially motivated against black Americans. Roof's website, *The Last Rhodesian*, was discovered just days after the shooting and details his white supremacist views and racist beliefs. In January 2017, Roof pleaded guilty to all charges and was sentenced to life in prison without the possibility of parole.

There are also terrorists who have been associated with the Jewish faith (Pedahzur and Perliger 2011). Rabbi Meir Kahane was an American cleric who was deeply offended by the violence against Jews in Israel. He created the Jewish Defense League as a way to ensure the survival of the Israeli state. His views claim that God made a covenant with Abraham to maintain biblical lands. Kahane's followers were involved in terrorist attacks in the United States in the 1960s. Although he was later assassinated, his son established a new organization (named Kahane Chai) to carry out similar missions. One member of his group killed 12 Muslims in a mosque in 1994.

The threat of religious terrorism has also come from radical Muslim extremists. The vast majority of Muslims are peaceful. However, many of the Islamic terrorists follow and support **Wahhabism**, a very stringent and legalistic religious movement that attempts to ensure the purity of the Muslim faith with no deviations whatsoever. There are thousands of fundamentalists who engage in terrorism under the banner of this doctrine (Cordesman 2017). For instance, Hezbollah means the "Party of God," while the Mujahideen are known as "holy fighters" in Afghanistan. Hamas is a name given for the Islamic Resistance Movement, and it is located in the Gaza Strip and in the West Bank. This group is an offshoot of the Muslim Brotherhood and was founded in the early to mid-1900s to denounce the national borders drawn up by colonial powers in Europe. Groups like Hamas and the Muslim Brotherhood oppose the creation of the Jewish state and desire to unify all Arabs under a pure Islamic government. These groups have been willing to kill anyone who supports peace.

The term "Jihad" is often used to describe terrorism that is influenced by Islamic fundamentalists. **Jihad** has a few different meanings. It is a term to describe an internal struggle to pursue righteousness or could be construed as a war of self-defense. However, terrorists, such as Osama bin Laden, have altered this term to encourage his followers to engage in a "holy war" against nonbelievers or infidels. He stated in his **fatwa** (a religious edict) that it is the individual duty of every Muslim to fight against Israel and murder Americans everywhere. Many individuals and organizations around the world have followed this religious call to arms, and it seems as if related attacks almost occur on a daily basis.

Terrorism is not confined to Christian, Jewish, or Islamic religions, however. Shoko Asahara, the leader of Japan's Aum Shinrikyo (Supreme Truth), stated he had been called as a messenger for God. After returning from a trip to the Himalayas, he prophesied that Armageddon would occur at the end of the 1900s and that only a divinely appointed race would survive in Japan. His religious views mixed Buddhism and Hinduism along with strong anti-American sentiment. On March 20, 1995, his followers used sarin gas to attack commuters on five subways in Tokyo. The attacks

In The Real World

The Emergence of the Irish Republican Army

During the Protestant Reformation, King Henry VIII separated himself from the Catholic Church and created the Church of England. At about the same time, Elizabeth, the king's daughter, sanctioned the creation of the Ulster Plantation in Ireland, which resulted in the displacement of many of the Catholic inhabitants. Thousands of Irish Catholics who lost their land eventually perished because of malnourishment. This, coupled with a new law passed in 1801 (the Act of Union) to join Ireland under the United Kingdom, created serious tensions between Irish Catholics and Protestants from England. Witnessing the plight of their kin, many of the Irish who had migrated to the United States created the Irish Republican Brotherhood (IRB). Although the IRB had the intention of helping poor relatives in their old country, it soon became apparent that Irish independence was desired. The IRB mutated into a revolutionary organization. Its members traveled back to Ireland and later became known as the Irish Republican Army (IRA). The founder, Michael Collins, used religion as a way to recruit members to fight the British Protestants. The IRA was responsible for many bombings in Ireland and England in the 1970s and 1980s. Although various peace accords have been signed between England and the opposition groups, attacks continued for some time. Luckily, terrorism does not have close ties to religion in Ireland today.

---

had the purpose of delaying police investigation into the groups' acquisition of deadly chemical weapons (Hoffman 1998, p. 126).

In each of these cases, terrorists use religion as a tool of violence and claim to act in God's name. Under these circumstances, "believers must identify with a deity and believe they are participating in a struggle to change history. They must also believe in cosmic consequences; that is, the outcome of the struggle will lead to a new relationship between good and evil. When they feel the struggle has reached a critical stage, violence may be endorsed and terrorism may result" (White 2002, p. 52).

Self-Check

1. According to Samuel Huntington, culture will be the major source of conflict in the future. True or false?
2. According to bin Laden, Jihad refers to:
   a. Internal struggle
   b. Holy war
   c. Fatwa
   d. Wahhabism
3. Do all Muslims support terrorist attacks?
4. What are examples of religious terrorism?

3.4 Ideology

Based on the previous discussion, it appears that historical grievances, poverty, foreign policy, political factors, culture, and religion are often regarded to be positively correlated with terrorism. Nevertheless, it is obvious that these variables cannot always be considered as the only motivating "causes" of terrorist phenomena. For instance, the overwhelming majority of poor people do not participate in terrorism. Ineffective political systems do not always produce citizens who engage in terrorism. Not all Muslims attack the United States, even if they disagree with Western foreign policy or culture. It consequently appears that there must be an additional reason as to why terrorism occurs. Gus Martin suggests that while "not all extremists become terrorists, some do cross the line to engage in terrorist violence. For them, terrorism is a calculated strategy. It is a specifically selected method that is used to further their cause" (2003, p. 56). Stated differently, ideology and the use of ideology as a tool are logical explanations for the occurrence of terrorism. In fact, Michael Chertoff has asserted that "Al Qaeda and like-minded organizations are inspired by a malignant ideology, one that is characterized by contempt for human dignity and freedom and a depraved disregard for human life" (2009, p. 22).

So, what is ideology? The word "ideology" is based on the prefix "idea." An **ideology** is consequently a set of beliefs related to values, attitudes, ways of thinking, and goals (Plamenatz 1970). In other words, ideologies are comprehensive theoretical viewpoints that often recommend certain types of political action. For this reason, ideologies are often referred to as "secular religions."

The notion of ideology has roots in early political philosophy from prominent intellectuals such as Socrates, Plato, and Aristotle. However, the term "ideology" did not appear until the late 1700s when a man named Destutt de Tracy attempted to discredit the political institutions in France. He claimed that the king was not divinely appointed and argued that the government did not represent the interests of the people. Destutt de Tracy's ideology opened up the possibility of political change in France. This was one of the first examples of how ideology is used for broad political purposes.

3.4.1 The Nature of Ideologies

Ideologies cover a broad range of subjects and may be related to numerous questions (Macridis and Hulliung 1996):

- What is truth, and how can it be pursued?
- What makes political authority or a political system legitimate?
- Should individual rights be promoted or should the majority rule?
- Is it important to espouse political freedoms or economic equality?
- Is the acquisition of wealth justified, or is protecting the environment more vital?
- Should you follow the laws of men or God's commandments?

As can be seen, ideologies therefore serve several purposes (Macridis and Hulliung 1996). They help promote understanding and simplify a complex world. They facilitate communication among individuals and generate identity and emotional fulfillment in groups. They are also used as tools by leaders and provide guidelines for the behavior of followers. An ideology not only endorses certain viewpoints (while rejecting others) but also helps to mobilize people in the accomplishment of goals and priorities. Ideologies can therefore be very powerful motivators in the lives of their adherents.

Although all ideologies share many similarities, they differ in dramatic ways. Ideologies may discourage change, promote revolution, or accept incremental reform (Macridis and Hulliung 1996). Ideologies may unite some groups and cause divisions with others. Roy Macridis and Mark Hulliung (1996, pp. 16–17) suggest that ideologies may be assessed on five bases:

1. **Scope**. What subjects does the ideology cover? Does it focus on human rights or environmental issues?

2. **Coherence**. Does the internal logic make sense? In other words, does the ideology contradict itself?

3. **Pervasiveness**. How long has it been in existence? Is it a relatively new ideology like feminism, or has it been prevalent for centuries?

4. **Extensiveness**. How many people share that particular belief? What is the number of people that support it?

5. **Intensiveness**. What is the strength of attachment? Are people casually associated with the ideology, or are they fully committed to their particular beliefs?

3.4.2 Ideological Dimensions of Terrorism

According to Mostafa Rejai, an Indian scholar, there are five dimensions that are integral to any ideology (1991). Rejai states that all ideologies have a **cognitive dimension**. This refers to the knowledge and beliefs of the ideology. Ideologies are also related to **affect dimension**—specific feelings or emotions that are generated in conjunction with beliefs. The **valuation dimension** deals with the norms and judgments of the ideology. The **programs dimension** conjures up the plans and actions to support goals. The **social base dimension** refers to the individuals or groups that espouse the ideology.

These dimensions are useful for understanding the ideologies associated with terrorism. Terrorists favoring radical Islam have unique attitudes about the world (e.g., they see the United States as a problem and assert that its influence needs to be countered or eliminated). Such groups have strong sentiments about a particular issue (e.g., a desire to retain a pure version of Islam). They are passionate about values and preferences (e.g., they are willing to injure others or even kill themselves if that is required). Radical Islamic terrorists also develop methods to accomplish their objectives (e.g., training and operational planning are needed to carry out attacks). Organizations are clearly identified to support this ideology (e.g., Al-Qaeda and ISIS members endorse violence, while most Muslims prefer peace).

Of course, it is important to recognize that there is no shortage of ideologies that may be related to terrorism (Miller 2017). Some conservative ideologies have been associated with political regimes that engage in violence against the masses to maintain the power of the ruling elite. In contrast, a liberal ideology may endorse terrorism as a means to expand the rights of individuals and citizens. Some terrorists also favor nationalistic movements, thereby attempting to promote a group with a particular culture, ethnicity, or language. Terrorism may have a relation to Marxist ideology. In this case, some people denounce capitalism and see a violent revolution as a way to promote economic equality. Other ideologies, like fascism and Nazism, sanction terrorism to expand a nation or promote racial superiority. At times, extreme conservationists engage in terrorism to avert global environmental degradation. Religious beliefs may be used by some people to endorse terrorism and counter the advances of secular society. Others, like anarchists and post-modernists, question all types of authority structures and accept violence as a way to protect individual freedoms. When you understand these ideological sources of conflict, you are in a better position to start dealing with the causes and consequences of terrorism.

In The Real World

Terrorist Ideology and Abortions

People have at times engaged in terrorist activities to halt abortions in the United States (Wirken et al. 2023). The ideology of these individuals and groups has all of the elements identified by Mostafa Rejai. The cognitive dimension of antiabortionists is their belief that abortion is wrong and that this practice has increased dramatically since it was legalized. A feeling of deep sorrow for the loss of innocent life is the affect dimension of this ideology. The preference to protect unborn children is the value dimension. With this in mind, some terrorists have relied on a program of violence against doctors and their clinics to halt abortions. The social base of this ideology is composed of certain fundamentalist Christians. While many of those adhering to Christianity oppose abortions, they are not willing to kill others to stop them from participating in this practice. Being able to detect which people are willing to participate in terrorism is one reason that it is necessary to have a solid understanding of how ideology may be used to engage in terrorist activity.

Self-Check

1. Ideology is not related to beliefs, ideas, or attitudes. True or false?
2. Marxism has been used as an ideology to support terrorism. True or false?
3. Which is not one of the five dimensions of ideologies suggested by Mostafa Rejai?
   a. Program
   b. Cognitive
   c. Affect
   d. Coherence
4. Think of the terrorist attacks that have occurred in recent history. Can you identify the ideology of the responsible terrorists in each case?

Summary

If you are to reduce the probability of terrorism, it is imperative that you understand what motivates people to participate in this type of behavior. In this chapter, you discovered that historical conflicts, mistakes in foreign policy, and extreme levels of poverty may impel some to engage in terrorist attacks. It was also illustrated that political systems, political functions, and political structures are often associated with the cause of terrorist activity. Diverse cultures and extreme religious beliefs may likewise be used by terrorists to justify violent or disruptive activity. While each of these variables may influence terrorist behavior, it is likely that ideology is a better predictor of terrorism. People's beliefs seem to play a large role in terrorism. Once you recognize this fact, you can begin taking steps to identify terrorists and implement other measures to prevent future attacks.

Assess Your Understanding

Understand: What Have You Learned

Go to **www.wiley.com/go/mcentire/homelandsecurity3e** to assess your knowledge of the causes of terrorism.

Summary Questions

1. There have been continual conflicts in the Middle East. True or false?
2. Absolute poverty implies that everyone is poor or lacks the desired resources. True or false?
3. There is considerable disagreement about what constitutes good government. True or false?
4. The economic class model is not related to terrorism. True or false?
5. Language is an aspect of culture but not moral systems. True or false?
6. Terrorism has only been undertaken by those professing the Islamic faith. True or false?
7. The intensiveness of an ideology may have a direct bearing on terrorism. True or false?
8. The wars committed by Christians against the Muslims were known as:
    a. Crusades
    b. Wahhabism
    c. Fatwas
    d. Holocaust
9. An environment that has inputs, decision processes, outputs, and feedback is known as a(n):
    a. Maintenance
    b. Structure
    c. Political system
    d. Corporatist model
10. What is the religious movement that adheres to a strict interpretation of Islam?
    a. Fatwa
    b. Wahhabism
    c. Marxism
    d. *The Turner Diaries*

11. The cognitive dimension of an ideology refers to:
    a. Feelings
    b. Norms and judgments
    c. Plans and actions
    d. Knowledge and beliefs

12. According to bin Laden, Al-Qaeda is justified in attacking America because the United States:
    a. Supports Israel
    b. Distrusts Israel
    c. Distrusts puppet governments
    d. Is opposed to colonialism

Applying This Chapter

1. You are the public information officer for the Department of Homeland Security. The media wants to know why terrorism occurs. List at least five causes of terrorism and evaluate which explanation makes the most sense.

2. If you were in charge of U.S. foreign policy, what causes of terrorism should you be aware of?

3. You are a scholar with expertise in terrorism. While appearing on the news show *60 Minutes*, you are questioned about the political systems and terrorism. Diagram the components of a political system and explain how the feedback loop may or may not relate to terrorism.

4. As an FBI analyst, your job is to identify potential terrorists. Predict why some people may use Christian, Jewish, and Islamic faiths to justify terrorism.

5. While discussing terrorism, a friend questions your knowledge about the characteristics of ideologies. Contrast the difference between pervasiveness and extensiveness in terrorist ideologies.

Be a Homeland Security Professional

Briefing the President

You are a political advisor with expertise in homeland security. The president has asked you to brief him/her on the causes of terrorism. What would you say? What are the differing perspectives on this matter? Which one(s) is(are) most important?

Promoting Cultural Understanding

You are the Secretary of State, and it is your job to improve relations among groups and nations internationally. An important aspect of your position is to foster cultural understanding. Explain how this might ease the occurrence of terrorism around the world. Do you think it would be possible to eliminate culture as a cause of terrorism? Why or why not?

Assignment: Understanding Ideology

Write a paper about the impact of ideology on terrorism. Be sure to discuss what an ideology is, what the common characteristics of ideologies are, and the types of ideologies that may lead to violence.

Key Terms

Absolute poverty   A situation in which people lack so many resources that they cannot even meet basic necessities such as food, clothing, and shelter

Affect dimension   Specific feelings or emotions that are generated in conjunction with an ideology

Cognitive dimension   Refers to the knowledge and beliefs of the ideology

Coherence   The internal logic of an ideology

Corporatist model   A model that stresses the integration of various components of society into the state government (e.g., close ties to business, churches, and clubs)

Crusades   Wars endorsed by the pope to recapture the Holy Land of Jerusalem from the control of Muslims

Culture   The lifestyle of groups, including their shared history, language, religion, and moral system

Economic class model   A model that suggests a division of society based on the amount of wealth one possesses (e.g., bourgeoisie, middle class, proletariat)

Extensiveness   How many people share a particular ideology

Fatwa   A religious edict

Group competition model   A model of politics that asserts that interest groups interact with or counteract one another in their attempt to sway government policy

Holocaust   The extermination of approximately six million Jews by the Nazi regime during World War II

Ideology   A set of beliefs related to values, attitudes, ways of thinking, and goals

Input functions   Activities that influence priorities in the political system

Intensiveness   The strength of attachment to an ideology

Jihad   An internal struggle to pursue righteousness or a war of self-defense but has been used by terrorists to denote an offensive attack

Maintenance   The feedback function of the political system

Muslims   Those following Prophet Muhammad and adhering to the religion of Islam

Output functions   Activities emanating from the political system

Pervasiveness   How long an ideology has been in existence

Political elite model   A model in which the leaders are ruling over the masses

Political system   A government that operates in a self-contained environment (e.g., a national territory)

Politics   Refers to the authoritative allocation of values in society

Programs dimension   The plans and actions to support goals

Relative poverty   Situation in which people are less wealthy than their fellow citizens

Religion   The beliefs and practices espoused by those sharing a common spiritual faith

Scope   The subjects covered by an ideology

Social base dimension   The individuals or groups that espouse an ideology

Structure   Refers to the organizational relationships within a political system in reference to the ability to acquire and wield power

System requirements   Functions that must be performed to maintain the operation of a political system

Theocracy   A government run by clerics in the name of God

Valuation dimension   The norms and judgments of an ideology

Wahhabism   A very stringent and legalistic religious movement that attempts to ensure the purity of the Muslim faith with no deviations whatsoever

References

Almond, G.A. and Coleman, J.S. (1960). *The Politics of the Developing Areas*. Princeton, NJ: Princeton University Press.

Badey, T.J. (2002). The role of religion in international terrorism. *Sociological Focus* 35 (1): 81–86.

Bakkour, S. and R. Sahtout. (2023). The dimensions and attributes of state failure in Syria. Journal of Balkan and Near Easter Studies. DOI:10.1080/19448953.2023.2167337 (Accessed September 27, 2023).

Barber, B. (1992). Jihad vs. McWorld. *The Atlantic* (March). http://www. theatlatntic.com/doc/print/199203/barber (accessed 21 September 2017).

Barber, B. (1996). *Jihad vs. McWorld: How Globalism and Tribalism Are Reshaping the World*. New York: Ballantine Books.

Chertoff, M. (2009). *Homeland Security*. Philadelphia, PA: University of Pennsylvania Press.

Cordesman, A.H. (2017). Islam and the Patterns in Terrorism and Violent Extremism. Center for Strategic Studies. **https://www.csis.org/analysis/islam-and-patterns-terrorism-and-violent-extremism** (Accessed September 27, 2023).

Crenshaw, M. (2010). *Terrorism in Context*. State College, PA: Penn State Press.

Crenshaw, M. (1981). The causes of terrorism. *Comparative Politics* 13 (4): 379–399.

Easton, D. (1953). *The Political System*. New York: Alfred P. Knopf.

Falk, A., Khun, A., and Zweimuller, J. (2001). Unemployment and right-wing extremist crime. *The Scandinavian Journal of Economics* 113 (2): 260–285.

Francis, D.R. (2002). Poverty and low education don't cause terrorism. *The NBER Digest* (September), pp. 1–2.

Hoffman, B. (1995). Holy terror: the implications of terrorism motivated by a religious imperative. *Studies in Conflict and Terrorism* 18: 271–284.

Hoffman, B. (2003). The logic of suicide terrorism. *The Atlantic Monthly*. **https://www.rand.org/pubs/reprings/RP1187.html** (Accessed September 28, 2023).

Hoffman, B. (1998). *Inside Terrorism*. New York: Columbia University Press.

Holmberg, S. and B. Rothstein. (2012). *Good Government: The Relevance of Political Science*. Cheltenham, UK: Edward Elgar.

Huntington, S.P. (1993). The clash of civilizations. *Foreign Affairs* 72 (3): 22–49.

Jager, A. (2018). Does poverty cause terrorism? International Institute for Counter-Terrorism. **https://www.ict.org.il/images/Does%20Poverty%20Cause%20Terrorism.pdf** (Accessed September 27, 2023).

Juergensmeyer, M. (1993). *The New Cold War? Religious Nationalism Confronts the Secular State*. Berkeley, CA: University of California Press.

Kluch, S.P. and A. Vaux. (2015). Culture and terrorism: The role of cultural factors in worldwide terrorism (1970-2013). *Terrorism and Political Violence*. 29(2): 323–341.

Krueger, A.B. (2007). *What Makes a Terrorist?* Princeton, NJ: Princeton University Press.

Lesser, I.O. (1999). Countering the new terrorism: implications for strategy. In: *Countering the New Terrorism* (ed. I.O.Lesser, B.Hoffman, J.Arguilla, et al.). Santa Monica, CA: Rand Corporation.

Lozada, C. (2005). Does poverty cause terrorism? *The NBER Digest* (May), p. 1.

Madigon, N. (2003). Cries of activism and terrorism in SUV torching. **https://www.nytimes.com/2003/08/31/us/cries-of-activism-and-terrorism-in-suv-torching.html** (Accessed September 27, 2023).

Martin, G. (2003). *Understanding Terrorism: Challenges, Perspectives, and Issues*. Thousand Oaks, CA: Sage Publications.

Macridis, R.C. and Hulliung, M.L. (1996). *Contemporary Political Ideologies: Movements and Regimes*. New York: HarperCollins.

Miller, E. (2017). Ideological motivations of terrorism in the United States, 1970–2016. National Consortium for the Study of Terrorism and Responses to Terrorism. **https:// www.start.umd.edu/pubs/START_IdeologicalMotivationsOfTerrorismInUS_ Nov2017.pdf** (Accessed September 27, 2023).

Milton, D.J. (2012). Foreign Policy and TransnationalTerrorism. Dissertation. Florida State University. **https://diginole.lib.fsu.edu/islandora/object/fsu:183018/datastream/ PDF/view** (Accessed September 27, 2023).

Pasagic, A. (2020). Failed states and terrorism: Justifiability of transnational interventions from a counterterrorism perspective. *Perspectives on Terrorism* 14 (3): 19–28.

Pedahzur, A. and A. Perliger. (2011). *Jewish Terrorism in Israel.* New York: Columbia University Press.

Pierce, W. and (aka Andrew MacDonald) (1978). *The Turner Diaries.* Hillsboro, WV: National Vanguard Books.

Pillar, P.R. (2001). *Terrorism and US Foreign Policy.* Washington, DC: Brookings Institution Press.

Plamenatz, J. (1970). *Ideology.* New York: Praeger.

Rejai, M. (1991). *Political Ideologies: A Comparative Approach.* New York: M.E. Sharpe.

Schmid, A.P. and A.J. Jongman. (2005). *Political Terrorism: A New Guide to Actors, Authors, Concepts, Data Bases, Theories, and Literature.* New York: Routledge.

Sizoo, B, Strijbos, D., and G. Glas. (2022). Grievance-fueled violence can be better understood using an enactive approach. *Frontiers in Psychology.* Doi:10.3389/fpsyg.2022.997121 (Accessed September 27, 2023).

Stanton, Z. (2021). It's time to talk about violent Christian extremism. *Politico Magazine.* February 4. **https://www.politico.com/news/magazine/2021/02/04/qanon-christian- extremism-nationalism-violence-466034** (Accessed September 27, 2023).

Sultalan, Z. (2008). The causes of terrorism. *Organizational and Psychological Aspects of Terrorism* 3 (1): 1–11.

United Nations Economic and Social Commission for Asia and the Pacific. (no date). What is Good Governance? **https://www.unescap.org/sites/default/files/good-governance. pdf** (Accessed September 27, 2023).

Weatherby, J.N., Cuikshanks, R.L., Evans, E.B. Jr. et al. (2000). *The Other World: Issues and Politics of the Developing World.* New York: Longman.

White, J.R. (2002). *Terrorism: An Introduction.* Belmont, CA: Wadsworth.

Wirken, B., Barten, D., De Cauwer, H., Mortelmans, L., Tin, D., and G. Ciottone. (2023). Terrorist attacks against health care targets that provide abortion services. *Prehospital and Disaster Medicine* 38 (3): 409–414.

CHAPTER 4

Comprehending Terrorists and their Behavior

Who They Are and What They Do

DO YOU ALREADY KNOW?

- How to define terrorism based on intentions
- The common attitudes and personality traits of terrorists
- Similarities and differences among terrorist organizations
- The tactics used by terrorists

For additional questions to assess your current knowledge of terrorists and their behavior, go to **www.wiley.com/go/mcentire/homelandsecurity3e**

WHAT YOU WILL LEARN

4.1 Examples of individuals, groups, and states involved in terrorism

4.2 The cultural and personal traits of terrorists

4.3 How terrorist organizations are similar and unique

4.4 The ways terrorists operate

WHAT YOU WILL BE ABLE TO DO

- Classify terrorists based on their intentions
- Critique stereotypical views about terrorists
- Predict common terrorist behaviors
- Anticipate how terrorists plan and carry out attacks

Introduction

If you are to increase your ability to deal with the threat and consequences of terrorism, it will be imperative that you comprehend who terrorists are and how they operate (Yayla et al. 2007). In this chapter, you will examine terrorists—whether they are acting alone or in conjunction with others. You will be able to categorize terrorists based on their goals and personality traits. In addition, you learn how to evaluate the similarities and differences among the individuals and organizations that are involved in terrorism. By assessing the behavior of terrorists, you will be able to recognize how they recruit and train members, support their activities financially, and plan and carry out attacks. Thus, the following chapter will enable you to understand the characteristics of terrorists are and how you can counter what they do.

4.1 Terrorists and Terrorist Organizations

Who is a terrorist? What groups are involved in terrorism? Are nations also engaged in terrorism? These are important questions that must be addressed by those working in homeland security. As will be seen, some terrorists are well known, while others are not. Certain terrorists are new to the scene, and others have been engaged in violence for years and decades. Specific proponents of terrorism have given up their violent activities, while different organizations remain heavily involved at the current time. Numerous examples of terrorists can be given, and the list seems to grow each day. Terrorists often have many similarities, but there are significant differences as well.

Anders Behring Breivik is one example of a terrorist. He espoused far-right ideology and conducted one of the worst terrorist attacks in Norway. He decried Islam, criticized feminism, and rejected the cultural Marxism that was being witnessed in Europe. To enforce his ideology, Breivik detonated a bomb in a van, which killed eight people on July 22, 2011. He then traveled to a summer camp on the island of Utøya where he shot and murdered another 69 individuals.

Other examples of individual terrorists include Richard Baumhammers (a neo-Nazi who attacked people in Pennsylvania in April 2000) and Ramzi Yousef (the ringleader of the World Trade Center [WTC] bombing in 1993). **Theodore "Ted" Kaczynski** also comes to mind when discussing an individual terrorist. Known as the Unabomber (because he attacked universities and major airlines), he opposed technology and even wrote a manifesto decrying advances in this area. His attacks were aimed mainly at university professors and corporate leaders. He was responsible for at least 15 bombings around the United States that killed 3 people and injured 22 others. The individuals who act alone are sometimes described as **lone-wolf terrorists**. They are often the most difficult terrorists to identify and capture because there is no communication with others so people are not aware of their intentions.

In The Real World

Excerpt of the Unabomber Manifesto

Theodore "Ted" Kaczynski wrote a manifesto to share his concerns about technology. He threatened attacks to reverse the modern advances he decried. Here are a few of his observations:

1. The Industrial Revolution and its consequences have been a disaster for the human race. They have greatly increased the life expectancy of those of us who live in "advanced" countries, but they have destabilized society, have made life unfulfilling, have subjected human beings to indignities, have led to widespread psychological suffering (in the Third World to physical suffering as well), and have inflicted severe damage on the natural world. The continued development of technology will worsen the situation. It will certainly subject human beings to greater indignities and inflict greater damage on the natural world, it will probably lead to greater social disruption and psychological suffering, and it may lead to increased physical suffering even in "advanced" countries.

2. The industrial-technological system may survive or break down. If it survives, it may eventually achieve a low level of physical and psychological suffering but only after passing through a long and very painful period of adjustment and only at the cost of permanently reducing human beings and many other living organisms to engineered products and mere cogs in the social machine. Furthermore, if the system survives, the consequences will be inevitable: there is no way of reforming or modifying the system to prevent it from depriving people of dignity and autonomy.

3. If the system breaks down, the consequences will still be very painful. But the bigger the system grows, the more disastrous the results of its breakdown will be, so if it is to break down it had best break down sooner rather than later.

4. We therefore advocate a revolution against the industrial system. This revolution may or may not make use of violence: it may be sudden or it may be a relatively gradual process spanning a few decades. We can't predict any of that. But we do outline in a very general way the measures that those who hate the industrial system should take to prepare the way for a revolution against that form of society. This is not to be a POLITICAL revolution. Its object will be to overthrow not governments but the economic and technological basis of the present society.

Besides individual terrorists, there are literally scores and scores of terrorist groups and organizations around the world. Examples include Abu Nidal, the Armed Forces of National Liberation, the Democratic Front for the Liberation of Palestine, the Earth Liberation Front, the German Red Army Faction, Hamas, the Irish National Liberation Army, the Liberation Tigers of Tamil Eelam, and the Revolutionary Armed Forces of Colombia. There are literally too many terrorist groups to mention, and new organizations appear continually.

The **Japanese Red Army** (JRA) was one such terrorist organization that emerged in the 1960s. Its members protested the presence of the U.S. military in Japan after World War II, disapproved of the Vietnam War, and rejected capitalism. The first operation of the JRA was in 1970 when it hijacked an airplane headed to North Korea. This extreme left-wing organization has been involved in attacks in the Middle East, Italy, and Japan, although its activities have disappeared since 2001.

Abu Sayyaf is another terrorist organization. Abu Sayyaf is a jihadist separatist group in the Philippines that desires an independent Islamic state in Mindanao. As a splinter group of the Moro National Liberation Front, it opposes any type of colonialism or foreign involvement in the Philippines. Abu Sayyaf has been involved in kidnappings and has used its hostages to acquire lucrative ransoms from family members or large corporations that are associated with their captured prisoners. Its membership has declined dramatically from about 1,200 in 2000 to about 20 members today.

In addition to the various groups involved in violent activity, there are many states that have participated in terrorism against their own people or have supported terrorist groups that attack other nations. Individual country reports about these threats can be accessed through the U.S. Department of State (2016).

Syria is a specific nation that has actively terrorized those who opposed this corrupt regime. In an attempt to maintain power, the al-Assad regime was involved in major episodes of ethnic cleansing within and around its borders. Its counterinsurgency operations included intense bombings of entire towns and key economic and medical facilities to weaken opposition parties or instill fear in the population. Various terrorist groups—including Al-Qaeda, ISIS, and Hezbollah—have also fought one another as they have vied for power during the Syrian Civil War since it started in 2011.

Libya is a country in Africa that was sympathetic to the Palestinian cause in Israel in the 1970s and 1980s. Under the direction of Muammar Qadhafi, this nation was involved in the 1986 La Belle discotheque bombing in Berlin. It was also responsible for the Pan Am Flight 103 bombing over Lockerbie, Scotland, in 1988. This particular attack killed 270 people. Because of sanctions imposed by the United Nations and the threat of retaliation from the United States after 9/11, Libya dramatically reduced its involvement in terrorism. It was subsequently taken off the list of nations supporting terrorism.

States have sponsored terrorism in other ways. For instance, the Sudan harbored Osama bin Laden until strong international pressure was put on this country, which resulted in his expulsion in 1996. At this point, bin Laden traveled to Afghanistan and received support from the Taliban.

While many states have provided safe haven, funding, arms, and diplomatic assistance to terrorists in the past, **Iran** is the state in the Middle East that has a long-standing history of participation in terrorism. After it underwent a revolution sponsored by Ayatollah Khomeini, this country increasingly denounced the United States and promoted anti-Western propaganda. Since this time, it has supported two terrorist organizations—namely, Hamas and Hezbollah. In addition, the Shia government in Iran had a role in attacks on U.S. and French embassies and on the Marine barracks in Beirut during the 1980s. Evidence suggests that Iran has provided and continues to provide missiles, improvised explosive devices, and other weapons to terrorists to destabilize the situation in Iraq or to eliminate the state of Israel. It is also highly likely that Iran has been the sponsor of over 100 attacks against US troops in the Middle East since the October 2023 eruption of conflict between Hamas and Israel. Both the United States and the United Nations are concerned that Iran is currently trying to develop nuclear weapons.

Terrorists—whether individuals, groups, or states—may have unique forms of organization and hierarchy. At times, terrorists may be under the direction of a single individual. For instance, many terrorists in Iraq support the ideology and vision promoted by **Abu Musab al-Zarqawi**. Al-Zarqawi was a Sunni terrorist who was responsible for many atrocities in Iraq, including the beheading of an American businessman named Nicholas Berg. Terrorist groups may also have a central headquarters. As an example, **Al-Qaeda** means the "base," a term that refers to the location from which the Afghan war against the Soviet Union was coordinated. At other times, terrorists have **cells** (or branches with members) around the world. It is believed that Khifa had units in New York, Chicago, Pittsburgh, and Tucson at one time. In contrast, Jemaah Islamiyah has a presence in Malaysia, Singapore, and Indonesia.

Terrorist organizations also interact one with another. The Red Army Faction was trained by other terrorists in the Middle East. Al-Qaeda has also consulted with Jemaah Islamiyah. Terrorists are increasingly decentralized and transform themselves to avoid detection. At times it is hard to tell exactly who belongs to certain terrorist organizations. There are also many splinter groups with similar ideologies. Both the Irish Republican Army (IRA) and Palestinian Liberation Organization (PLO) have fragmented over time, especially when disagreements arose regarding the organization's stance (e.g., hard line or willingness to negotiate and compromise).

4.1.1 Terrorist Classification

Regardless of their individual attributes or organizational structure, terrorists may be classified in one of three ways. Frederick Hacker (1976), a doctor who later became an expert in hostage negotiation, suggests that terrorists may be labeled as criminals, crusaders, or crazies. A terrorist who is a **criminal** seeks personal gain. Abu Sayyaf is an example of a criminal terrorist organization. This organization seems to be more interested in making money through extortion and less motivated by its initial founding goals of national independence. A **crusader** is a terrorist who promotes lofty moral goals (see Figure 4-1). Examples of this type of terrorist include left-wing groups seeking economic or political equality, right-wing groups desiring to limit government interference in their lives, and environmental groups attempting to protect the earth's resources. Nationalist groups advocating for separatism or religious groups trying to promote their beliefs and impose them on others are also examples of this category. A terrorist who is **crazy** is regarded to be psychologically disturbed. Ted Kaczynski might be regarded as falling into this type of terrorist since the reasoning for his attacks is not logical to others. Of course, it is important to recognize that some terrorists may exhibit elements of all three classifications. It is not always easy to distinguish among criminals, crusaders, and crazies. "The categories are not mutually exclusive; any terrorist group could contain a variety of these ... types" (White 2002, p. 25).

In The Real World

Saddam Hussein's Al-Anfal Attacks on the Kurds

When Saddam Hussein reigned with his Ba'ath Party in Iraq, he implemented state terrorism to rid his nation of the Kurdish people (Martin 2003, p. 104). He dropped chemical weapons on villages in northern Iraq to halt their desire for political separatism and autonomy. The mustard gas and nerve agents produced a yellow cloud that smelled like onions. When people inhaled the smoke, they would fall to the ground, and blood would spew from their mouths. Between 50,000 and 100,000 died, and 2.5 million Kurds were displaced as a result. This series of attacks, which occurred in 1988, was named Al-Anfal. The attack is reported to have been in reference to revelations given by Prophet Mohammed after his first great victory. The scripture advised the prophet to kill unbelievers when necessary. It is an example of a crusader-type attitude that is common among many terrorists today.

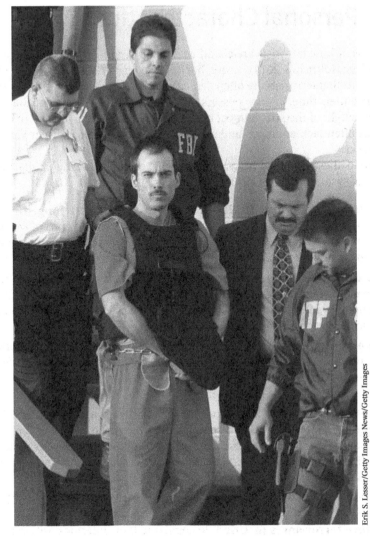

Erik S. Lesser/Getty Images News/Getty Images

FIGURE 4-1   Some individuals, like Eric Rudolph pictured here, use terrorist tactics to achieve moral objectives.

Self-Check

1. Terrorists always act alone. True or false?
2. Some terrorists are described as crusaders since they promote broad goals to change society or the world. True or false?
3. Which country has been the largest supporter of terrorist attacks in Iraq?
    a. Sudan
    b. Syria
    c. Iran
    d. Sri Lanka
4. How can you classify terrorists?

4.2   Personal Characteristics

Regardless of how terrorists are classified, it is apparent that they share remarkable similarities (Holtmann 2014). In his book, *Language of Violence*, Edgar O'Balance (1979) states that terrorists are often dedicated, brave, and stoic. Terrorists are willing to sacrifice their time, energy, possessions, and lives for what they believe in. They are calm in light of the possibility of being subjected to prison, torture, or even death. Terrorists often lack emotions and show no pity or remorse for their victims or for their attacks.

Beyond these characteristics, terrorists have other striking similarities across their belief systems (Combs 2000, p. 39). Terrorists:

1. **See the world simplistically in terms of right and wrong.** They may not accept the complexity of issues or recognize the pros and cons of certain points of view. For instance, they might have a negative perception of capitalism and fail to acknowledge any possible benefits from this type of economic system.

2. **Are disturbed by the current situation.** They are disappointed by existing societal problems and consequently identify what should be changed. As an example, Islamic fundamentalists assert that Western culture is responsible for the secularization of Arab societies.

3. **Have a unique image of themselves.** They feel that they are in a morally superior position to others. Terrorists are self-regarded to be the means for change and improvement. For instance, a terrorist who opposes abortions may view himself or herself as a messenger or servant of God.

4. **Identify the enemy and have strong feelings against them.** Terrorists frequently place blame and discredit those they regard to be at fault. Extreme nationalists dehumanize those of other ethnic groups as a way to mobilize action and limit guilt. They view others as enemies, collateral damage, or instruments of change. As a case in point, it is alleged that Michael Collins (IRA founder) stated after killing 14 men: "They were undesirables by whose destruction the very air is made sweeter" (Taylor 1958, p. 17).

5. **Believe terrorism is justified.** Recognizing their inferior position in terms of numbers or capacity as well as their inability to work through democratic processes, terrorists are willing to promote their goals through illegal means. In the view of terrorists, violence brings attention to a certain plight and speeds up the resolution of problems. To those participating in violence for religious reasons, terrorism may even bring salvation. Before being executed for killing an abortion doctor, a terrorist named Paul Hill stated that he expected a great reward in heaven once he was fatally punished.

4.2.1   Distinct Differences

Although terrorists may have several common personality traits, there are major differences among them as well. This divergence can be seen in terms of age, gender, education, and economic status.

Most terrorists are younger adults in their twenties (see Figure 4-2). This trend is seen throughout the world since young people have high ideals and possess more agile physical capabilities than older adults. During the 1970s many of the terrorists

FIGURE 4-2 Many, but not all, terrorists are young. During a march in Pakistan on October 5, 2001, a boy holds a pistol. Thousands of demonstrators shouted anti-American sentiment during this protest. *Source: © Chris Hondros/Getty Images News/Getty Images.*

captured in Ireland were as young as 12 years old. Arab terrorist groups today also include a large number of teenagers in their ranks. There are even young children who engage in terrorism today. Nevertheless, older terrorists contradict these expectations. Age is not always an accurate predictor of who will engage in terrorism. Osama bin Laden was in his mid-forties when the 9/11 attacks occurred. Thus, terrorists cannot be defined by a particular age. What we can state with some degree of certainty is that the leaders tend to be older and those who implement attacks are likely to be younger.

The typical terrorist is regarded to be male. Men have traditionally been involved in the dangerous activities of planning and carrying out terrorist attacks. In fact, terrorism has frequently been a male-dominated enterprise. However, women have always played a significant support role in terrorism, and some females have even participated in spying and actual operations (Griset and Mahan 2003; Ness 2008). Many women were involved in terrorism in Germany (i.e., Baader-Meinhof Gang and Red Army Faction). About half of the terrorists in Sri Lanka (i.e., Tamil Tigers) were girls or women, and they have been particularly deadly in their craft. While it is true that most terrorists are men, you should not expect that this will always be the case.

The educational status of terrorists is also in question. Many terrorists are well educated, obtaining degrees in difficult subjects such as engineering, chemistry, or computer science. In fact, some of the leaders of Al-Qaeda earned college degrees in the United States and elsewhere. Not all terrorists are highly educated though. Many of the terrorists in the Middle East receive basic education in schools known as **Madrasahs**. These institutions may indoctrinate orphan students in extreme Islamic thought as occurs in some remote areas of Pakistan. Other terrorists may lack any formal type of education whatsoever.

It is also difficult at times to predict terrorists based on economic class. Some terrorists, like Abu Nidal, were born into the lap of luxury. These and other terrorists rejected their parents' middle-class lifestyle in the 1960s to fight for economic equality. Osama bin Laden is another interesting example of wealth. He was an extremely rich terrorist who acquired millions of dollars from several oil and construction industries his family owned. However, today, many of the terrorists around the world are poor and lack the basic necessities of life. Thus, "it remains true that to generalize about the 'typical' terrorist can be very difficult with any degree of accuracy" (Combs 2000, p. 51).

In The Real World

Rearing Terrorists in the Middle East

Socialization into the world of terrorism is a common practice among many of the terrorists in the Middle East today. Children are brought up to support terrorist organizations such as Hamas and Hezbollah. They attend parades that celebrate militancy, and they are taught at an early age to fire guns and use knives. Children dress in camouflage uniforms and proudly wave the colors or flags of their terrorist organization. Cartoons illustrate the blessings of violence, and songs extol the virtue of terrorist heroes. Under these circumstances, it is no wonder that the ultimate goal in life for children is to die as a martyr. Other children have even executed enemies when asked to do so. This is one of the many reasons why terrorism is proving to be the enduring problem of our time.

Self-Check

1. Terrorists are not willing to sacrifice for their cause. True or false?
2. Terrorists often see themselves as the means for change. True or false?
3. Terrorists:
   a. Are always younger
   b. Are not always male
   c. Are never rich
   d. Are always educated
4. How are terrorists similar?

4.3 The Behavior and Tactics of Terrorists

To a certain extent, terrorist activities follow discernable patterns (Clutterbuck and Warnes 2011). Terrorists are involved in propaganda, recruiting, financing, and training. They also acquire false documents, travel, seek safe haven, use code words and secret communications, plan attacks, and search for weapons. Eventually, terrorists will launch attacks against their enemies. They also learn from their prior experience to plan more deadly attacks in the future.

4.3.1 Propaganda

Terrorists actively promote their ideological goals (Martin 2003). For instance, the Palestinian Islamic Jihad has a "manifesto." In its guiding documents, this organization rejects any peaceful solution to the Palestinian cause and lays out the goal of destroying Israel and ending Western influence in the Middle East. This document has been circulated widely among countries in the area. The spreading of religion and other ideologies by Hamas has also resulted a strong sense of "heroic resistance" and has been accompanied by "the legitimization of political violence in the Israeli-Palestinian conflict" (Singh 2012, p. 529). It is common to hear individuals in the Middle East say, "From the river to the sea, Palestine will be free." This statement, which now appears on mugs and T-shirts, essentially calls for the elimination of Jews and the state of Israel.

The media is often used to publicize the interests of terrorists. **Al Jazeera** (the most widely viewed TV station based in Qatar) often serves as the vehicle to disseminate information. Terrorists use it to call for the end of Western occupation in the Middle East and encourage viewers to "cut the head off of the snake" (i.e., destroy the United States). News material often shows videos of actual or recreated attacks as well as training activities. In other cases, terrorists use the media to denounce their enemies in very visible ways. For instance, the torturing of prisoners in the Abu Ghraib prison from 2003 to 2004 horrified the Western world. These disturbing actions by U.S. soldiers were publicized widely by terrorists to spread hatred against America. In different situations, members of Al-Qaeda or ISIS swear loyalty to religious or political leaders and film a martyrdom statement. Videos of these terrorists are then mass-produced and given to people in Saudi Arabia, Yemen, and elsewhere. Young men and women are told about the benefits of terrorism. For example, a 16-year-old terrorist apprehended in Israel said, "A river of honey, a river of wine and 72 virgins. Since I have been studying the Quran, I know about the sweet life that waits there" (Daraghmeh 2004, p. 27a). In other cases, groups like ISIS published a newsletter named Rumiyah from 2016 to 2017. This magazine called for the overthrow of democratic governments and provides recommendations on ways to attack the enemies of Islam. For instance, ISIS encouraged its supporters to kill infidels with anything, including vehicles. As a result, on March 22, 2017, a terrorist in England ran over several people and then stabbed a police officer. During the attack, six people died, and nearly fifty people were injured before the perpetrator was shot and killed.

4.3.2 Recruiting

Terrorists also attempt to recruit new members to their cause (Combs 2000). They may seek out other individuals or groups that espouse similar ideologies (e.g., anti-government groups or anti-American groups). In this situation, the terrorist may use well-developed websites to share their views with others. In different cases, terrorists may seek out students at universities who often have idealistic views of the world. Those advancing their education are aware of the world's problems, have specific technical knowledge sets, and desire change to make things better. Military personnel are often in high demand among terrorist organizations because they have an understanding of how to use weapons or manufacture explosives. Crooked law enforcement personnel may also prove useful for terrorists, especially if these corrupt officers can provide intelligence or help the terrorists to operate with impunity.

Terrorists may also be recruited at prisons (Dunleavy 2011). One New York Commissioner of Prisons said that "our prisons are stuffed full of people who have a hatred to the prison administration, a hatred toward America and have nothing but time to seethe about it" (Associated Press 2016). An example is José Padilla (a.k.a. Abdullah al-Muhajir). Padilla was thrown in jail after pulling a gun on another driver in the

United States. Padilla converted to Islam while in prison and traveled to Afghanistan and Pakistan in the late 1990s. He later talked to Osama bin Laden and Abu Zubaydah about plans to build and detonate a dirty bomb. Fortunately, Padilla was arrested before he could carry out the plot.

Those who are recruited into terrorist organizations often join the group due to unique psychological dynamics (Kruglanski et al. 2021) such as filling unmet emotional needs (Gerwehr and Daly 2006, p. 85). It has been illustrated that individuals who are prone to adhere to terrorist ideology are dissatisfied, are disillusioned, and lack strong family ties. Before recruitment, they often have no value system and desire clarity of purpose instead of ambiguity. For instance, it has been demonstrated that the radicalization of active or former military members in the United States may experience failed relationships, unemployment, and high rates of post-traumatic stress disorder (Jensen et al. 2022, p. 2). These factors may cause them to join terrorist organizations.

Once the terrorists are recruited, they are questioned extensively. For instance, terrorists who arrived in Afghanistan during the Al-Qaeda era were asked: What brought you to our country? How did you hear about us? What attracted you to the cause? How did you travel here? What is your educational or professional background? Such inquiries not only help to filter out spies but also determine the usefulness of new recruits.

In The Real World

The Process of Radicalization

Research by Karl Kaltenthaler and Arie Kruglanski has explored reasons why some individuals turn to terrorism (2021). These professors assert that radicalization is the process whereby people internalize and accept an extremist ideology. According to these authors, this transition is most likely to occur when three conditions—the "3 Ns"—are met. This includes needs, narratives, and networks. Needs refers to the desire people have to feel significant and be regarded as one that matters. Narratives are the messages people receive about how they have been mistreated, who is responsible for this injustice, and what can be done to rectify the situation. Networks are groups of family and friends that meet needs, reinforce narratives, and provide a supportive environment for further action (which could include violent attacks). Therefore, to prevent or minimize radicalization, Kaltenthaler and Kruglanski assert that it is vital to "understand how the need for significance drives so much of human behavior. By conceptualizing how the need for significance, networks and narratives work together to influence behavior, information operations [or communications to sway opinions] have a better chance of success" (2021, p. 12).

4.3.3 Financing

Terrorist groups also seek to obtain financial resources to support their activities (Emerson 2006). At times, terrorist organizations may appear to resemble large corporations (Napoleoni 2005). Cindy Combs has researched these issues, and she concludes that terrorism can be a "big business" (Combs 2000). Two organizations seem to illustrate her point very well.

Irish Northern Aid (NORAID) was established by Michael Flannery in 1969. This organization had headquarters in the Bronx (New York City) at 273rd East and 194th Street. Its goal was to assist the IRA through fundraising. On several occasions, it sponsored dinners and gathered between $20,000 and $30,000 in cities around the United States. James Adams, one expert on terrorism, asserted:

"From the onset of modern terrorism in Northern Ireland in 1969, the United States has played a key role in its support. The enormous Irish-American population has always felt a strong sentimental attachment to the 'old country,' and this has been translated into a steady stream of cash and guns to the IRA, which has, in part, enabled them to survive"

(in Combs 2000, p. 95). Fortunately, this is less of a problem today.

Another example is the Palestine Liberation Organization (PLO), which was led by Yasser Arafat. Its initial office was in Damascus where it has a bank of computers as well as accountants and other employees from MIT and Harvard. Much of its funds were acquired through dairy and poultry farms as well as cattle ranches or duty-free stores in airports. The PLO also invested in stocks on Wall Street. Money acquired from these transactions has been stored in bank accounts in Switzerland and Germany or passed through the Arab Bank for Economic Development. Combs asserted that this organization would have been on the Fortune 500 list if it were a company (Combs 2000, p. 90). In 1985 the PLO was worth an estimated $5 billion. The PLO was considered to be a terrorist organization up until 1991. Mahmoud Abbas took over the organization when Arafat died in 2004. The organization has experienced political turmoil during its history, but it still promotes the cause of the Palestinian people today (although perhaps in a less violent way).

A more recent example of the funding given to terrorists by states is Iran. According to the U.S. State Department, Iran funnels roughly $100 million to Hamas and $700 million to Hezbollah each year. This may include the transfer of currency as well as in-kind donations (e.g., the provision of thousands of explosives and rockets that have been used to attack U.S. troops in Iraq and the nation of Israel). The resources Iran shares with Hamas and Hezbollah pose serious threats to the Jewish state.

Terrorists may also obtain money from their involvement in the drug trade. There is often a very close connection between terrorism and narcotics. In fact, the word "assassin" comes from the Arabic term "hashashin"—one who eats or is addicted to hashish (Combs 2000, p. 18). Centuries ago, Muslim leaders gave drugs to terrorists who acted against the crusaders. Marco Polo noted this in his travels to the Middle East. During the 1980s it was reported that "Lebanese hashish helps to pay for everything from hijacking and bombing spectaculars in Europe and the Middle East to a simmering revolt by Muslim insurgents in the Philippines" (U.S. News & World Report 1987, pp. 36–37). In December 2003, U.S. investigators found 54 bags of hashish weighing 70 pounds each. In another case, 600 kg of the heroine was traded to buy stinger missiles in a case that included suspects from Chicago and New York. The Sendero Luminoso also made large sums of money from cocaine production. Al-Qaeda operatives were caught with drugs in the Persian Gulf. Al-Qaeda also sold opium to finance their terrorist operations. For these reasons, narco-terrorism is described by *U.S. News & World Report* as "the unholiest of alliances, a malevolent marriage between two of the most feared and destructive forces on modern society—terror and drugs" (U.S. News & World Report 1987).

Money for terrorism may also come from other illegal activities. The Red Army Faction in Germany robbed banks and engaged in the theft of automobiles. Ahmed Ressam, the LAX bomber, stole the chemicals that were used in his explosives. Al-Qaeda has relied on African clans' production of conflict diamonds to raise money. Gemstones are also used to buy weapons for terrorists in Liberia. Children trafficking supports Jemaah Islamiyah and Abu Sayyaf. Corrupt charities, such as the Haramain Islamic Foundation, have been used to support terrorism. Counterfeiting perfume, software, and music CDs has also been undertaken to raise money. In the United States, cigarette smuggling (from states with low taxes/prices to states with higher taxes/prices) resulted in $8 million in profits. Nearly $100,000 of this was sent to Hezbollah to purchase stun guns, night vision equipment, and other devices (Emerson 2006, p. 217).

Terrorists may also get their financial resources from individuals or governments. Osama bin Laden received a large inheritance from his father. He also owned several construction companies. A significant portion of Al-Qaeda's operating budget ($30 million annually) came from bin Laden, and he used some of this to pay off the Taliban for safe harbor. Shoko Asahara also played a large financing role in Japan. His terrorist organization, Aum Shinrikyo, obtained its money from the sale of books and computers. In other cases, states have sponsored terrorism. During the 1980s, Libya frequently provided terrorists with monetary resources. As mentioned earlier, Iran is a major financier of terrorist groups like Hamas and Hezbollah today and it has worked diligently to depose Western leadership in Iraq and Syria. Terrorists therefore seem to be able to raise significant amounts of funds for their operations.

4.3.4 Training

Terrorists seek to educate themselves and others about their deadly craft. For instance, in 1986, there was an international terrorist congress in Germany. Five hundred people attended including terrorists from France, Ireland, Portugal, Spain, and Latin America (Combs 2000, p. 89). Similar conferences have taken place in the United States. One event was held in 1990 in Dallas, Texas. It is reported to have hosted more radical Muslims under one roof in America than any other event up to that point in history (Reeve 2002, p. 224). Such events seek to reiterate the need for terrorist activities and discuss the best methods for accomplishing violent goals.

Training camps are also created and utilized by terrorists. There have been terrorist training camps in Cuba, Bulgaria, Czechoslovakia, East Germany, Lebanon, Libya, North Korea, South Korea, and Syria. The Mes Aynak training camp was located in an abandoned copper mine in Afghanistan. In secluded areas such as this, terrorists may participate in advanced commando courses. They may be taught how to raise money, recruit others, gather tactical information, disguise themselves, travel discretely, fit into a foreign culture, read maps, and understand cryptology. At other camps they may be taught to speak English or German and be introduced to weapons manufacturing. Some training locations help terrorists use knives while butchering sheep or implement gas attacks on dogs. According to the Federal Bureau of Investigation (FBI), there have been a number of similar training sites in the United States in isolated locations. It is argued that 22 jihadist camps have been identified in America and were operated by Jamaat al-Fuqra. Video footage has demonstrated the use of automatic weapons and hand-to-hand combat in isolated compounds. The extent of this threat has been disputed, but the situation does produce concern among citizens and law enforcement officials.

In many situations, terrorists will be given a kit, manual, or literature to educate them further (Griset and Mahan 2003, p. 196). Ramzi Yousef—the person responsible for the 1993 WTC bombing—possessed bomb-making instructions, false identification documents, videos denouncing the United States, and other documents with operational guidance. Other materials found at training camps include sections on how to acquire and make chemical explosives as well as the best location to place charges on bridges and overpasses. An Al-Qaeda manual illustrated how terrorists may enter the United States through Canada. These and other documents also suggested that terrorists should avoid hanging out at radical mosques, use cash only for expenditures, and rely on public Internet access to send e-mails. Terrorists have also advised in manuals to use pay phones to communicate or throw away cell phones once they are used.

4.3.5 False Documents, Travel, and Safe Haven

According to the 9/11 Commission (2003), terrorists often seek to obtain fraudulent documents. They also travel frequently and seek safe haven among sympathizers. For instance, Al-Qaeda had an office in the Kandahar airport. Personnel in this facility helped some of its members obtain Yemeni passports. Terrorist organizations are able to substitute photos and add or erase entry stamps. They also can submit false paperwork in an attempt to acquire visas. For instance, Mohammed Atta told U.S. officials that he lost his old passport. Because he was worried about visas from Pakistan, he ordered a new one. This allowed him to enter the United States with greater ease prior to 9/11.

As terrorists go about their activities, they may travel extensively. Khalid Sheikh Mohammed journeyed to India, Indonesia, and Malaysia. Others have flown frequently on commercial airlines between Afghanistan and the United States. Terrorists have traveled through Iran, which neglected to stamp passports. As terrorists move about to conduct their business, they will seek a safe haven. Both Sudan and Afghanistan provided bases for Osama bin Laden. Weak countries with remote areas that have little government presence are attractive to terrorist organizations. When terrorists enter enemy territory, they are picked up by associates and taken to homes or apartments where food, shelter, and other necessities of life can be met.

4.3.6 Code Words and Secret Communications

To avoid detection, terrorists are very cautious about their communications. When discussing the timing to pick up a bomb, one terrorist told his colleague that they were in the "ninth month of pregnancy." In another situation, a bin Laden supporter in Yemen said his impending "marriage" would be a "surprise." This was in reference to an attack that was being planned. However, leaders of terrorist organizations do not always tell their subordinates everything about operations. For instance, bin Laden wanted to limit the information his recruits possessed in case they were caught. Terrorists may also use the Internet to communicate. In March 2001, Al-Qaeda used the hidden scripts on web pages. Known as steganography, this technique allows terrorists to hide maps, photos, and letters in chat rooms or pornographic sites. Louis Freeh, a former FBI director, says, "uncrack-able encryption is allowing terrorists—Hamas, Hezbollah, Al-Qaeda and others—to communicate about their criminal intentions without fear of outside intrusion" (Thetford 2001, p. 252).

4.3.7 Planning

Terrorists spend a great deal of time planning attacks (9/11 Commission 2003). In researching possible targets, terrorists may rely on the Internet and open-source records (e.g., library or government documents), insider information (e.g., knowledge from employees), electronic equipment (e.g., police scanners or phone taps), or physical surveillance (e.g., casing a location) (Pluchinsky 2006).

Terrorists also plan how to implement their attacks (see Figure 4-3). In 1994, Khalid Sheikh Mohammed designed the Manila air plot. He and other terrorists intended to bomb 12 U.S. jets over the Pacific Ocean during a two-day span. They started casing flights and purchased timers and nitrocellulose to make bombs. Fortunately, the Philippine government uncovered the plot and was able to thwart it. Terrorists involved in the embassy bombings in Africa used state-of-the-art video cameras for reconnaissance purposes. They obtained this equipment from dealers in China and Germany. They were interested in traffic patterns, building construction, and occupancy rates to inflict maximum casualties. The 9/11 plot was discussed seven or eight years before it actually occurred. It was initially supposed to have about 10 planes but was scaled back due to complexity. Nevertheless, the terrorists involved in the 9/11 attacks cased flights frequently to determine the best time to hijack the planes. These terrorists also went to flight school in the United States so they could maneuver the aircraft.

CAREER OPPORTUNITY | Federal Protective Service

Knowing what terrorists and criminals do is only one of many components of the homeland security enterprise. There is also a need to ensure a safe and secure work environment in government facilities, which have been and could be the target of terrorist attacks. This is one of the goals of the Federal Protective Service (FPS).

Employees of the FPS provide security at the estimated 9,500 federal facilities around the nation. This includes locations such as the Congressional Building, federal courthouses, and department and agency regional offices.

FPS officers and agents assess vulnerabilities, install entry control systems, educate facility tenants, contract with private security companies, gather intelligence, monitor security breaches, provide police response, perform hazardous materials operations, and coordinate other activities (e.g., emergency services in high profile events and after natural disasters). The FPS also has "Mega Centers" in Pennsylvania, Michigan, and Colorado where personnel monitor the videos and alarms at government facilities and dispatch personnel during any threat or emergency.

The FPS often hires veterans and others who have worked in law enforcement in the past. Applicants can apply for positions on the USAJOBS website. Like other jobs in homeland security, applicants must not have any felony convictions and must be able to pass a thorough background investigation and drug screening. Once hired, new employees will be trained at the Federal Law Enforcement Training Center in Glynco, Georgia.

For further information, see Adapted from **https://www.dhs.gov/homeland -security-careeers/careers-federal-protective-service.**

An important function of planning is to find a symbolic date or location so terrorist attacks can occur and have maximum impact (Martin 2003). The bombing of the Murrah Federal Building took place on the anniversary of Waco. Timothy McVeigh was frustrated with the way the Bureau of Alcohol, Tobacco and Firearms and FBI dealt with the Branch Davidians, and he wanted to get even. Similarly, the targets on 9/11 were not selected by chance. Their preferred airlines—American Airlines and United Airlines—had reference to America and the United States. The WTC was chosen because it is the heart of the Western capitalist economy. The Pentagon was selected as a target because it signified the strength of the US military. Although Flight 93 never arrived at its intended location, many believe it was headed for the White House or the Capitol Building (which included the political leadership of the country). The meticulous planning and choice of targets adds to the death, damage, social disruption, and visibility of the event.

FIGURE 4-3 Terrorists spend a lot of time planning attacks, as was the case in the twin African Embassy bombings. *Source: © FBI.*

In The Real World

Nice Truck Attack

On July 14, 2016, 31-year-old Mohamed Lahouaiej Bouhlel carried out a terrorist attack while people were celebrating Bastille Day in the French city of Nice. Bastille Day is the national independence holiday in France, and it commemorates the beginning of the French Revolution with the storming of the Bastille prison on July 14, 1789.

The perpetrator was born in Tunisia but lived in Nice and worked as a delivery truck driver. His attack began when he purposely drove a large, rented truck (weighing about 19 tons) into crowds of people along the Promenade des Anglais. Bouhlel also fired a gun at French police officers while driving the truck. The violent assault ended when police officers shot into the vehicle, killing him. However, prior to his death, Bouhlel was responsible for killing 84 people and injuring over 200 others in this horrific event.

Bouhlel was identified after the attack, and his belongings and apartment were investigated by law enforcement. The discovered that Bouhlel had a previous criminal record but was not on intelligence services' radar. He also had a history of violence toward his family, and, in 2004, a psychiatrist described him as edging on psychopathy. Bouhlel was apparently depressed after his wife divorced him and likely became radicalized very quickly.

Evidence shows Bouhlel was in contact with Islamic radicals and planned the terrorist attack for over a year with several accomplices who have since been taken into custody (none of whom were previously known to have any connections to Islamic radicals). The Islamic state claimed responsibility for the incident shortly after it occurred, but Bouhlel may have acted independently with the same radicalized ideology.

4.3.8 Weapons

Sometime before, during, or after individuals and groups plan terrorist attacks, they will also acquire weapons (Martin 2003). These may range from simple devices to advanced military armaments. Such weapons may impact one or a few people, or even hundreds and thousands. For instance, mace, pepper spray, or box cutters were used against some of the passengers and crew on 9/11. Knives have been used by ISIS to cut off the ears of those who desert their cause. In other cases, terrorists will obtain and use pistols and light artillery. Guns and rocket-propelled grenades are also commonly used by terrorists. Incendiary devices (e.g., Molotov cocktails that start fires) are used when destruction of property is desired.

Terrorists have been increasingly interested in weapons of mass destruction (**WMDs**), which will be discussed in depth in Chapter 14. Common acronyms for WMDs are NBC or CBRNE. **NBC** stands for nuclear, biological, and chemical weapons. CBRNE is a more recognized acronym since it is more comprehensive. **CBRNE** implies chemical, biological, radiological, nuclear, or explosive devices.

Chemical weapons may include typical household cleaning agents that have been mixed into hazardous or unstable combinations. Terrorists may also try to find military-grade weapons such as mustard gas or nerve agents. Shoko Asahara spent an estimated $30 million to make anthrax.

Biological agents include pathogens, toxins, or viruses. Larry Wayne Harris was a member of the Aryan Nations. After illegally ordering the plague from a research facility, he was arrested in Arizona and caught with the deadly agent in the glove box of his car.

Radiological weapons are devices that contain and emit alpha-, beta-, or gamma-emitting material. Radiological material has been reported lost or stolen at several industrial or medical facilities. A nuclear weapon is far different from a radiological weapon. Nuclear weapons are large bombs created by fusion or fusion processes. They can decimate a single city or an entire region. Although nuclear weapons have generally been obtained by states, there are concerns that terrorists may seek them. At one point, it was reported that bin Laden was working with a Sudanese military officer to purchase weapons-grade uranium.

Explosive devices are other names for conventional bombs. Some "dirty bombs" can be laced with radiological materials. Improvised explosive devices (**IEDs**) were used extensively by insurgents against U.S. troops in Iraq and Afghanistan. Pipe bombs are examples of IEDs. They can pierce armor and kill or maim soldiers and police officers.

4.3.9 Acts of Civil Disorder or Terrorism

Once terrorists plan attacks and obtain weapons, they are ready to initiate and engage in violent activity (Pluchinsky 2006). Types of attacks can vary dramatically. There have been numerous cases where individuals have mailed letters with white powdery substances. These hoaxes and actual attacks instill fear and are intended to disrupt society. Responding to such events requires large numbers of personnel and can shut down businesses and government facilities for hours and even days. At other times, other forms of sabotage are preferred. This includes unique attacks that are geared toward the accomplishment of ideological goals. In the Northwest, the Rajneeshee group sprayed *Salmonella* on a salad bar at a restaurant in Dalles, Oregon. Seven hundred people became ill as the group tried to influence the outcome of local elections.

Terrorists also hijack planes or seagoing vessels. The 9/11 attacks that resulted in so much destruction in New York and Virginia began with the takeover of aircraft. There have been other attacks on ships in the Philippines. Hostage-taking and kidnapping are also frequent ploys of terrorists. Some hostages are used for ransom or killed if the demands of terrorists are not met. For instance, Nicholas Berg was a contractor who was captured and later beheaded in Iraq. Other people are murdered through assassinations and ambushes. Yitzhak Rabin, a prime minister of Israel, was shot to death on November 4, 1995, by a person who opposed the Oslo Peace Accords. In another case, Al-Qaeda terrorists pretended to be reporters. They met with Ahmed Shah Massoud (a Northern Alliance leader) and killed him with a bomb on September 9, 2001.

Arson is a frequent choice of terrorists, and it has been used to intimidate members of black religious congregations among other groups. In addition, there is mounting evidence that Islamic terrorists have promoted, planned, and perhaps even carried out arson attacks on wildland-urban interface areas in the United States (NCTC 2022). Al-Qaeda's *Inspire Magazine* and ISIS's *Believers* video advocate this type of arson and provide instructions on how to make timed incendiary devices. On July 20, 2019, an ISIS supporter was arrested in California for his plans to set fire to the Berkeley Hills mountain range.

In The Real World

Boko Haram

The Jama'atu Ahlis Sunna Lidda'awati wal-Jihad is an Islamist terrorist group that emerged in 2002 in Nigeria. In Arabic, this official name means "People Committed to the Propagation of the Prophet's Teachings and Jihad." However, the organization is better known by the name Boko Haram, the loose translation of which sheds light on their ideology to forbid Western education. Mohammed Yusuf united his followers to form the group. Under his leadership, Boko Haram created an Islamic school and took action to establish an Islamic state. It was categorized by a campaign of violence, and the group's headquarters were seized in 2009 by Nigeria's security forces. Many of Boko Haram's members were captured, and Yusuf was killed.

Boko Haram regrouped under Abubakar Shekau and continued their insurgency with assassinations, bombings, raids, and other attacks. Perhaps one of the most widely publicized of these incidences was the 2014 abduction of 276 schoolgirls from Chibok in Borno (northeastern Nigeria). The girls were between 16 and 18 years old. About 50 of the girls were able to escape soon after being abducted. Shekau threatened to sell the remainder of the girls into slavery, and it is likely they have been brutalized and forced to convert to Islam.

The kidnapping gained international attention and spurred a discussion on social media with the hashtag #BringBackOurGirls, which many powerful celebrities and high-ranking officials participated in, including Michelle Obama. Boko Haram continues to abduct women and children as they seek to control more of Nigeria and the surrounding regions. Some victims have since escaped or been rescued from captivity, and speculation has arisen on whether any of the Chibok girls were among those recovered. Despite the wide media coverage and government efforts in Nigeria, the United States, and other countries, over 200 Chibok girls are still missing.

At one point, Boko Haram was a very active threat to the region with a force estimated to be around 9,000 men. The group may have had alliances with foreign groups and extremists like Al-Qaeda. During this period, Boko Haram attempted to obtain the release of their peers who were taken prisoner by the government and desired to build an Islamic state. The poverty-stricken and chaotic state of the Nigerian government fueled the group's insurgency. However, with the death of its leader, Boko Haram has splintered into various factions. But the followers are still of concern.

Bombings are the most prevalent type of attack, accounting for nearly 75% of violent actions committed by terrorists. Bombings have taken place at abortion clinics, the WTC (in 1993), the Marriott Hotel in Indonesia, the African embassies, and the USS Cole. In the Middle East, such attacks are frequently initiated by suicide bombers. These death squads wear or carry explosives that they detonate around others. It may be a matter of time before these occur in the United States.

While WMDs have not been used often, terrorists did use sarin in an attack on the subway in Tokyo. Terrorists have used chemicals such as chlorine in their bombings in Iraq. Syria launched gas attacks against certain villages to "cleanse" the area of ethnic groups of which they disapprove. There is fear that these types of attacks will be more commonplace in the future. If terrorists survive attacks, they will quickly flee the scene. For instance, after the first WTC bombing, Ramzi Yousef traveled to Pakistan. He remained there for two years until he was caught by the special agents from the FBI. The acquisition and use of WMD is just one of the many threats terrorists pose to the United States today.

In The Real World

Operations of the Tamil Tigers

The Tamil Tigers seek independence in northeastern Sri Lanka. When India gained autonomy at the end of World War II, both the Sinhala and Tamil ethnic groups shared political authority. In 1955, Sinhala-only policies were implemented, and the Tamils became disgruntled due to their loss of power. After assassinating the Sinhala leader in 1959, the Tamil Tigers emerged as a serious terrorist threat. They recruited many members to join their cause. Funding was obtained through bank robberies. Training camps were founded in India, and members received instruction from terrorists in the Middle East. The Tamil Tigers built a navy to attack Indian fleets. In 1991 they killed the Indian prime minister. The Tamils prefer bombings, and suicide attacks are common to this day.

Self-Check

1. Terrorists want to recruit others, but they are also careful who they allow to join with them. True or false?
2. Training does not appear to be a major priority of terrorists. True or false?
3. Terrorists:
    a. Often obtain false documents
    b. Do not travel frequently
    c. Spend little time on planning attacks
    d. Do not prefer explosives
4. Why do terrorists select some targets over others?

Summary

Understanding who terrorists are and how they operate is extremely important if you are to reduce both the number and the extent of attacks. In this chapter, you have assessed the nature of individual terrorists and those associated with organizations and nation-states. You have recognized common personality traits as well as significant differences among terrorists. Your review of terrorist behavior has enabled you to identify how they finance operations, communicate with secret codes, and carry out attacks. Such knowledge is imperative if you are to work successfully in homeland security.

Assess Your Understanding

UNDERSTAND: WHAT HAVE YOU LEARNED

Go to **www.wiley.com/go/mcentire/homelandsecurity3e** to assess your knowledge of terrorists and their behavior.

Summary Questions

1. The JRA is a terrorist organization that supports capitalism. True or false?
2. Leaders of Libya were responsible for the bombing of Pan Am Flight 103. True or false?
3. Individuals who commit terrorist acts are sometimes referred to as lone-wolf terrorists. True or false?
4. Abu Sayyaf is an Islamic separatist group that wants foreign involvement in the Philippines. True or false?
5. Terrorist groups may be led by a single individual, a headquarters station, or several cells throughout the world. True or false?
6. Terrorists often see the world as a complicated place with many different interpretations of what is right and wrong. True or false?
7. Terrorist groups avoid using the Internet as a means of communication. True or false?
8. It is possible that terrorists have set forests on fire with timed incendiary devices. True or false?
9. Which of the following answers does not accurately describe Meir Kahane?
   a. He was a Jewish rabbi.
   b. He founded the Jewish Defense League.
   c. He founded the Palestinian Jihad.
   d. He was assassinated in 1990.
10. Theodore "Ted" Kaczynski was known for his:
    a. Opposition to technological advancements in society
    b. Loyalty to professors and corporate leaders
    c. Collaboration with terrorist organizations
    d. Efforts to combat terrorist activity
11. All of the following nation-states have recent involvement in sponsoring terrorism except:
    a. Iran
    b. Syria
    c. Sudan
    d. France
12. Left-wing groups seeking economic or political equality through terrorist acts are an example of what kind of terrorist group?
    a. Criminals
    b. Careless
    c. Crusaders
    d. Crazies

13. Terrorist groups use all of the following techniques to raise funds except:
    a. Drug sales and other illegal activities
    b. Mass mail-outs to households
    c. Using organized "big business" techniques
    d. Obtaining funding from individuals and governments
14. CBRNE stands for:
    a. Chemical, biological, radiological, nuclear, or explosive devices
    b. Chemical, biological, radiological, nuclear, or explosive disasters
    c. Catastrophic, biological, radiological, nuclear, or explosive devices
    d. Catastrophic, biological, radiological, nuclear, or expensive devices
15. Bombings account for what percentage of terrorist attacks?
    a. 25
    b. 50
    c. 75
    d. 80

Applying This Chapter

1. Pick a terrorist attack not mentioned in this chapter and describe it based on the following criteria:
   a. What was the terrorist act?
   b. Was it performed by an individual or group?
   c. Would the terrorist(s) be classified as a criminal, crusader, or crazy? Why?
2. Using the same terrorist attack you picked for question one, describe what personal characteristics the person or group involved in the act has in common with other terrorists described in this chapter. Are there differences? If so, what?
3. Terrorist groups often recruit new members to grow their organization and increase their influence. What individuals and groups of people do terrorists recruit? What are some reasons why people join terrorist groups? What happens once they are recruited?
4. Why is it important for us to understand the various behaviors and tactics of terrorist groups? Using the information in the chapter, describe how we can use information on terrorist groups to help monitor and suppress terrorist activity.

Be a Homeland Security Professional

Common Terrorist Characteristics

You are an analyst with the FBI. Your boss would like you to write a report on common terrorist profiles. How could you describe terrorists? In what ways could you classify them?

Similarities and Differences Among Terrorists

Write a three-page paper about the personal characteristics of terrorists. Be sure to discuss if terrorists are similar or different from one another. Mention why it is important to understand who terrorists are.

Looking for Terrorists

You work as a member of a terrorist task force in law enforcement in local government. Your job is to identify potential terrorists before they launch attacks. What type of behavior are you looking for? Explain at least five activities that you would want to consider.

Key Terms

Abu Musab al-Zarqawi   A Sunni terrorist responsible for many atrocities in Iraq, including the beheading of an American businessman named Nicholas Berg

Abu Sayyaf   An Islamic separatist group in the Philippines that desires an independent state in Mindanao

Al Jazeera   A TV station based in Qatar that is popular in the Middle East and used to disseminate terrorist information

Al-Qaeda   A well-known terrorist organization whose name refers to the "base"— the location from which its supporters attacked the Soviet Union to free Afghanistan

Anders Behring Breivik   A terrorist who espoused far-right ideology and conducted one of the worst terrorist attacks in Norway

CBRNE   An acronym for chemical, biological, radiological, nuclear, or explosive devices

Cell   A terrorist branch or unit operating in locations away from the organization's headquarters

Crazy   A terrorist who is regarded to be psychologically disturbed (e.g., Ted Kaczynski)

Criminal   A terrorist who seeks personal gain through illegal means (e.g., drugs or crime)

Crusader   A terrorist who promotes high moral goals (e.g., Islamic fundamentalists)

Japanese Red Army   A left-wing terrorist organization that emerged in the 1960s to protest U.S. military presence in Japan after World War II, the war in Vietnam, and capitalism

IED   Improvised explosive devices

Iran   A state in the Middle East that denounces the United States and Israel, promotes anti-Western propaganda, and has a long history of participation in terrorism

Libya   A country in Northern Africa that supported terrorism heavily in the 1980s

Lone-wolf terrorists   Individual terrorists who act alone

Madrasahs   Schools in Pakistan and elsewhere that at times indoctrinate students in extreme Islamic thought

Theodore "Ted" Kaczynski   A terrorist known as the "Unabomber" who opposed technological advances

WMDs   Acronym for weapons of mass destruction

References

9/11 Commission (2003). *The 9/11 Commission Report: Final Report of the National Commission on Terrorist Attacks Upon the United States*. New York: W.W. Norton and Company.

Clutterbuck, L. and Warnes, R. (2011). *Exploring Patterns of Behavior in Violent Jihadist Terrorists: An Analysis of Six Significant Terrorist Conspiracies in the UK*. Santa Monica, CA: Rand Corporation.

Combs, C.C. (2000). *Terrorism in the Twenty-First Century*. Upper Saddle River, NJ: Prentice Hall.

Daraghmeh, A. (2004). Family: teen exploited for bomb mission Dallas Morning News (Friday, March 26), p. 27a.

Dunleavy, P.T. (2011). *The Fertile Soil of Jihad: Terrorism's Prison Connection*. Washington, DC: Potomac Books.

Emerson, S. (2006). *Jihad Incorporated: A Guide to Militant Islam in the U.S.* New York: Prometheus Books.

Gerwehr, S. and Daly, S. (2006). Al Qaida: terrorist selection and recruitment. In: *The McGraw-Hill Homeland Security Handbook* (ed. D.G.Kamien), 73–89. New York: McGraw-Hill.

Griset, P.L. and Mahan, S. (2003). *Terrorism in Perspective*. Thousand Oaks, CA: Sage.

Hacker, F.J. (1976). *Crusaders, Criminals, Crazies: Terror and Terrorism in Our Time*. New York: Norton.

Hickey, Jennifer. 2016. Ripe for radicalization: federal prisons "breeding ground" for terrorists, say experts. *Fox News* (January 5). **http://www.foxnews.com/us/2016/01/05/ripe-for-radicalization-federal-prisons-breeding-ground-for-terroriss-say-experts. html**.

Holtmann, P. (2014). Terrorism and jihad: differences and similarities. *Perspectives on Terrorism* 8 (3): 140.

Jensen, M.A., Yates, E., and S. Kane. (2022). Radicalization in the Ranks: An Assessment of the Scope and Nature of Criminal Extremism in the United States Military. January 17. **https://www.start.umd.edu/sites/default/files/publications/local_attachments/Radicalization%20in%20the%20Ranks_April%202022.pdf**. (Accessed October 11, 2023).

Kaltenthaler, K. and A. Kruglanski. (2021). The Three N Approach to Strategies for United States Information Operations. May. https://nsiteam.com/the-three-n-approach-to-strategies-for-united-states-government-information-operations/. (Accessed October 11, 2023).

Kruglanski, A., Kopetz, C., and E. Szumaowska. (2021). *The Psychology of Extremism: A Motivational Perspective*. New York: Routledge.

Martin, G. (2003). *Understanding Terrorism: Challenges, Perspective, and Issues*. Thousand Oaks, CA: Sage.

Napoleoni, L. (2005). *Terror Incorporated: Tracing the Dollars Behind the Terror Networks*. New York: Seven Stories Press.

NCTC. (2022). Mitigating the threat of terrorist-initiated arson attacks on wildland-urban interface areas. FIRSTRESPONDERSTOOLBOX. January 27. **https://www.dni.gov/files/NCTC/documents/jcat/firstresponderstoolbox/124s_-_Mitigating_the_Threat_of_Terrorist-Initiated_Arson_Attacks_on_Wildland-Urban_Interface_Areas.pdf**. (Accessed October 11, 2023).

Ness, C.D. (2008). *Female Terrorism and Militancy: Agency, Utility and Organization*. New York: Routledge.

O'Balance, E.O. (1979). *The Language of Violence: The Blood Politics of Terrorism*. Presidio: San Rafael, CA.

Pluchinsky, D.A. (2006). A typology and anatomy of terrorist operations. In: *The McGraw-Hill Homeland Security Handbook* (ed. D.G.Kamien), 365–390. New York: McGraw-Hill.

Reeve, S. (2002). *The New Jackals: Ramzi, Yousef, Osama bin Laden, and the Future of Terrorism*. Lebanon, NH: Northeastern University Press.

Singh, R. (2012). The discourse and practice of 'heroic resistance' in the Israeli-Palestinian conflict: The case of Hamas. *Politics, Religion & Ideology* 13 (4): 529–545.

Taylor, R. (1958). *Michael Collins*. London: Hutchinson.

Thetford, R.T. (2001). The challenge of cyberterrorism. In: *Terrorism: Defensive Strategies for Individuals, Companies and Governments* (ed. L.J.Hogan), 239–257. Frederick, MD: Amlex, Inc.

US Department of State (2016). Country Report on Terrorism. **https://www.state.gov/j/ct/rls/crt/2016/**.

U.S. News & World Report (1987). Narcotics: terror's new ally. US News & World Report (4 May), pp. 36–37.

White, J. (2002). *Terrorism: An Introduction*. Belmont, CA: Wadsworth.

Yayla, A.S., Ekici, S., Durmaz, H., and Sevinc, B. (2007). *Understanding and Responding to Terrorism*, NATO Security Through Sciences Series. Amsterdam: IOS Press.

CHAPTER 5

Uncovering the Dynamic Nature of Terrorism

History of Violence and Change over Time

DO YOU ALREADY KNOW?

- The early history of terrorism

- How terrorism has evolved in other countries

- Ways terrorism has affected the United States

- The differences between traditional and modern terrorism

For additional questions to assess your current knowledge of the dynamic nature of terrorism, go to **www.wiley.com/go/mcentire/homelandsecurity3e**

WHAT YOU WILL LEARN

5.1 When terrorism appeared

5.2 The manifestation of terrorism Hinternationally

5.3 The prevalence of domestic terrorism

5.4 Differentiation between modern and traditional terrorist attacks

WHAT YOU WILL BE ABLE TO DO

- Synthesize the factors that influenced the emergence of terrorism

- Recount the evolution of terrorism abroad

- Evaluate the impact of terrorism in the United States

- Judge how terrorism has changed over time

Introduction

One of the best ways to decrease the probabilities and consequences of terrorism is to understand how it has evolved over time. Terrorism has likely existed throughout history in one form or another, and it has not remained static in terms of its various manifestations. In this chapter, you will learn how the use of ideology, violence, and fear has influenced the emergence of terrorism. You will then acquire an understanding of the dynamic nature of terrorism around the world and be able to describe the history of terrorism in the United States. Finally, this chapter will help you defend the assertion that terrorism is different today than it was in prior decades. Understanding the fluid nature of terrorism is imperative for anyone working in homeland security. Recognizing how terrorism has changed over time may help you anticipate what to expect in the future.

5.1 The Appearance of Terrorism

Violence with an eye toward creating fear among an enemy has occurred repeatedly throughout history (Chalian and Blin 2007, p. 5). The examples of warriors painting their faces, brandishing weapons, and yelling loudly at their opponents are too numerous to mention. In such cases, this activity could be regarded as an effort to intimidate and dissuade the violent intent of the adversary. Averting conflict by having the enemy give up was the goal of engaging in such behavior.

If hostilities did break out, ruthlessness was made painfully evident to illustrate the consequences of engaging in combat. Those from the opposing tribe were killed, and their lifeless bodies were displayed visibly for everyone to see. For instance, corpses were hung in trees and skulls were placed on poles. In addition, homes were burned, property and cattle were taken, women were raped, and children were enslaved. These types of actions were a vivid message regarding the strength of the prevailing party. In this sense, the fear exhibited in war is somewhat similar to that associated with terrorism. However, the two cannot be considered synonymous.

Violent activities undertaken with the goal of obtaining ideological objectives—which is an integral aspect of terrorism—can be traced back to the Roman and Greek republics (Combs 2000, p. 18). Aristotle was one of the famous Greek philosophers who lived from 384 to 322 BC. He asserted that killing despotic rulers could be justified in certain situations. If such leaders failed to serve the interests of the people, Aristotle acknowledged that their elimination could be warranted. Others felt similarly. For example, when Brutus killed Roman emperor Julius Caesar, Cicero observed: "There can be no such thing as fellowship with tyrants, nothing but a bitter feud is possible ... [Such] monsters ... should be severed from the common body of humanity" (in Griset and Mahen 2003, p. 2). Thus, people have rationalized violence based on their perception of the behavior of others, suggesting that murder may be "just" and necessary.

In conjunction with these viewpoints, assassinations of political leaders have correspondingly been a prevalent form of terrorism throughout history. For example, in 1605, Guy Fawkes was caught in an attempt to murder King James I and other leaders of the British Parliament (Griset and Mahan 2003). The cause of his **"Gunpowder Plot"** was a disagreement between King Henry VIII and Pope Clement VII. King Henry desired an annulment of his marriage to Catherine so he could wed Anne Boleyn. Because the pope disapproved of this request, King Henry denounced Catholicism and created the Church of England. Monasteries were closed, and cathedrals were taken over by those loyal to the king. Catholics were suppressed in England and lost their ability to worship as they saw fit.

Fawkes and others were angry with how this religious disagreement was unfolding. They desired to reinstall the pope as the religious head of England by blowing up the Palace of Westminster. When the king was notified of the scheme, Fawkes was immediately captured. He and the other conspirators were drawn and quartered by galloping horses in front of the public that gathered at Westminster.

While Fawkes' attempt to kill the king may be regarded as an act of terrorism, the word "terrorism" did not appear until the Enlightenment. The **Enlightenment** was a period in history when a new way of looking at the world emerged. For instance, advances in scientific knowledge offered an alternative to the religious explanations of nature during that period. Improvements in communications and travel promoted the sharing of ideas within and across nations. As these and other changes took place, the divine appointment of kings was subsequently questioned. Common people started to demand political rights and freedoms. Democratic governments began to emerge as a result. The feudal economic system gave way to capitalism, and sociopolitical structures were altered. This transformation is evidenced by such events as the movement for independence in the United States (1775–1783) and the French Revolution (1789–1795).

The Revolutionary War in North America resulted from the questioning of the religious authority of the king as well as taxation without representation. This war of independence was successful in that it only required a transfer of power from the King in England to the elite in America. The ambitions in France were loftier in that the middle classes desired to oust the reigning nobility. However, those seeking "liberty, equality, and fraternity" in France could not accomplish their goals. The Jacobin party acquired control of this country, and with their power they began to repel the revolutionary movement. During this period, an estimated 20,000 persons were killed by France's Committee of Public Safety. The state-sponsored violence against its own citizens is what Edmund Burke and others called the "**Reign of Terror**." Marie Antoinette was one of the most notable casualties in this conflict. She was beheaded on October 16, 1793.

Self-Check

1. In many ways, violence, fear, and terrorism have always existed. True or false?
2. When did the word "terrorism" appear?
   a. During Aristotle's time
   b. In medieval times
   c. During the Enlightenment
   d. Long after the American Revolutionary War
3. Discuss why Guy Fawkes wanted to kill King James I.

5.2 The Evolution of Terrorism Abroad

Terrorism has changed dramatically since the Enlightenment. However, the history of terrorism is not simple or linear. Instead, the evolution of terrorism is complex, and numerous manifestations of violent activity have coexisted at times (Chalian and Blin 2007). This history is best understood when considered as an unfolding transformation.

Terrorism took on distinct forms as the nineteenth century got underway. In the late 1700s, Napoleon Bonaparte gained control of France and he began to spread his empire to the rest of Europe. Bonaparte's armies were meticulously trained, sufficiently staffed, and well equipped. Nevertheless, the Spanish attempted to repel French invasions in 1808. Hit-and-run tactics were common in this war for liberation. Spanish resistance was aided by the support of the British military.

Nationalist movements like this have been common in many parts of the world. The associated attacks of these political initiatives have had significant and far-reaching impacts. For instance, World War I was triggered on June 28, 1914. A 19-year-old member of the Black Hand (a Serbian terrorist organization) opposed the Austro-Hungarian Empire. He was therefore recruited to assassinate Archduke Franz Ferdinand. Such terrorist attacks with nationalistic origins have been evident in many countries ever since. Numerous terrorist acts have occurred in the Middle East, Ireland, and India. Countless and cruel acts have also been witnessed in the Balkan region (i.e., Yugoslavia) and African nations (e.g., Rwanda) during the 1990s.

Anarchists—those opposing specific governments or all governments—also launched numerous terrorist attacks in the 1800s. In Russia, men like Michail Bakunin sought to dismantle the czarist state. He and others disapproved of the government and desired a major transformation of society. As a means to accomplish this goal, Bakunin advocated the selective killing of Russian officials. Sergei Nechaev, another Russian, also advocated that terrorism be used in his publication, *Revolutionary Catechism*. He asserted that the anarchist must "have one single thought, one single purpose: merciless destruction. With this aim in view, tirelessly and in cold blood, he must always be prepared to kill with his own hands anyone who stands in the way of achieving his goals" (Venturi 1966, p. 366). In 1879, Zemlya i Volya (a terrorist organization known as the Will of the People) was created in Russia. This group frequently used terrorism as a form of political protest against the ruling elite (Combs 2000, p. 25).

In the early and mid-1900s, fascist governments were involved in terrorist attacks against their own citizens. **Fascism** is an ideology that promotes the uniting of citizens in support of the state. It is often an extreme movement aimed at protecting government and broad societal interests. Fascist regimes existed in Spain under Francisco Franco and Italy under Benito Mussolini. Of course, the most egregious use of force against the populous occurred in Germany. Adolf Hitler killed an estimated six million Jews during the Holocaust. Sadly, many political associations - including the Aryan Nations today - have been influenced by Hitler's hatred toward those of Jewish descent or others from diverse races.

In the mid-to-late 1900s, **communism** and reactions against it were the driving force of terrorism (see Figure 5-1). Communism is an ideology that sympathizes with the poor and downtrodden. It attempts to do away with private property and exploitative class relations. Communists have commonly been involved in terrorism in Latin America. During the Cold War, the United States supported many military dictators in an attempt to halt Russian political influence in the area. As a result, many peasants and intellectuals began to attack the governments they saw as illegitimate in the 1960s, 1970s, and 1980s. This impelled those in power to squash opposition movements.

One vivid example is from El Salvador. Máximo Hernández ruled with an iron fist and persecuted anyone suspected of having ties with Russia. In some cases, he forced alleged communists to dig ditches that were then filled with their bodies after these individuals were executed. Some historians assert that the U.S. Marines were

FIGURE 5-1 Many leaders in Latin America have employed terrorist tactics in the name of communism. *Source: © U.S. Department of Justice.*

stationed offshore in case their assistance in the effort was needed. One of the prominent leftists of the time, Farabundo Marti, was put to death during the purge known as the Matanza (the killing). This resulted in the creation of the Farabundo Marti Liberation Front (FMLF). The FMLF was involved in guerilla-type activities in El Salvador until a peace accord was signed with the government in 1992.

There have been many other terrorist organizations that have fought for economic justice including the Zapatista National Liberation Army in Mexico and the Rebel Armed Forces in Guatemala. The Morazanist Patriotic Front, Sandinistas, and Revolutionary Armed Forces were other left-wing terrorist organizations in Honduras, Nicaragua, and Colombia, respectively. Bolivia, Chile, Uruguay, Argentina, and Brazil also witnessed many terrorist attacks in the name of communism. The Shining Path and Tupac Amaru Revolutionary Movement (TARM) had similar ideologies in Peru. These groups desired to rid their country of outside imperial influence. For this reason, TARM assassinated General Enrique López Albújar eight years after the organization was founded.

In the 1970s and 1980s, state-sponsored terrorism was common in the Middle East (Combs 2000). For instance, Iran and Syria supported the suicide bomber who attacked the Marine barracks facility in Lebanon in 1983. These nations facilitated the terrorist who delivered a 2,000 lb bomb on a flatbed truck against U.S. forces in the lobby of the headquarters building, killing 241 soldiers. Syria was also involved in bombings in France and plane hijackings. This nation frequently provided training and a safe haven for terrorists operating in this region of the world. In addition to Syria, Sudan has supported many terrorist groups. In the 1990s, it was the home of Osama bin Laden and his followers. Because of international pressure, however, Al-Qaeda was forced to leave this country, and this terrorist organization took up its presence in Afghanistan. The Taliban (the ruling government in Afghanistan) then allowed bin Laden to use the country as the operating base of this terrorist organization. In return, bin Laden paid these leaders to permit the group's presence and command post in this country. Terrorists have often benefited from the states that support their ideological objectives. In these cases, both terrorists and various states in the Middle East oppose Israel and the United States for political and religious reasons.

Today, the major threat of international terrorism comes from radical **Islamic fundamentalists**. Israel has been dealing with such violence for decades, but the groups opposing the Jewish people and the United States have often grown in number, political strength, and operational sophistication. Russia has likewise been dealing with attacks from similar groups who desire independence in Chechnya. There are also many other Islamic fundamentalist sympathizers and organizations throughout Africa, Europe, and nations such as the Philippines and Indonesia. Arab and Muslim extremists are often viewed as the most dangerous terrorists globally because of events like 9/11, the bombing of trains in Spain in 2004, and the attacks on buses and subways in England in 2005. Such extremists appear to have no reservations about dying as martyrs for their cause.

In The Real World

The Brutality of ISIS

The Islamic State of Iraq and Syria (ISIS) is a radical militant group well known for its strict opposition to the Western world and its ruthless efforts to establish a caliphate based on fundamental Sunni Islamic principles. ISIS is also referred to as the Islamic State of Iraq and the Levant (ISIL) or simply the Islamic state. The group originally emanated from Al-Qaeda in 1999, and Abu Musab al-Zarqawi was its first leader. Al-Zarqawi professed his allegiance to Osama bin Laden in 2004 and was later killed by U.S. forces. Abu Bakr al-Baghdadi was the subsequent leader of ISIS, and he continued the legacy of carnage until he was killed in 2019. Other leaders have taken over and been eliminated in counterterrorism operations up to this point in time.

Regardless of the turnover in leadership, ISIS has been notorious for its brutality toward others who do not share their beliefs and ideologies. Mass bombings, shootings, beheadings, and additional types of executions are common and are frequently recorded to be shared publicly. For example, in June 2014, the group shared a series of tweets and photos on social media claiming to have mass-executed 1,700 Iraqi soldiers. In January 2015, the terror cell imprisoned a pilot from Jordan, trapped him in a cage, doused him with gasoline, and burned him alive while the entire incident was captured on video. In June of that same year, ISIS released a video showing a cage full of spies slowly being lowered into a swimming pool to drown the victims. In August 2016, ISIS publicly executed nine youths by tying them to a pole and then slicing them in half with an electric chainsaw.

On many occasions, members of ISIS have thrown men off tall buildings because of their sexual orientation. ISIS has also been known to stone homosexuals and adulterers to death. Other forms of torture include everything from starving prisoners or boiling them alive. When taking over new territory, ISIS often takes females and uses them as sex slaves. In other situations, children are kidnapped, forced to learn Islamic beliefs, and used as informants or soldiers. The terrorist organization has also used sulfur mustard, sarin nerve, and chlorine gases as chemical weapons against its enemies. In another instance, ISIS put "infidels" in a car and then launched a rocket-propelled grenade at it. The video was uploaded to the Internet for all to see. A recent bombing attack by ISIS against Iran in January 2024 killed nearly over 80 people and injured hundreds of others. These horrendous tactics are utilized with the intent to incite terror and cause as many casualties as possible. This has led the international community to attack ISIS (see Figure 5-2).

FIGURE 5-2 Coalition forces attacking ISIS—one of the most brutal terrorist organizations in the world. *Source: © Orlok/Shutterstock.com*

This brief history of terrorism abroad generates at least two lessons for those involved in homeland security. First, terrorism has been manifested in different ways over time. On some occasions, citizens rose against their own governments. There have been countless cases where political leaders have been assassinated by those who felt mistreated. Later on, states engaged in terrorism against their own people. Russia, under Joseph Stalin, had a deplorable history of state-sponsored terrorism. The government killed millions of its citizens. So did Cambodia. From 1975 to 1978, it is believed that Pol Pot (the leader of the Khmer Rouge in Cambodia) executed as many as two million people. He accomplished this through executions, starvation, and the grueling conditions of forced labor (Simonsen and Spindlove 2000, p. 227). Terrorism has also been related to nationalistic and communistic movements. States have sponsored terrorist groups over time, and a major threat today emanates from radical Islamic fundamentalists around the world.

In The Real World

Chechen Terrorists in Russia

When the Soviet Union disintegrated at the end of the Cold War, pent-up ethnic tensions and rivalries began to re-emerge. Many Chechens disliked Russian involvement in the political decisions that affected them and consequently desired increased independence. In addition, because Chechens share a greater affinity for Islamic tradition than their Russian counterparts, this has been a notable source of conflict among the two groups of people. Because of the continued presence and influence of Russia over the affairs of this nation, Chechens have launched numerous terrorist attacks against them. On August 24, 2004, two passenger aircraft departing from Moscow were blown up in coordinated terrorist attacks. The explosive hexogen was used to bring down the planes. The bombings resulted in the death of over 85 passengers and crew members. It is believed that two females from the Islambouli Brigades and a Chechen field commander were responsible for the attacks.

A second point to be drawn from history is that terrorism often breeds violent counterterrorism activity. Terrorism in France, Spain, Russia, Latin America, and other parts of the world often generated long-standing conflicts among those promoting and opposing this violent behavior. For instance, the civil war in El Salvador lasted from 1980 to 1992. It resulted in the death of 75,000 citizens and displaced millions of people. The perception that violence is an effective solution to ideological disagreements is not always supported with evidence. In most cases, violence has only brought more problems to those seeking a better way of life. Terrorism creates a cruel cycle of viciousness (Combs 2000, p. 24).

Self-Check

1. World War I was initiated in part due to a terrorist attack. True or false?
2. Which ideology was not involved in the early development of terrorism?
   a. National liberation
   b. Anarchism
   c. Environmentalism
   d. Communism
3. What are some of the major lessons we learn from terrorism in other nations?

5.3   Terrorism and The United States

It is imperative that those involved in homeland security realize that terrorism is not just a foreign problem. In fact, certain historians believe that U.S. citizens and its government have participated in terrorism. For instance, and as mentioned earlier in this book, it is sometimes asserted that the tactics used against the British during the American Revolution bear close relation to those of terrorists. Others suggest that the U.S. government sponsored terrorism abroad while it fought communism during the Cold War. As an example, it is alleged that the United States spent $7 million to finance opposition groups to destabilize the leftist government in Chile. The CIA is even reported to have been involved—directly or indirectly—in an attack on Rene Schneider, a commander of the Chilean Army. Evidence also suggests that the United States tried to assassinate Fidel Castro from 1961 to 1962. **Operation Mongoose**, as it was known, involved an attempt to kill the Cuban leader with a poisoned cigar. It is believed that the United States tried to hire Castro's former girlfriend to kill him on another occasion. This plot was also unsuccessful. American politicians at the time claimed that Castro was threatening the security of the United States. Cuba had close ties to the Soviet Union in this era, and Castro accepted the placement of Russian nuclear missiles on this Caribbean Island. For these reasons, other scholars argue that the United States was not engaged in terrorism but was instead involved in violence for geopolitical purposes. This case reiterates that it is not always easy to define terrorism.

The United States may or may not have been involved in terrorism as a participant (depending on your political view or how one defines terrorism and war).

However, this country has definitely experienced terrorism throughout its history. For instance, the **Molly Maguires** was one of the many groups involved in terrorism in this nation after it was founded. The Mollies, as the organization was commonly known, was a collection of citizens who originally joined together in Ireland to dispute the treatment of tenants by landlords. At times, their protests were violent in nature. In the early 1800s, many members of this group migrated to the United States and settled in Pennsylvania. Feeling that they were again being treated unfairly in the coal mines of this area, they began to engage in acts of violence. The Mollies killed many of the mine superintendents in protest. The situation became so bad in the 1870s that Franklin Gowen, president of the Philadelphia and Reading Railroad, told detectives that they were "to remain in the field until every cut-throat has paid with his life for the lives so cruelly taken" (Lejeune 2001, p. 208). Nineteen members of the Molly Maguires were subsequently killed in the late 1800s.

Disputes over workers' rights and compensation also led to other terrorist acts in large cities across the United States (Lejeune 2001, p. 208). In 1886, labor unions requested an eight-hour workday. These collective bargaining organizations believed that the existing and strenuous work hours were unreasonable. As strikes and picketing took place in Chicago, skirmishes broke out among police and rioters. One person was shot and killed in the riot. A few days later, a protest was held to denounce the death. When police tried to disperse the crowd, a bomb was detonated. Eight police officers died in the blast. Explosives were also set off to protest political and economic conditions elsewhere. In New York City, a horse-drawn carriage was taken near J.P. Morgan's house. Its deadly cargo killed 35 people and injured hundreds more. The perpetrators of the bombing, who were believed to have Bolshevist or anarchist leanings, were never apprehended.

The **Ku Klux Klan (KKK)**, a white supremacist group, has also been involved in terrorism in the United States (see Figure 5-3). The KKK emerged after the Civil War era (Simonsen and Spindlove 2000, p. 40). It is alleged that six Confederate veterans were talking around a fireplace one night in Tennessee in the late 1800s and decided to create a secret society. After coming to an agreement on a name to represent the group, these individuals disguised themselves in sheets and rode through the town on horseback. The regalia they wore gained so much attention that the KKK adopted white-hooded clothing as a symbol of the organization. In time, more people joined, and the KKK began to threaten African Americans. This abhorrent behavior was evident during the civil rights movement and again in the early and mid-1990s. Hate speech soon turned to terrorism, and the KKK was involved in "hanging, acid branding, tar-and-feathering, torture, shooting, stabbing, clubbing, fire branding, castration and other forms of mutilation" (Simonsen and Spindlove 2003, pp. 41–42). Thousands of African Americans were killed or attacked by members of the KKK. Fortunately, such violent activities have waned dramatically in the past few decades. But this type of activity has yet to be completely eliminated. And, in fact, there are many neo-Nazi groups in the United States that share much of the ideology of the KKK. Examples include the American National Socialist Party, Aryan Nations, Creativity Alliance, Folks Front, the National Alliance, the White Aryan Resistance, and Vinland Rebels among several others. They desire to establish a white-controlled government in the United States. This racist ideology has been accompanied by violence in May 2022, 10 black people were killed by a white supremacist at a Tops supermarket in Buffalo, New York.

FIGURE 5-3 The United States has had to deal with terrorist organizations, like the Ku Klux Klan in Madison County, Mississippi. *Source: © FBI.*

During the 1950s, 1960s, and 1970s, terrorists in the United States also came from the political left. Individuals and groups like Bill Ayers and the Weather Underground Organization decried class struggles and desired equal rights akin to Marxism. The terrorists planted bombs in public areas, police stations, and military bases (Griset and Mahan 2003, p. 87). They also protested the Vietnam War. Others, like the Symbionese Liberation Army (SLA), had utopian visions for society. This organization desired a unified populous established on cooperative economic institutions. The SLA was responsible for several bank robberies, kidnappings, and assassinations.

The **Black Panthers**, in contrast, was a left-wing group that was more concerned about racial injustice. The Black Panthers sought to oppose the mistreatment of African Americans by the KKK (History.com Editors 2017). Their visibility

In The Real World

Protests, Riots, or Terrorism?

The actions of various individuals and groups on both the left and right have been extremely violent in recent years. Starting on May 26, 2023, between 15 and 25 million people took to the streets to protest the death of George Floyd in response to police brutality in Minneapolis. While most individuals were peaceful, several people attacked police officers, killed 25 people, and caused an estimated $2 billion in damages. One Black Lives Matter leader stated that they will continue to "burn down the system" if change does not happen (Moore 2020). On January 6, 2021, thousands of supporters gathered at the U.S. Capitol building to protest the 2020 election results. Some of these individuals were carrying weapons of various kinds and acted violently against Capitol police officers and destroyed property at this historic building. Some people even brought a gallows to intimidate Vice President Pence, who was certifying election results. More recently, groups supporting Palestinian liberation have voiced their animosity toward the Jews and they have participated in a "Day of Rage" protest. A few days later, one Jewish leader was found stabbed to death in Detroit, Michigan. Many synagogues have also been vandalized. Although arguments could be made to include or exclude these as acts of terrorism, it is clear that our political environment today has become far more violent than in the past.

---

has in some ways become less notable over time, although there was a resurgence of activity at the end of the twentieth century. It was also asserted that a few members of the Black Panthers may have intimidated voters at one polling place in Philadelphia during the 2008 elections (von Spakovsky 2010). A man was carrying and waving a billy club outside the doors of the building, and it has been documented that several people did not vote at that time as a result. This was not interpreted as terrorism, however, and the charges by the Department of Justice were eventually dropped.

In the 1970s and early 1980s, much of the terrorist activity in the United States came from Puerto Rican terrorists. The **Armed Forces of National Liberation (FALN)** emerged in 1974. The goal of this insurgency was to obtain liberation of Puerto Rico, and its supporters were willing to utilize violent means if necessary. The FALN was responsible for at least 100 bombings in the United States in the mid-to-late 1970s. Since that time, many other Puerto Rican groups formed and launched attacks against the United States. "Between 1982 and 1994, approximately 44 percent of terrorist incidents committed in the United States and its territories are attributed to Puerto Rican terrorist groups" (Lejeune 2001, p. 211).

In the 1980s and 1990s, terrorism was associated with many different advocacy issues, and it became one of the biggest threats to the United States. Some **right-wing terrorist organizations** opposed the government. They disliked taxes as well as U.S. involvement in the United Nations. Other conservative groups wanted to protect individual rights to own firearms, or they discriminated against Jews and homosexuals. Some people and groups have been involved in terrorism for these reasons and to stop abortions. For instance, Eric Rudolph bombed abortion clinics and the Olympic Park in Atlanta during the 1996 Summer Games. Numerous people were killed or injured as a result. Some of these right-wing groups of this era included the National Alliance and the World Church of the Creator.

CAREER OPPORTUNITY | Private Security Specialists

The history of terrorism often suggests that public officials and government buildings are often preferred targets of attacks. While there is truth to this statement, terrorism and crime also impact private sector facilities in significant ways. Critical infrastructure, which is often owned and operated by businesses, is especially vulnerable.

For these reasons, many companies hire private security officers to help ensure safety and deter illegal activity at corporate headquarters, retail stores, industrial plants, and other office buildings. These private security officers often have a visible presence at major entrances, and they may or may not carry a firearm or work with k-9 units.

Private security guards receive on-the-job training and make on average about $35,000 per year. While projections show little change in the employment situation over the next decade, there are still about 150,000 openings each year. The advertised positions are posted to replace those who retire and others who seek employment elsewhere.

For further information, see **https://www.bls.gov/ooh/protective-service/security-guards.htm**.

There are also terrorists who used violence in the 1990s to promote animal rights or environmental conservation. Examples include the **Animal Liberation Front** and the **Earth Liberation Front**. For instance, terrorists attacked research facilities that conduct tests on animals. The perpetrators are concerned about cruel procedures in meat packing facilities as well. New neighborhood developments have likewise been vandalized or set ablaze to discourage the destruction of forests (for wood) or the expansion of urban sprawl. According to one study, there were a total of 239 arsons and bombings in the United States between 1995 and 2010 (National Consortium for the Study of Terrorism and Responses to Terrorism 2013). Sadly, "Earth Now, one militant environmental group in the United States rationalized that, if it was necessary to kill people to save the trees, then they would be justified in killing people" (Combs 2000, p. 43).

All of these cases illustrate that the United States has been affected by terrorism long before the first World Trade Center attack in 1993, the bombing of the Murrah Federal Building in 1995, or the 9/11 attacks in 2001. It is also evident that terrorists in the United States come from diverse ideological interests as well. These facts illustrate the complicated nature of terrorism in our modern era.

Self-Check

1. The United States is not believed to have been involved in committing terrorist attacks. True or false?

2. The Black Panther Movement likely emerged due to the racist activities of the KKK. True or false?

3. Which terrorist organization emerged due to concerns over workers' rights?

   a. Molly Maguires

   b. The KKK

   c. The Armed Forces of National Liberation

   d. Animal Liberation Front

4. Explain how terrorism has manifested itself in different ways over time in the United States.

5.4 Terrorism Today

Another important requisite to effectively deal with terrorism is to be aware of how different current terrorist activity is as compared with that of the past (Rubin and Rubin 2008). In the past few decades, notable alterations in the manifestation of terrorism have become evident. Scholars such as Veness (2001), Hoffman (2001), Kegley (2003), and Jenkins (2008) have explored how terrorism has changed over time. They cite numerous examples of significant transformations, including the following:

- In the past, terrorists were organized hierarchically. Organizations such as the Palestinian Liberation Organization often had a clear leader and an identified chain of command. *More recently, terrorists around the world (like Al-Qaeda) organize in a diffuse manner and are cell oriented.* A cell is a semiautonomous terrorist organization that may be loosely affiliated with other terrorist groups. Current terrorism therefore resembles a network of many groups that operate in a loosely coordinated fashion. In other words, direction among terrorists may or may not always emanate from a central headquarters.

- Terrorists were frequently supported by states in prior years. Governments like Libya or Syria provided sanctuary and financial assistance to those engaging in terrorism. Many terrorists, such as Hamas and Hezbollah, continue to work closely with Iran now. However, *today's terrorist organizations may launch attacks with or without approval from governments even if these states sanction such behavior.* What is more, several terrorist organizations have become more independent in terms of raising funds. For instance, Aum Shinrikyo, the infamous Japanese terrorist group, acquired money through religious donations and legitimate business activities. It was not reliant on any type of state support.

- In former years, it was also easier to distinguish between crime and terrorism. The differences between theft and assassinations were more readily apparent. *Currently, it is more challenging to make a distinction between criminal and terrorist activities.* Many terrorist groups rely on crime to raise money, and some organizations that were founded on terrorist aspirations now promote extortion for financial gain (e.g., left-wing groups or drug cartels in Latin America or the Communist Party of the Philippines).

- In the past, citizens of the United States typically regarded terrorism to be a problem in a very limited geographic area. The feeling was that terrorism was traditionally believed to be a feature of the Middle East or, to a lesser extent, Europe. Nonetheless, terrorism has been prevalent in the United States throughout its history, and 9/11 and recent attacks brought the reality of international terrorism home. Terrorism is also making significant headway in South Asia and even Australia. No place appears to be immune.

- Historically, terrorists desired goals such as liberty, national independence, or political stability. They used violence as a way to promote freedom or strengthen government control over the masses. *At the moment, the mission of terrorists is more messianic in orientation.* There is a desire to "save" the world from abortions, environmental degradation, or the evils of Western civilization. Extreme Christian groups and the Environmental Liberation Front come to mind as examples of these types of terrorist organizations. Of course, each of these groups has their own strategies and tactics to obtain their messianic objectives. As an example, the approach of ISIS today is based firmly on providing inspiration to those who may join their cause. Right-wing terrorists believe their actions are needed to preserve and maintain the Constitution of the United States of America.

- Twenty years ago, terrorists may have shared their ideological goals and messages with the public at large. The goal was to instill fear into the minds of individuals and governments to change their behavior. *More recently, terrorists such as Hamas or Hezbollah give greater attention to sharing their message in detail with their sympathizers.* Their goal is to improve recruitment for sustained operations. Terrorists recognize that they need continual backing if they are to be successful in the future.

- Since its birth, terrorism has largely been related to the assassination of specific leaders for clearly defined purposes. By "taking out" individual politicians (e.g., Yitzhak Rabin), the hope was to advance a particular political or ideological agenda. *Today, terrorists are increasingly willing to target citizens.* The belief is that citizens will cave into terrorists' demands due to fear, which will put pressure on governments and allow terrorists to pursue their desires and ambitions with limited opposition.

- Terrorists have always been willing to kill others when it advances their cause. Violence was seen as a "logical" means of goal attainment. While terrorists have periodically engaged in suicide attacks, the practice has been given more legitimacy in the late 1900s and early 2000s. *Terrorists from the Middle East and elsewhere are increasingly willing to kill themselves in the process of taking the lives of others.*

- Terrorist activities in the past were relatively small and isolated. Attacks killing more than a handful of individuals were, for the most part, rare. *Terrorism, especially international terrorism, is more deadly today.* Many attacks kill dozens of people, and there are others that have extinguished the lives of hundreds and even thousands. Terrorism in Africa, the 9/11 attacks in the United States, and events in Spain, Russia, England, and France all resulted in more deaths as compared with the past.

- Terrorist attacks a few decades ago were relatively simple and limited in terms of impact. A conventional explosive was used to kill the occupants of a single building. In contrast, *terrorists today employ novel techniques in attacks, and their objectives are more grandiose.* Planes are used as missiles (e.g., 9/11), and vehicles are employed to run over innocent bystanders (e.g., England and the United States). Terrorists also intentionally disrupt urban infrastructure and love to produce adverse effects on the economy.

- Terrorists were somewhat cautious about their statements in the past and did not always want to take responsibility for their harsh words or activities. They may have tried to avoid being ridiculed or held accountable for their actions. *Terrorists have more vocal and defiant attitudes today.* While all terrorists are not alike, many publicly denounce their enemies and are eager to take credit for their violent and disruptive behavior. ISIS is one of many organizations that publicly denounce the United States as the "great Satan." This group has not hesitated to acknowledge its role in violent behavior.

- Terrorists in earlier decades were relatively easy to track down and capture. They were not always careful about their communications and were less than meticulous in leaving evidence that could be used for prosecution. *Terrorists are now more elusive.* They utilize technology (e.g., calling cards or the Internet) to minimize detection. They blend into the community or hide in countries with weak governments and in isolated areas that cannot be reached effortlessly. As an example, Osama bin Laden operated from remote locations in Afghanistan and Pakistan. Terrorists' obscure activities create serious challenges for those trying to thwart terrorism.

The Distinct Waves of Modern Terrorism

According to certain scholars who study terrorism, several distinct phases of this type of violent phenomenon have been experienced over time.

David Rapoport wrote an important work (2011) in which he asserted that modern terrorism has witnessed four waves throughout history. This includes the anarchist, anticolonial, new left, and religious episodes. He asserted that each wave has unique aspects related to ideology, strategies, tactics, accounting (finance), recruiting, and targets. For instance:

1. **Anarchism.** This phase began with the assassination of Czar Alexander II and ended with World War I. Strong antigovernment sentiments were the central feature of this wave. This period was manifest in numerous assassinations of political leaders. The supporters of this type of terrorism desired freedom and therefore targeted totalitarian rulers.

2. **Anticolonialism.** This phase ended at the conclusion of World War II. The objective here was to push back on imperialists who entered and controlled countries throughout the world. Hit-and-run tactics were utilized by nationalists who disapproved of the police and military personnel who supported puppet governments that were established by foreign entities.

3. **New Left.** This phase ended as Russia pulled out of Afghanistan. Marxist thought was the major characteristic of this particular wave. To delegitimize capitalism, those supporting this perspective engaged in kidnappings, hijackings, and hostage-taking as common terrorist strategies. The theatrical symbolism of targets during this era sent a strong message that U.S. involvement in the political and economic affairs of others was not to be tolerated.

4. **Religious.** This phase has been one of the most recent manifestations of terrorism. The ideology in this wave was extremely concerned about secularism and it had a strong desire to rid society of Western influence. Suicide bombings were supported by Islamic state sponsors who wanted to push back against the West to establish and strengthen Islamic governments.

Other scholars—Hess, Dolan, and Falkenstein (2020)—invite us to consider if we are now entering another wave which is a combination of all of the prior grievances in a continuous phase of turmoil and perpetual violence. In this fifth wave, many people and groups are engaging in terrorism for diverse reasons. There is ample opposition to rules and a yearning to promote transformational change. Political leaders, governments, and prominent religious leaders are targeted through bombings and other types of attacks. The current phase began with the disruptive COVID-19 pandemic and is ongoing. Recruiting and the promotion of ideological causes will be spread through the use of the Internet, and attacks will likely be exacerbated by modern technology. For this reason, Hess and his colleagues assert that additional research is needed to understand terrorists and find ways to counter such activities.

As a participant in homeland security, you should be aware of these changes and their potential impact on your profession. It is imperative that you recognize that terrorism is not static but is instead ever changing and likely to morph in unique ways in the future. You must anticipate unprecedented and more consequential attacks from diverse terrorist organizations. Uncertainty and complexity are the characteristics of terrorism today.

In The Real World

Terrorism and Technology

Terrorists increasingly exploit modern technology to carry out their attacks (Thetford 2001). In 1998, the Tamil Tigers sent repeated e-mails to the Sri Lankan embassies to disrupt the activities of this government. During the Gulf War in 1990, "High Tech for Peace," a group of Dutch hackers, asked Iraqi agents if they would pay $1 million to disrupt communications between bases in the United States and military units in Saudi Arabia. There have been other instances of foreigners or high school students accessing sensitive computers (e.g., 911 systems, NASA, and the Pentagon). Many terrorist groups also use encryption techniques on the Internet to hide their secret communications. James Kallstrom, chief of engineering at the FBI laboratory in Virginia, states that "In the old days ... Fort Knox was the symbol of how we protected things of great value: we put them in buildings with thick walls and concrete. We put armed guards at the doors, with sophisticated multiple locks and locking bars. We could even build a moat and fill it with alligators ... [Today] we are not equipped to deal with ... [cyber-terrorism], both in the government and [in] private industry" (in Thetford 2001, pp. 243–244).

Self-Check

1. Terrorists today are always organized hierarchically. True or false?
2. Terrorism is a problem in the Middle East only. True or false?
3. Which is not a feature of modern terrorism?
    a. Terrorists are individually sponsored.
    b. Terrorists attack key individuals only (e.g., politicians).
    c. Terrorists work harder to recruit members.
    d. Terrorism is more complex today than in the past.
4. List three ways that terrorism has changed in recent years.

Summary

In this chapter, you have been exposed to the dynamic history of terrorism. You have gained a better comprehension of why terrorism emerged initially. The chapter illustrated how terrorism evolved in other nations and how it has been manifested in the United States as well. In addition, you have increased your knowledge about the differences between terrorism in the current century and the attacks of the past. The major lesson of this chapter is that terrorism has changed over time and will continue to be manifested in diverse ways going forward. You must therefore anticipate unique future attacks if you are to be successful in working toward the achievement of goals in homeland security.

Assess Your Understanding

Understand: What Have You Learned

Go to **www.wiley.com/go/mcentire/homelandsecurity3e** to assess your knowledge of the dynamic nature of terrorism.

Summary Questions

1. The fear created by warriors painting their faces, brandishing weapons, and yelling loudly at their opponents is somewhat similar to the creation of fear associated with terrorism. True or false?

2. The word "terrorism" appeared during the Enlightenment. True or false?

3. Communism is an ideology that sympathizes with the poor and downtrodden. True or false?

4. The KKK was formed in Tennessee. True or false?

5. The Black Panther movement arose due to the violence of the KKK. True or false?

6. Operation Mongoose was an attempt to assassinate Saddam Hussein. True or false?

7. A cell is a completely autonomous terrorist group with no affiliation to other terrorist groups. True or false?

8. Anarchists are those people who:
    a. Want a communist nation
    b. Oppose specific governments or all governments
    c. Fight for a democratic state
    d. Think that the government is not doing enough

9. Which of the following individuals was not in some way associated with the "Gunpowder Plot?"
    a. Pope Clement VII
    b. King Henry VIII
    c. Sarah Elizabeth
    d. Anne Boleyn

10. The following groups or individuals were a part of the "Reign of Terror" except:
    a. The Jacobin party
    b. Marie Antoinette
    c. France's Committee of Public Safety
    d. The Spaniards

11. The Molly Maguires were:
    a. A Scottish group opposed to the religious state
    b. An American group opposed to immigration
    c. An Irish group supporting the elite of Ireland
    d. An Irish group that disputed the treatment of tenants by landlords

12. Syria is known for all of the following except:
    a. Being sympathetic to U.S. Marines
    b. Supporting the bombing of barracks in Lebanon
    c. Bombings in France
    d. Plane hijackings
13. Which of the following groups set fire to ski resorts in Colorado?
    a. Earth Liberation Front
    b. Black Panthers
    c. Molly Maguires
    d. The Armed Forces of National Liberation
14. All of the following are characteristics of today's terrorist groups except:
    a. Some rely on crime for money.
    b. Terrorism is a feature of the Middle East only.
    c. The practice of suicide attacks is seen as more legitimate by many terrorists.
    d. Terrorist activity is carried out on a larger scale.

Applying This Chapter

1. How did the Enlightenment change the way people saw their role in society? How did it alter the way citizens viewed their leaders? What is one reason for the increased violence and the creation of the term "terrorism" during the Enlightenment?
2. Describe the difference between the violence mentioned in the first section of the chapter and the terrorist attacks we see today.
3. The chapter makes clear the fact that terrorism is employed by groups acting against established public organizations and is also implemented by government organizations themselves. Why is it important that we understand both kinds of terrorism? Provide one example of a terrorist group rising against an established organization or government and one example of a terrorist group supported by a government organization.
4. There are several examples of different terrorist groups that have acted within the United States. What common threads run throughout these groups? Why is it important to understand these commonalities?
5. This chapter lists 12 ways in which terrorism has changed and evolved. Why is it important for those working in homeland security to understand these changes? What does the evolution of terrorism tell us about terrorism in the future?

Be a Homeland Security Professional

The Emergence of Terrorism

You are a historian who has been hired as a consultant by the CIA. Your job is to write a two-page brief about the birth of terrorism. What could you say about the emergence of terrorism and its manifestations in the 1700s and 1800s?

History of Terrorism Abroad or in the United States

Write a three-page paper describing the evolution of terrorism abroad or in the United States. Be sure to describe how it has changed over time.

Differences Between Modern and Traditional Terrorism

As a local emergency manager or employee in homeland security, your job is to anticipate terrorist attacks in your community. Could the knowledge about the differences between modern and traditional terrorism help you in your job? How so?

Key Terms

Anarchists   Those opposing specific governments or all governments

Animal Liberation Front   A terrorist organization that opposes cruelty to animals

Armed Forces of National Liberation (FALN)   A Puerto Rican terrorist organization seeking the liberation of Puerto Rico from the United States

Black Panthers   An organization composed of African Americans to counter the actions of the KKK and other white supremacists

Communism   An ideology that sympathizes with the poor and downtrodden and attempts to do away with private property

Earth Liberation Front   A terrorist organization that opposes environmental degradation

Enlightenment   A period in history when a new way of looking at social, political, and economic structures emerged

Fascism   An ideology that promotes the uniting of citizens in support of the state

Gunpowder Plot   An attempted terrorist attack in 1605 against King James I and other leaders of Parliament to reinstate Catholic involvement in England

Islamic fundamentalists   Individuals or groups of Muslims who violently oppose Israel and the United States

Ku Klux Klan (KKK)   A white supremacist group that has been involved in terrorism in the United States since the Civil War

Molly Maguires   A group of Irish citizens joined together to dispute the treatment of coal mine workers in the United States

Nationalist movements   Efforts on the part of a group or nation to obtain political independence and autonomy

Operation Mongoose   An attempt by the United States to kill Cuban leader Fidel Castro with a poisoned cigar

Reign of Terror   A period during the French Revolution when an estimated 20,000 persons were killed by France's Committee of Public Safety

Right-wing terrorist organizations   Conservative groups that promote gun rights, oppose abortions, question the value of the government and the United Nations, and discriminate against blacks, Jews, and homosexuals

References

Chalian, G. and A. Blin. (2007). *The History of Terrorism: From Antiquity to Al Qaeda*. Berkeley, CA: University of California Press.

Combs, C.C. (2000). *Terrorism in the Twenty-First Century*. Upper Saddle River, NJ: Prentice Hall.

Griset, P.L. and S. Mahan. (2003). *Terrorism in Perspective*. Thousand Oaks, CA: Sage Publications.

Hess, J.H., Dolan, J.P., and Falkenstein, P.A. (2020). The fifth wave of modern terrorism: perpetual grievances. *American Intelligence Journal* 37 (2): 128–138.

History.com Editors. (2017). Black Panthers. November 3. https://www.history.com/topics/black-history/black-panthers. (Accessed October 2, 2023).

Hoffman, B. (2001). Change and continuity in terrorism. *Studies in Conflict and Terrorism* 24: 417–428.

Jenkins, B.M. (2008). The new age of terrorism. In: *Weapons of Mass Destruction and Terrorism* (ed. R.D. Howard and J.J.F. Forest), 23–31. New York: McGraw Hill.

Kegley, C.W. Jr. (2003). *The New Global Terrorism: Characteristics, Causes, Control*. Upper Saddle River, NJ: Prentice Hall.

Lejeune, P.H.B. (2001). History and anatomy of terrorism. In: *Terrorism: Defensive Strategies for Individuals, Companies, and Governments* (ed. L.J.Hogan), 203–238. Frederick, MD: Amlex, Inc.

Moore, M. (2020). BLM leader: If change doesn't happen, then "we will burn down this system." *New York Post*. June 25. https://nypost.com/2020/06/25/blm-leader-if-change-doesnt-happen-we-will-burn-down-this-system/. (Accessed October 2, 2023).

National Consortium for the Study of Terrorism and Responses to Terrorism. (2013). An overview of bombing and arson by environmental and animal rights extremists in the United States, 1995–2010. https://www.dhs.gov/sites/default/files/publications/OPSR_TP_TEVUS_Bombing-Arson-Attacks_Environmental-Animal%20Rights-Extremists_1309-508.pdf. (Accessed October 2, 2023).

Rapoport, D.C. (2011). The four waves of modern terrorism. https://international.ucla.edumedifilesrapoport-four-waves-of-modern-terrorism.pdf. (Accessed October 2, 2023).

Rubin, B.M. and J.C. Rubin. (2008). *Chronologies of Modern Terrorism*. Armonk, NY: Routledge.

Simonsen, C.E. and J.R. Spindlove. (2000). *Terrorism Today: The Past, the Players, the Future*. Upper Saddle River, NJ: Prentice Hall.

Simonsen, C.E. and J.R. Spindlove. (2003). *Terrorism Today: The Past, the Players, the Future*. Upper Saddle River, NJ: Prentice Hall.

Thetford, R.T. (2001). The challenge of cyberterrorism. In: *Terrorism: Defensive Strategies for Individuals, Companies, and Governments* (ed. L.J.Hogan), 239–257. Frederick, MD: Amlex Inc.

Veness, D. (2001). Terrorism and counterterrorism: an international perspective. *Studies in Conflict and Terrorism* 24: 407–416.

Venturi, F. (1966). *Roots of Revolution: A History of the Populist and Socialist Movement in the Nineteenth Century*. New York: Norton.

von Spakovsky, H.A. (2010). Voter intimidation New Black Panther Style. April 27. https://www.heritage.org/commentary/voter-intimidation-new-black-panther-style. (Accessed October 2, 2023).

CHAPTER 6

Evaluating a Major Dilemma

Terrorism, the Media, and Censorship

DO YOU ALREADY KNOW?

- How the media has changed over time
- Why terrorists rely on the media
- If the media benefits from terrorism
- Why the portrayal of terrorism is important to government officials
- The drawbacks of media censorship

For additional questions to assess your current knowledge of the dilemmas associated with terrorism, the media, and censorship, go to **www.wiley.com/go/mcentire/homelandsecurity3e**

WHAT YOU WILL LEARN

6.1 **How reporting is different today than in the past**

6.2 **Why the media is utilized by terrorists**

6.3 **What the media gains from terrorist activity**

6.4 **If the government is concerned about the reporting of terrorist attacks**

6.5 **Why censorship is to be avoided**

WHAT YOU WILL BE ABLE TO DO

- Comprehend the features of modern news coverage
- Predict how terrorists take advantage of the media
- Evaluate reasons why publicity of terrorist attacks is in the interest of the media
- Assess why reporting about terrorism is of concern to government officials
- Critique the possibility of media self-censorship in the United States

Introduction

An important way for you to reduce the impact of terrorism is to comprehend the role of the media in publicizing and perhaps even aggravating this type of violent activity. Those involved in homeland security activities must grasp how the media has changed over time and be conscious of the complicated relationships between terrorists and news organizations. It is imperative that you recognize what the government desires from the media in addition to the drawbacks of censorship or self-censorship. Finding appropriate ways to work with the media before and after terrorist attacks is a notable challenge and opportunity you and others will confront in homeland security.

6.1   Changes in The Media Over Time

To adequately envisage the nature of terrorism today, it is imperative that you first recognize how the media has changed over time. In many ways, there were no unofficial outlets for news up until the mid- or late 1700s. Information was distributed (and sometimes even created) by the king or political party in power. The crafting and custody of information resulted in a great deal of control over people's perspectives. In addition, news traveled slowly as it was transmitted by horse, rail, or telegraph. The dissemination of official statements was thereby severely limited due to geographical distance. The lack of education around the world also constrained the demand for information. Furthermore, people could not always read or understand what was being shared with them or decipher if something was accurate or the implications of what was being relayed.

Although authoritarian governments still control the production and flow of information in certain countries around the world today (e.g., China, Iran, North Korea, Russia), things have changed dramatically with the creation of the media over the past few centuries and the ease of transmission that has become apparent in recent decades (Hoffman 1998, p. 136). For instance, the establishment of democratic governments and the rise of capitalism have resulted in the appearance of additional news sources. Media organizations now determine what the news is and convey it to an increasingly educated public citizenry. Furthermore, technological innovations have dramatically reduced the time it takes to send information to national audiences and around the world. Television, cable TV, satellite dishes, the Internet, e-mail access, and cell phones with cameras have revolutionized the media. Getting news to the public took months, weeks, or days in the past. Now, the sharing of information is measured in hours, minutes, and even seconds (see Figure 6-1). News that was only given to people in large urban areas and developed communities is now accessible to anyone around the world. With the advent of the Internet, ordinary people are able to become online journalists (Mahan and Griset 2013). They can utilize social media applications like Instagram, Twitter, and Facebook to relay news events around the world at a dizzying pace. Such changes have had dramatic consequences. On the positive side, we are more aware of what is going on, and we can share information quickly with one another. People are knowledgeable about current and unfolding events. On the negative side, new technologies and the media can be fully exploited by terrorists. The same tools that can benefit individuals and organizations can be employed to maximize the impact of terrorist attacks on people far from the impact zone.

FIGURE 6-1   News can be quickly broadcast around the world with modern technology.
Source: Gorodenkoff/Adobe Stock.

Self-Check

1. Prior to the 1800s, most news came from the king or ruling elite. True or false?
2. Information traditionally traveled slowly around the world because of limitations in travel or communications. True or false?
3. What type of technology has not transformed the media in recent years?
   a. Television
   b. The typewriter
   c. Satellite dishes
   d. The Internet
4. What are the pros and cons of the modern media?

6.2 Terrorists and The Media

It is increasingly accepted as fact that terrorism and the media have an extremely close relationship (Mahan and Griset 2013; Nacos 2016). Indeed, terrorists are heavily dependent on the media in our modern era. To achieve their ideological goals, terrorists want to promote legitimacy for their cause, shock as many people as possible, and instill widespread fear. Terrorists also want to augment their own visibility (for monetary support and recruiting purposes) and destabilize the enemy (by illustrating that the government cannot cope effectively with attacks). In the mind of terrorists, an excellent way for them to reach their desired objectives is to rely on the assistance of reporters.

One of the most notorious examples of this occurred during the 1972 Munich Olympics (see Figure 6-2). In an attempt to reverse real or perceived injustices against Palestinians, **Black September**—an operational unit of the Al-Fatah terrorism organization—launched an attack against Israeli athletes who were participating

FIGURE 6-2   The media was used extensively by terrorists in the 1972 Olympics. *Source:*
© Popperfoto/Getty Images. Reproduced with permission of Getty Images.

in the international games (Simonsen and Spindlove 2000, p. 140). Decrying the
eviction of the Palestinian Liberation Organization from Jordanian training camps
in September 1970, Abu Iyad (the head of the Palestinian intelligence network) and
his followers entered the Olympic Village in Germany and immediately killed two
members of the Israeli team. Using the threat of AK-47 assault rifles, the terrorists
gained control over the remaining athletes and took them hostage. With this bargaining chip in their pocket, the terrorists then demanded the release of 234 Palestinians who were held in Israeli jails. The large global audience that was initially
tuned into the high-level competition of the Olympics soon became consumed
with the unfolding drama. Terrorists thought the broadcasting of the hostage situation would put intense pressure on the Israeli government and cause them to
acquiesce to their demands. Deep sympathy was manifested toward the Jewish people as international viewers were given updates about the situation. But the state of
Israel decided not to release the prisoners. Nevertheless, Israeli and German leaders did allow the terrorists to fly by helicopter to Furstenfeldbruck Airport where
they were told they could transfer to a plane and fly to Cairo, Egypt. Within a short
time, most of the terrorists were killed by snipers, but not before the remaining
hostages were killed. Because this was "the first time a terrorist incident reached
a global audience during a live broadcast" (Doubek 2022), the event illustrated the
potential impact of the media in issuing reports about unfolding hostage situations
and terrorist attacks.

Since this time, the media has increasingly been seen as an important tool
to be used at the disposal of terrorists. For instance, after taking hostages at the
headquarters of the Organization of Petroleum Exporting Countries (OPEC) in

Vienna in 1975, Carlos "the Jackal" waited for the media to arrive before exiting the building with the oil ministers he kidnapped (Hickey 1976, p. 6). Another case is also illustrative. When 52 American hostages were being held at the American Embassy in Tehran in 1979, the mob outside only began to shake their fists, burn flags, and denounce President Carter when a Canadian broadcasting company showed up at the scene (Hoffman 1998, p. 142). To increase public relations, the Irish Republican Army historically promoted "a close relationships with the print and broadcast media" (Martin 2003, p. 281). Ted Kaczynski likewise sent his 35,000-word manifesto against technology to the *Washington Post* and The *New York Times*. This document denouncing economic development and the U.S. government was published in 1995 (Griset and Mahan 2003, p. 132). Thus, these terrorists wanted to bring attention to their cause and saw the media as a vehicle to make that happen.

Osama bin Laden also used the media to promote Al-Qaeda's grandiose wishes for a new world order in which he and other Muslim leaders would establish and strengthen Islamic governments around the globe. In the late 1990s, bin Laden was interviewed by Western reporters. He told them that America would be attacked if it did not withdraw its troops from the Middle East and stop supporting puppet governments in the region. Because this ultimatum was not met, bin Laden's suicide pilots took over four planes on 9/11 and began their campaign of death. Al-Qaeda relied on the morning news broadcasts and large international media organizations in New York to publicize the lengths it was willing to go to force the United States and other nations to cave into its call to depart Muslim countries. Later on, bin Laden and ISIS leaders encouraged sympathizers to acquire weapons of mass destruction and use them against Americans or their interests, regardless of their location around the world.

It is apparent that terrorists have relied on the media to get their message across. But they are increasingly turning to social media as well (Thompson and Isaac 2023). It is true that Hamas has been banned by Facebook, Google, Instagram, TikTok, and X (formerly Twitter). However, on October 7, 2023, Hamas broadcast its gruesome attack on Israel via Telegram (a global social media app that was originally created in Russia). Hamas shared posts showing Hamas fighters killing Israeli soldiers, attacking those attending a music festival, shooting dogs, and taking hostages. Hamas used the app to justify its attack on Israel, invite sympathizers to join Hamas, and call for additional violence against the Jews. Many of the subscribers to this platform have given a "thumbs up" to the posts distributed by Hamas. The number of Hamas accounts tripled and even quadrupled after the attacks on Israel. Thus, it is evident that the media and social media are viewed as useful avenue through which terrorist may achieve their aims and objectives.

Understanding this unique relationship, Brian Jenkins—a well-known RAND Corporation researcher—stated, "terrorists want a lot of people watching and a lot of people listening. ... Terrorists choreograph dramatic incidents to achieve maximum publicity, and in that sense, terrorism is theater" (1985, p. A4). Jenkins also asserted, "Terrorism is violence for effect—not primarily, and sometimes not at all for the physical effect on the actual target, but rather for its dramatic impact on an audience" (Jenkins 1985, p. 101). Others agree with this assessment. For instance, Walter Laqueur believes the "media are a terrorist's best friend." Frederick Hacker emphatically declares that if the mass media organizations did not exist, "terrorists would have to invent them!" (see Combs 2000, p. 128).

In The Real World

Terrorism on TV

On June 14, 1985, three Lebanese terrorists hijacked TWA Flight 847 from Athens to Rome. The plane was initially flown to Beirut, and it then made several trips between Lebanon and Algeria. During the 16-day ordeal, nearly 500 reports on nightly news programs (CBS, ABC, NBC) were devoted to the event. The terrorists skillfully used the media to publicize their cause, and they even attempted to charge networks for a tour of the plane and to conduct interviews with them. The broadcasts ended up creating enough pressure on the U.S. government that it gave in and negotiated with the terrorists. Everyone on board the plane was released when 756 Shi'a terrorists were freed from prison in Israel. Gus Martin believes "the hijackers masterfully manipulated the world's media" in a way to get what they wanted (2003, p. 294).

Self-Check

1. The media helps give terrorists visibility. True or false?
2. Osama bin Laden did not speak with Western media because he regarded them to be infidels. True or false?
3. Which terrorist attack was among the first to exploit international media?
   a. Tehran hostage crisis
   b. The OPEC hostage crisis
   c. The Munich Olympic massacre
   d. 9/11
4. Why is the media regarded to be the terrorist's "best friend"?

6.3 The Media and Terrorism

While terrorists highly value the publicity generated by reporters, the media has a very complicated relationship with terrorism (Eid 2013). On the one hand, the leaders of news organizations may not always want to report the cause or impact of terrorist attacks to shape the political narrative regarding this phenomenon. In a 2017 speech to the military at MacDill Air Force Base, President Donald Trump decried the "fake news" and asserted that the dishonest press does not want to relay to the public what is actually happening. Trump's argument was that the media is typically liberal and wants to avoid social discrimination, harsh interrogation practices, and covert military operations. In fact, after Trump's speech, the White House released a list of 78 attacks carried out by ISIS that were underreported by the media. While the media probably downplayed these terrorist attacks to protect Muslims and promote the interests of the Democratic Party (e.g., lax immigration policies), the media's stance on reporting is not always consistent.

In fact, the media can be obsessed with terrorism at times, and this could have an enormous impact on terrorist behavior in the future. According to former British

Prime Minister Margaret Thatcher, the media provides "the oxygen of publicity on which [terrorists] depend" (in Hoffman 1998, p. 142). The media also stands to gain much from terrorists and their ideologically motivated attacks. In particular, the media desires to make "the news entertaining enough to 'sell'" (White 2002, p. 257).

In other words, and fortunately for the media, terrorist attacks offer a dramatic presentation that keeps the public engaged in the story at hand. The attention given to terrorism is owing to the violent attacks, heroic responses, unfolding outcomes, and momentous consequences. These are only a few of the features that keep people glued to their television, encourage them to watch TV and cable coverage, listen to radio broadcasts, and remain riveted to social media posts. Ratings are also important to the media, and, therefore, reporters exploit terrorism to their advantage. In other words, the media broadcasts terrorist attacks to maintain or increase their standings in the competitive news world.

Of course, it is necessary to acknowledge that the media does present a timely portrayal of the news to keep society informed of important events (Perl 1997, p. 7). The media may also play a positive role in the resolution of some terrorist conflicts. For instance, in one case, hostages felt that reporters kept politicians interested in their plight until their situation could be peacefully resolved (Hoffman 1998, p. 147). The media can increase awareness and bring publicity to potential or actual victims of various terrorist attacks.

Media involvement in terrorism is not without drawbacks, however. One of the major problems created by the media revolves around their interference in counter-terrorism operations. For instance, as the Bureau of Alcohol, Tabaco, and Firearms (ATF) was making final preparations to raid the Branch Davidian compound in Waco, Texas, because of illegal weapons violations, the media relayed information that made its way to David Koresh of the government's impending plans. This tip given by the media intensified the Branch Davidian's fear of the government and may have triggered the violence that broke out on February 28, 1993. Ten people (both ATF agents and Branch Davidians) died during the exchange of gunfire, and this violent shootout resulted in a 51-day standoff between the two opposing parties. Eventually, 79 Branch Davidians died when federal law enforcement agencies launched tear gas into the building and David Koresh ordered fires to be lit in the compound to ensure no one survived in a mass suicide operation (see Figure 6-3).

Another excellent case in point concerns the Hanafi siege that occurred in Washington, D.C., from March 9 to March 11, 1977 (Combs 2000, pp. 135–136). Twelve Islamic activist gunmen took over three buildings in an attempt to force the release of imprisoned convicted murderers and protect the name of the Prophet Mohammad. During the ordeal, 2 people were killed and another 149 were taken hostage. Throughout the crisis, the media complicated government negotiations with the terrorists. One reporter who communicated with the terrorists said the police were preparing to attack. Another member of the press called the terrorist leader and stated that the police were trying to trick him. This terrorist then selected 10 hostages for execution. The assailant did not carry through with the threat, however, because the police agreed to defuse the situation by removing sharpshooters from nearby buildings. Nevertheless, the participation of the media slowed down and even reversed negotiations. One of the hostages (who also happened to be a reporter) said this about the media's behavior and mistakes during the takeover:

> As hostages, many of us felt that the Hanafi takeover was a ... high impact
> propaganda exercise programmed for the TV screen, and for the front pages

FIGURE 6-3   Aerial view of Branch Davidian compound near Waco, Texas, in flames on April 19, 1993, following a 51-day siege by the FBI and law enforcement. *Source: © FBI.*

of newspapers around the world. Beneath the resentment and the anger of my fellow hostages toward the press is a conviction gained that the news media and terrorism feed on each other, that the news media and particularly TV, create a thirst for fame and recognition. Reporters do not simply report the news. They help create it. They are not objective observers, but subjective participants (Schmid and de Graff 1982, p. 172).

Besides complicating negotiations and rescue operations, the media may exacerbate terrorism or attacks in other ways. For instance, by giving terrorists airtime, reporters may help the terrorists appear more powerful than they actually are. In addition, continued coverage of terrorist attacks could create emotional problems for viewers. As an example, numerous children suffered from nightmares after repeatedly witnessing the World Trade Center towers collapse in media reports after 9/11. TV stations soon stopped airing this footage as a result. Furthermore, the media may use terrorist attacks to advance a political agenda. Reporters frequently criticized George W. Bush for his aggressive foreign policy in the Middle East. Whether justified or not, the media seemed to be very critical of the decisions made under President Bush and insinuated that his actions were aggravating the ongoing conflict in that part of the world.

More recently, critics have argued that the media is not reporting facts and may even be sympathetic to terrorists. For instance, some in the media have refused to label Hamas as terrorists after the attack on against Israel October 7, 2023. In addition, over 10 state attorneys around the nation sent a letter to those in charge of the New York Times and Reuters in which they expressed concern that these news organizations may have knowingly individuals with known ties to Hamas. Of course, others assert that the media has not adequately covered Israeli actions that may foment terrorism. The media has also asserted that it is not providing material support to terrorists.

Alex Schmid offers other perspectives about the negative impact of the media on the occurrence of terrorist attacks. Four are worthy of mention here:

- **Social learning theory**. Observing terrorist attacks in the news may be the first step in generating similar types of behavior among others (e.g., people learn from what they see).
- **Disinhibition hypothesis**. Violence portrayed by the media may weaken the reticence of others to participate in terrorism (e.g., people become immune to the violence of terrorism).
- **Arousal hypothesis**. Media reports on terrorism can increase people's interest in acting aggressively (e.g., terrorism may be seen as a legitimate way to accomplish goals).
- **Built-in escalation hypothesis**. More deadly and visible attacks in the future are required to get the same amount of media coverage as in the past (in Combs 2000).

The point to remember about the foregoing discussion is that the media has a complicated relationship with terrorism and that attacks can be magnified if reporting occurs in a careless way.

In The Real World

The *Achille Lauro* Incident

In 1985, the Palestine Liberation Front hijacked an Italian cruise ship, the *Achille Lauro*. The goal of this organization was to exchange passengers for 50 Palestinian terrorists imprisoned in Israel. During the incident, terrorists killed a man who was confined to a wheelchair and dumped his body into the Mediterranean Sea. In response, the United States announced a $250,000 reward for anyone who could provide information that would lead to the arrest of the leader Abul Abbas. The government struggled in its attempt to find Abbas and bring him to justice. However, the media was able to contact him for an exclusive interview on NBC's news program. Although some viewed the interview as tasteless, others claimed it was indeed of interest to the American people. The *Achille Lauro* incident was an excellent case of the symbiotic relationship between terrorists and the media (Hoffman 1998, p. 144).

Self-Check

1. The Hanafi hostage crisis illustrates how the media worked to effectively deal with terrorism. True or false?
2. Social learning theory implies that people will mimic the behavior they see on TV. True or false?
3. Which of the following concepts describes why people become complacent about terrorism in the media?
   a. Social learning theory
   b. Disinhibition hypothesis
   c. Arousal hypothesis
   d. Built-in escalation hypothesis
4. Explain why terrorism is appealing to the media and those interested in the news.

6.4 Government and The Media

According to an editorial in the journal *Terrorism and Political Violence*:

> The relationship between publicity and terror is indeed paradoxical and complicated. Publicity focuses attention on a group, strengthening its morale and helping to attract recruits and sympathizers. But publicity is pernicious to the terrorist groups as well. It helps an outraged public to mobilize its vast resources, and produces information that the public needs to pierce the veil of secrecy all terrorist groups require (Rapoport 1996, p. viii).

This complex relationship is convoluted further because the government also has an interest in using the media to counter terrorism and its impacts.

The government may take special interest in how the media reports the news of terrorist attacks so it can portray itself in the best possible light to the public. This was probably the case with President Barack Obama and some of his senior leadership during their time in office. Three situations revealed that the U.S. foreign policy and the war on terrorism were not as successful as was being implied. First, the **Arab Spring**—a series of uprisings and armed rebellions that spread across the Middle East in 2010 and 2011—revealed that violent extremism was becoming more commonplace in that part of the world. This change in behavior seemed to be downplayed in the media. Second, after four Americans were killed in Benghazi, Libya, Susan Rice (a National Security Advisor) went on five Sunday morning news programs to claim that the deaths were a spontaneous demonstration against a hateful video posted on YouTube. However, Libya president Mohamed Magariaf claimed the event was preplanned and he mentioned the fact that the perpetrators had rocket launchers. This called into question the assumption about a spontaneous demonstration to some degree.

Third, the rising number of terrorist attacks abroad and at home indicated that not enough was being done to counter such violent behavior. Instead of acknowledging and addressing this problem, President Obama implied that the situation was improving, and he even called ISIS the "JV Team." This caused many Americans to question the administration's policies and perspective. As a matter of fact, after a string of attacks occurred in the United States, Michelle Malkin wrote a scathing article entitled "Where Was President Obama?" This political commentator asserted that instead of tackling the problem, the president was "Sabotaging our borders, restricting our gun rights, working to free Gitmo Jihadists, decrying Islamophobia, demonizing conservatives, welcoming jihad sympathizers to the White House, and putting politics over national security" (Malkin 2014). These complex cases indicate that the media and government may try to spin what is happening, although other reporters may question these types of assumptions and assertions.

There are other, more recent, examples of how government influences the narrative about terrorism. On January 3, 2020, Donald Trump ordered a drone strike on an Iranian major general near the Baghdad International Airport. Trump wanted to take out this leader of the Quds Force due to his participation in attacks that killed over 500 U.S. troops in the Middle East. While some in the media criticized the president for assassinating a foreign leader, Trump noted in a speech that "Soleimani was plotting imminent and sinister attacks on American diplomats and military personnel, but we caught him in the act." Trump also shared his foreign policy with the media: "To terrorists who harm or intend to harm any American, we will find you; we will eliminate you."

In The Real World

Fort Hood Shooting

Nidal Hasan carried out a mass shooting against military personnel at Fort Hood, Texas, on November 5, 2009. Hasan shouted "Allahu Akbar!" (meaning "Allah is great!" in Arabic) as he opened fire at the Soldier Readiness Center. The shooting killed 13 people and injured 32 others. Hasan was shot in the incident and ended up becoming paralyzed from the waist down. Evidence later revealed that Hassan corresponded with an Al-Qaeda supporter and became radicalized. Hasan was later convicted in the trial and sentenced to death.

President Barack Obama originally claimed the shooting was workplace violence, which caused public outrage. People believed the president cast the event this way to maintain a perception that he was effective in his efforts to deal with terrorism. Because of this stance, victims were not awarded Purple Heart medals (a military decoration given to those wounded or killed while serving in the military). In November 2011, survivors and other family members of the Fort Hood shooting filed a lawsuit against the U.S. government. The victims and relatives reprimanded the government for not taking sufficient actions to prevent the attack and demanded that it be classified correctly as terrorism.

As a result of this heated disagreement, Congress issued new legislation in the National Defense Authorization Act in December 2014. The new provision now treats military personnel who are killed or wounded in domestic terrorist attacks in the same manner as those who are targeted during foreign military campaigns. The Fort Hood tragedy was also reclassified as domestic terrorism, and victims were subsequently awarded either Purple Hearts or the civilian equivalent (i.e., Defense of Freedom medals). President Obama began to publicly recognize the shooting as an act of terrorism five years later. However, when President Obama gave his farewell address in January 2017, he again reiterated that no attacks by terrorist organizations occurred in the United States during his time in office. This clear exaggeration is a striking example of how government leaders may use the media to advance their policy agenda or attempt to protect their political legacy.

After the events on January 6, 2021, Donald Trump issued several statements to the media to downplay what some people have labeled not just protests but acts of insurrection and even terrorism at the U.S. Capitol. He pointed out in his remarks that he had no role on that day in agitating the violence and asserted that those who attended were "peaceful people, these were great people." He also commented that House Speaker Nancy Pelosi and other democrats were the "ones responsible" for the outcome because they did not provide the necessary security for the event as requested by a Defense Department official on January 2, 2021. When the January 6 Report was issued in December 2023, Trump denounced it and said the investigation and resulting document "did not provide a shred of evidence." Justified or not, Trump was clearly using the media as a vehicle to get his side of the story across to the American people.

Another example is from the school shooting that occurred in Nashville, Tennessee, on March 27, 2023. Audrey Elizabeth Hale (a young transgender man who changed her name to Aiden) entered the Covenant School after sending an Instagram message stating that he was going to die that day. Aiden was armed with a pistol and two rifles and began to discharge 150 rounds at the people located in the building. Three students and three staff members were killed by the gunfire. The police arrived quickly, and within 14 minutes of the 911 call, Aiden was fatally shot. The event brought further attention to the prevalence of mass shootings. However, the incident became even more controversial because the FBI would not release documents provided by Aiden, which may have been a manifesto about his reasonings for targeting this particular Christian school. Several politicians—including Donald Trump, Senator Bill Hagerty, and House Representative Jeremy Falson—argued that the journal entries should be released so a similar tragedy could be prevented in the future. At the time of this writing, portions of the manifesto have been leaked and there is debate about the need to share the entire document with the public.

The October 7, 2023, attack on Israel provides an additional interesting case. Immediately after the violence occurred, Prime Minister Benjamin Netanyahu held a press briefing to give a statement on the atrocities committed against his nation and lay out his plans for retaliation. He acknowledged in his statement that "Hamas initiated a murderous surprise attack against the state of Israel and its citizens." He denounced the fighters as animals and vowed to wipe them off the face of the earth like ISIS. He also noted, "I am initiating an extensive mobilization of the reserves to fight back on a scale and intensity that the enemy has so far not experienced." Netanyahu also called upon the international community to take a stand against terrorism and support Israel in its unprecedented moment of crisis.

Meanwhile, President Joe Biden also relied on the media to share his assessment of the Hamas attack on the Jewish State and the subsequent military response by Israel. After traveling to Israel, he participated in a joint meeting with Israeli leaders and reporters. Biden sympathized that the "recent terrorist assault on the people of this nation has left a deep, deep wound." He said, "Hamas committed atrocities that recall the worst ravages of ISIS, unleashing pure unadulterated evil." The president then reiterated that it would stand by Israel and that the United States would make sure "you have what you need to protect your people, to defend your nation." He also noted in his conversations with reporters that he would do whatever it takes to bring American hostages home. Later on, President Biden provided additional remarks perhaps as a way to de-escalate the crisis abroad and limit Palestinian protests at home. He promised aid to the Palestinians in Gaza and asked Israel to delay a ground invasion. When Israel took over Gaza, the president pressured Israel to leave so peace could resume in the region.

The relationship between the government and the media is thus more complicated than simply controlling the narrative around terrorism. Raphael Perl, a specialist in foreign affairs, believes that the government has several requests from the media before, during, and after terrorist attacks (1997, pp. 5–6). Government leaders generally desire to do the following:

1. Gain information about possible terrorist attacks (if those plans exist).
2. Separate the terrorists from the media to deny them a public platform.
3. Obtain information about terrorists if the media is aware of the location where hostages are being kept.
4. Present terrorists in media reports as criminals who devalue life and ignore international or national law.
5. Diffuse the crisis through the accurate dissemination of public information, rather than contribute to an already tense situation.

6. Avoid emotional stories as media reports may place extreme pressure on officials to negotiate with terrorists.

7. Withhold information that may notify terrorists of pending counterterrorism or rescue operations.

8. Avert the sharing of details about intelligence gathering or successful operations.

9. Seek media cooperation in holding back evidence that could be used for future prosecution.

10. Boost the image of their agencies by controlling leaks to the press and avoiding undue criticism.

As can be seen, the government is at times concerned about the media's involvement in terrorism. This is because "terrorists learn their tactics and copy methods from the mass media. Media coverage also serves as a motivation for terrorism. The most serious outcome is that violence seems to increase during media coverage. The mass media have become the perfect instrument of violent communication" (White 2002, p. 262).

How can this major problem be addressed? Two choices are available, but neither is totally desirable nor achievable. The first is to have the government censor the media. The second is to have the media control the content of its own broadcasts. These will be addressed in the subsequent section.

CAREER OPPORTUNITY | Public Information Officers

Public information and media reporting have a significant impact on how terrorism is portrayed and how people react to terrorist attacks and events like mass shootings. As a result, public information officers (PIOs) are hired to help people understand what is happening as well as what measures they should take to protect themselves.

PIOs serve in an important role between the key leaders they report to and the public at large. Specifically, PIOs collect, confirm, and coordinate the dissemination of information to government decision makers and departments (both vertically and horizontally) and to the populace in relevant jurisdictions. The goal is to share correct facts and details with the right audience in a timely manner so appropriate actions can be taken to save lives, protect property, and improve official responses and general public reactions.

PIOs complete most of their work under nonemergency situations, whether that is networking with the media, writing plans, or practicing public relations skills. However, if a terrorist attack or disaster were to occur, PIOs must be able to sort through lots of information, decipher what is unfolding, develop clear and concise messages, and be able to answer questions quickly in person, over the phone, or in a live video interview. Critical thinking and strong communication skills are a must.

PIOs and others in public relations positions earn about $70,000 per year (median pay). Although there are a relatively limited number of positions, openings for PIOs do exist in local, state, and federal government agencies including the Department of Homeland Security. Training is also available through the Federal Emergency Management Agency.

For more information, please see Adapted from **https://www.bls.gov/ooh/media-and-communication/public-relations-specialists.htm**or**https://www.fema.gov/programs/empp/pio/.**

Self-Check

1. The relationship between terrorism and the media is straightforward. True or false?

2. Which is not a reason that the government is concerned about the media's reporting of terrorism?

   a. The government does not want the media to make money.

   b. The media could share evidence that might hurt the prosecution of terrorists.

   c. The media may have information about forthcoming terrorist attacks.

   d. Emotional stories could cause officials to react to terrorism prematurely.

3. Why would the government worry if the media discusses intelligence operations?

6.5 Censorship and Self-Censorship

One of the ways to limit the negative aspects of media coverage is to implement some form of censorship, and this is certainly a controversial measure for the government. **Censorship** is the withholding, banning, or altering of information the media shares with the public. Such censorship is forbidden by law in the United States. In fact, the First Amendment to the Constitution states that "Congress shall make no law ... abridging the freedom of speech, or of the press." Nevertheless, there have been cases where censorship has occurred in our nation's history and elsewhere around the world. For instance, during World War II, the amount and type of information about the conflict was controlled by the government. This grip on reporting had the purpose of maintaining unity at home and obfuscating Germans' understanding of U.S. war plans. Also, in the first Gulf War, the U.S. government denied immediate access of reporters in war zones (presumably for their own safety). The United States is not alone in these types of actions. Britain has been particularly adept at restraining media coverage of terrorism. On various occasions, the media in the United Kingdom was forbidden to broadcast interviews or statements by members of the Irish Republican Army.

While this type of censorship is decried and extremely rare, some scholars suggest that restrictions on news coverage may be necessary in emergency conditions. "Lives hang in the balance during hostage crises. ... Reporting and the freedom to report are not the only critical concerns in such crises: The impact of the media on the event, and media interference with police operations, have also become central issues" (White 2002, p. 263).

The proposal for censorship is far from being accepted, however. Besides conflicting with democratic ideals for society, censorship probably does nothing to stop terrorist activity. In fact, prohibiting the freedom of speech or of the press may actually augment terrorism. This is because people will not have a voice to express concerns about government policies or highly contentious debates. Put differently, citizens have shown a tendency over time to engage in violence if they feel their political values are not being heard or heeded by elected officials. Chechen rebels have argued, for example, that Russia has limited their freedom over the press. This has been one of many factors that have aggravated terrorism in that region of the world. Although censorship has been implemented in certain nations to some degree and at different times, it will probably never occur completely in the West because of the ongoing interest in a free society.

The second option for dealing with terrorism is to have the media take more stringent measures to control their own reporting of news to the public. This is known as **self-censorship**. In some ways, self-censorship is in the interest of news organizations. Terrorists often see reporters as pawns of their enemies. Terrorists therefore seek to silence any opposition to terrorist ideology and violent behavior. These violent extremists may kill reporters, which is an additional way that terrorists spread fear among the population.

Reporters consequently participate in a very dangerous profession (see Figure 6-4). The Committee to Protect Journalists says 110 reporters were killed around the world in 1994 (Marks 1995). While a portion of these deaths were due to accidents, most resulted from assassinations. Hot spots for attacks against journalists include countries such as Algeria, Nigeria, Russia, Turkey, Syria, and Zaire. Drug cartels in Mexico and Colombia are also quick to kill journalists who point out corruption in the government. Reporters and media organizations are also targeted in the United States. A week after the 9/11 attacks, a deranged individual sent letters laced with anthrax spores to Tom Brokaw as well as ABC News, CBS News, the *New York Post*, and the *National Enquirer*. Although Brokaw was not harmed, 5 people died and 17 others were treated for exposure to the hazardous bacteria.

One of the most notable cases of such attacks on professionals in the media involved **Daniel Pearl**. Pearl was a reporter with the *Wall Street Journal*. He was investigating Richard Reid (the shoe bomber) and his possible ties with Al-Qaeda and Pakistan's intelligence service. On January 23, 2002, Pearl went to interview a suspected terrorist. In the process, a group calling itself the National Movement for the Restoration of Pakistani Sovereignty kidnapped Pearl. They demanded, among other things, the release of terror detainees in Guantanamo Bay, Cuba. The terrorists also stated that they would kill Pearl if the United States did not meet their request. On February 1, 2002, Khalid Sheikh Mohammed decapitated Pearl in Karachi, Pakistan. Pearl's death was captured on video and released with the comment that such actions would occur repeatedly if the United States did not give in to the terrorists' demands. The media refused to air this brutal murder and the demands of the terrorists. In time, Ahmed Omar Saeed Sheikh and three other suspects were captured. They were charged with murder in March 2002.

FIGURE 6-4 Journalists are often killed or held hostage by terrorist organizations.
Source: © PRESSLAB/Shutterstock..

Unfortunately, most of the slayings of reporters remain unsolved crimes. In Latin America, it is estimated that "95 percent of the 155 killings, 1109 known beatings, 49 kidnappings, and 205 terrorist acts against media installations ... have gone unpunished" (Marks 1995, p. 9). Other media personnel are frequently threatened verbally or jailed for long periods of time. As an example, China has more journalists in prison than any other Asian country. Attacks against journalists continue in the Middle East to this day.

With the above in mind, one would assume that reporters and media organizations would want to limit their involvement in controversial subjects like terrorism. This is surprisingly not the case. The media wants desperately to protect citizens' right to public information. In addition, a failure to report the news results in lower ratings in a very competitive media market. Although committees among media organizations have been and can be formed to monitor news reporting, disagreement about what should or should not be reported is also frequent.

Reporters therefore do their job in spite of potential threats or disagreement about content. Horria Saihi, an Algerian TV producer, stated, "I know what awaits me in the end is a bullet in the head, but what kills me more is censorship. That would be symbolic death" (Marks 1995, p. 9). Others agree that any form of censorship is an attack on liberty. Cindy Combs states, "[w]hen a democratic society, in panic and anger, abandons one of the cherished principles of law that make it democratic, the society has inflicted on itself a greater wound than the terrorists could achieve, were they to bomb a hundred buildings" (2000, p. 131). Knowing what to do in regard to how the media should report on terrorism remains problematic at best.

In The Real World

The Case of Terry Anderson

Terry Anderson was a reporter stationed in Beirut in 1985. He was snatched from his car and held for seven years by Islamic fundamentalists. When Anderson was finally freed, he later stated, "[i]f there ever was a sense that journalists should be given some sort of immunity, it is long gone" (Marks 1995, p. 1). As Anderson's comments indicate, reporters recognize that they may be kidnapped, beaten, or killed for their comments about terrorism. Unfortunately, terrorists intentionally target the media to maximize publicity of their cause or change the way they are presented in the news.

Self-Check

1. Amendments to the constitution protect freedom of the press. True or false?
2. Censorship has never occurred in democratic countries. True or false?
3. Which reporter was kidnapped and killed in Pakistan?
   a. Richard Reid
   b. Daniel Pearl
   c. Tom Brokaw
   d. Dan Rather
4. Why would the media want to censor its reporting of terrorist activity?

Summary

One of the greatest dilemmas you will face in homeland security is in regard to the relationship between terrorism and the media. Because the media can now relay information around the world in a matter of minutes, you should be aware of the reasons why terrorists desire to publicize their activities through reporters. However, the media also benefits from terrorist attacks because of the nature of such stories and high viewer ratings. Meanwhile, the government desires to alter how terrorism is reported to discredit the perpetrators or try to alter how the public views so politicians can maintain their popularity. For these reasons, you should carefully monitor the information you give the media during and after a terrorist attack. You should also be aware of the drawbacks and limitations of censorship, so you do not aggravate incentives for violent behavior.

Assess Your Understanding

Understand: What Have You Learned

Go to **www.wiley.com/go/mcentire/homelandsecurity3e** to assess your knowledge of the dilemmas associated with terrorism, the media, and censorship.

Summary Questions

1. News that once traveled slowly can now be transmitted around the world in minutes or seconds. True or false?
2. One of the reasons why terrorists want to have media coverage is to show the inability of the government to deal with their attacks. True or false?
3. Red October is the name of the terrorist group that killed Israeli athletes in the 1972 Olympic Games in Germany. True or false?
4. Terrorism is regarded to be "theater" because it is directed toward an audience and not just actual victims. True or false?
5. The media has no interest in reporting information about terrorist attacks. True or false?
6. People are interested in terrorist attacks because of their drama and suspense. True or false?
7. The government wants the media to portray terrorists as criminals. True or false?
8. Censorship is the withholding, banning, or altering of information the media shares with the public. True or false?
9. We are likely to see full censorship of the media in the future in the United States. True or false?
10. Which terrorist sent a manifesto to the media to complain about modern technology?
    a. Osama bin Laden
    b. Timothy McVeigh
    c. Ted Kaczynski
    d. Khalid Sheikh Mohammed

11. Which of the following was not one of Osama bin Laden's demands issued through the media?

    a. Withdraw U.S. troops from the Middle East

    b. Limit pollutants in the environment

    c. Stop supporting corrupt puppet governments

    d. Have Muslims acquire and use nuclear weapons against the United States

12. If a terrorist is successful in reaching his or her goals with the help of the media, others may be likely to engage in similar behavior. This is known as:

    a. The arousal hypothesis

    b. The disincentive hypothesis

    c. The built-in escalation hypothesis

    d. The inflammatory hypothesis

13. Why is the government interested in how the media conveys news about terrorism?

    a. Because terrorists do not care about media reports about attacks

    b. Because the government can never provide any details about what is happening

    c. Because censorship is the best way to protect the security of the United States

    d. Because the media can contribute to an already tense situation

Applying This Chapter

1. Explain why or how terrorists are able to have an impact on countries far away from the actual attack. Be sure to list at least three reasons.

2. Many scholars, including Brian Jenkins, state that "terrorism is theater." What is meant by this comment? How do terrorists use "theater" to get their message across?

3. What were some of the problems created by the media during the Hanafi siege? How was this resolved?

4. Raphael Perl states that the government is interested in media reporting of terrorism for at least 10 reasons. List three of them.

5. Censorship of the media is a controversial strategy for dealing with terrorism. Explain why this is the case, and be sure to discuss the advantages and disadvantages of the curtailing of news information to the public.

Be a Homeland Security Professional

Covering a Terrorist Attack

You are a reporter for CNN. You have just been notified that a terrorist is holding hostages in a shopping mall in Phoenix, Arizona. The terrorist group contacted you and wants your organization to relay their grievances. The terrorists decry U.S. involvement in the United Nations and are threatening to kill people if international policies are not reconsidered. Would your boss want you to conduct the interview? Are there any drawbacks to doing so? If so, what are they?

Public Information Officer

As the PIO for the Department of Homeland Security, your job is to ensure that the government's perspective of terrorism is shared with the media. A bomb has just been detonated in California, and an environmental group is claiming responsibility for the destruction at a petrochemical plant. What are some of the issues or comments you would like the media to address as they cover this attack?

Your Views About Censorship

Write a two-page paper about censorship of the media in relation to terrorist activity. Be sure to discuss why some form of censorship might be considered, as well as its advantages and disadvantages.

Key Terms

Arab Spring   A series of uprisings and armed rebellions that spread across the Middle East in 2010 and 2011

Arousal hypothesis   Media reports on terrorism can increase people's interest in acting aggressively

Black September   An operational unit of the Al-Fatah terrorism organization that initiated the terrorist attacks on Israeli athletes at the Munich Olympic Games in 1972

Built-in escalation hypothesis   More deadly and visible attacks are required to get equal media coverage in the future

Censorship   The withholding, banning, or altering of information the media shares with the public

Daniel Pearl   A reporter with the *Wall Street Journal* who was killed by terrorists in Karachi, Pakistan

Disinhibition hypothesis   Violence portrayed by the media may weaken the inhibition of others to participate in terrorism

Self-censorship   Media control over their reporting of news to the public

Social learning theory   Observing terrorist attacks in the news may generate similar types of behavior among others

References

Combs, C.C. (2000). *Terrorism in the Twenty-First Century*. Prentice Hall, NJ: Upper Saddle River.

Doubek, J. (2022). 50 years ago, the Munich Olympics massacre change how we think about terrorism. NRP. September 4. https://www.npr.org/2022/09/04/1116641214/munich-olympics-massacre-hostage-terrorism-israel-germany. (Accessed October 3, 2023).

Eid, M. (2013). The new era of media and terrorism. *Studies in Conflict and Terrorism* 36 (7): 609–615.

Griset, P.L. and S. Mahan. (2003). *Terrorism in Perspective*. Thousand Oaks, CA: Sage.

Hickey, N. (1976). Terrorism and television. *TV Guide* (July 31). Radnor, PA.

Hoffman, B. (1998). *Inside Terrorism*. New York: Columbia University Press.

Jenkins, Brian. (1985). Terrorism found rising, now almost accepted. *Washington Post* (December 3), p. A4.

Mahan, S. and P.L. Griset (2013). *Terrorism in Perspective*. Thousand Oaks, CA: Sage.

Malkin, M. (2014). Where was President Obama. *Real Clear Politics* (October 24). http://www.realclearpolitics.com/articles/2014/10/24/where_was_president_obama_124411.html. (Accessed June 22, 2016).

Marks, A. (1995). Reporters at risk: war and lawlessness increasingly turn foreign correspondents into targets. *Christian Science Monitor* (November 10), pp. 1, 10–11.

Martin, G. (2003). *Understanding Terrorism: Challenges, Perspectives and Issues*. Thousand Oaks, CA: Sage.

Nacos, B.L. (2016). *Mass-Mediated Terrorism: Mainstream and Digital Media in Terrorism and Counterterrorism*. Lanham, MD: Rowman & Littlefield Publishers, Inc.

Perl, R.F. (1997). Terrorism, the media, and the government: perspectives, trends, and options for policy makers. Volume 97, Issue 960 of CRS Report for Congress. Library of Congress. Congressional Research Service, Library of Congress. Foreign Affairs and National Defense Division.

Rapoport, D. (1996). Editorial: the media and terrorism: implications of the unabomber case. *Terrorism and Political Violence* 8 (1): viii.

Schmid, A.P. and J. de Graff. (1982). *Violence as Communication: Insurgent Terrorism and the Western New Media*. Beverly Hills, CA: Sage.

Simonsen, C.E. and J.R. Spindlove. (2000). *Terrorism Today: The Past, the Players, the Future*. Prentice Hall: Upper Saddle River, NJ.

Thompson, S.A. and M. Isaac. (2023). Hamas is barred from social media. Its messages are still spreading. *New York Times*. October 18. https://www.nytimes.com/2023/10/18/technology/hamas-social-media-accounts.html. (Accessed October 25, 2023).

White, J.R. (2002). *Terrorism: An Introduction*. Thomson/Wadsworth: Belmont, CA.

CHAPTER 7

Contemplating a Quandary

Terrorism, Security, and Liberty

DO YOU ALREADY KNOW?

- The difference between terrorism and war
- Why security and freedom are important
- How terrorism prevention policies can impact liberty
- If rights can be protected in spite of terrorist threats

For additional questions to assess your current knowledge of the relationship of terrorism to security and liberty, go to **www.wiley.com/go/mcentire/ homelandsecurity3e**

WHAT YOU WILL LEARN	WHAT YOU WILL BE ABLE TO DO
7.1 How terrorism is unlike traditional combat operations	• Appraise how security and liberty may be at odds with one another
7.2 Why security and freedom are vital but competing priorities	• Argue why security is necessary
7.3 Ways that liberty is adversely impacted by efforts to prevent terrorist attacks	• Evaluate reasons why people are increasingly concerned about liberty
7.4 How to ensure rights while promoting homeland security	• Predict ways to secure the nation against terrorism while also maintaining rights

Introduction

The priority of homeland security is to protect the safety and physical well-being of citizens within the United States. One of the major difficulties you will have in achieving this goal concerns efforts to ensure the proper relationship between security and liberty. Specifically, you must be able to make and implement decisions that will minimize the probability and impact of terrorist attacks while also preserving values coveted by democratic societies such as the United States. It is vital therefore that you understand international law relating to conflict. You must also be able to critically evaluate the probability of terrorist attacks and estimate the negative impact that counterterrorism and antiterrorism policies may have on rights and freedoms. Above all, you must select policies that will limit the threat of terrorism but protect liberty as a cherished principle. These issues are addressed in the chapter that follows.

7.1 War, Terrorism, and Law

All countries want to ensure their security from external or internal threats. Such dangers may come from enemy nations, terrorist groups, or individuals within the country who would subvert the political system or attack political leaders. If the government is unable to successfully defend itself or protect its citizens from these aggressors, the nation could logically cease to exist. All other goals—such as education, healthcare, and economic prosperity—would be jeopardized. For this reason, government leaders want to identify potential threats and, if required, take measures to protect their people and way of life. Not all threats are alike, however. War and terrorism both involve violent behavior, but these types of conflict are distinct in fundamental ways.

Fighting in traditional conflicts (e.g., nation against nation) poses a difficult challenge, but the enemy is typically identified and known. In many cases, the governments issue a formal statement of war, the two sides engage in violent conflict, and soldiers are clearly marked with their respective matching uniforms. The nature and location of the battle is also distinguishable in that efforts are supposed to follow Geneva Convention rules. The **Geneva Conventions** are a set of internationally accepted laws pertaining to the conduct of war. They were initiated in Europe by Henry Dunant in 1863 after he witnessed the depressing lack of medical treatment for wounded soldiers at the Battle of Solferino in 1859. Dunant believed both sides of any conflict have a responsibility to care for the injured. He was instrumental for the initial development of international law pertaining to the humane treatment of soldiers.

Since this time, the conventions have undergone numerous changes. One of the most notable revisions of the Geneva Convention occurred after World War II. After seeing between 35 and 40 million people killed in the conflict, the United Nations determined that non-combatants should not be targeted in war because they have no way of protecting themselves (e.g., weapons, training, and armor). Signatory nations therefore established moral guidelines to limit the most disturbing aspects of war.

Nearly 200 countries have declared their support of the Geneva Convention, which prohibits attacks on civilians.

Terrorism is unlike the national conflicts mentioned earlier. First, the perpetrator of a terrorist attack is not always distinguishable like a military soldier. Terrorists may be foreigners or citizens of the targeted country. These terrorists do not typically wear uniforms that identify themselves and denote their intent to engage in conflict. Instead, they may be dressed as an ordinary person to carry out their attacks in a covert manner. Second, terrorists may or may not provide advanced notice of their intent to attack. They are far more likely to engage in violence and then claim responsibility afterward. Terrorists also disregard other aspects of the Geneva Conventions. Although terrorists may kill soldiers in attacks, they also prefer to target civilians to increase fear and publicity. It is this aspect that makes terrorism so deplorable.

In The Real World

War or Terrorism?

Generally speaking, it is fairly easy to distinguish between the nature of war and the phenomena of terrorism. However, the lines can be blurred at times. A visible example is the terrorist group, Hamas. Hamas soldiers have acted violently against Israel in prior decades and launched attacks on and after October 7, 2023. During an attack in 2014, Hamas soldiers wore uniforms of the Israeli Defense Force (IDF) to blend in with their enemy. This cover allowed them to approach IDF soldiers and easily kill four troops. Hamas personnel have also worn their own uniforms on various occasions to film propaganda videos. When Hamas attacked Israel on October 7, 2023, some (but not all) of its personnel were dressed in commando-style uniforms.

The Hamas attack almost appeared to be a war in other ways. Hamas carefully breached the Israeli border with drones, paragliders, and heavy equipment. Hamas also launched thousands of rockets and missiles into Israel in the coordinated attack, and its personnel were well-equipped with small arms and ammunition. In some respects, the violent assault looked like a well-planned war from a well-trained army.

In other ways, the attack was clearly a terrorist attack. Some Hamas personnel were dressed in civilian clothing. They targeted civilians on the street, at parties, and in their homes. Israeli citizens were gunned down, killed by mortars, set on fire, tortured, raped, and beheaded. No one—the elderly nor the infants—was immune to the violent attack. In addition, the entire operation was livestreamed on social media. In this sense, there is no doubt that this attack was the work of a terrorist organization.

The unique attributes of terrorism present other major challenges to democratic governments and the law under which they operate. One of the greatest struggles facing those involved in homeland security is to protect the United States from terrorism while also maintaining the rights we enjoy (Demmer 2004). It is argued that "the Constitution weighs heavily in both sides of the debate over national security and civil liberties" (Rosenzweig 2006, p. 1020) (See Figure 7-1). Thus, the president and Congress must take steps to protect the nation against terrorist attacks. In addition, national leaders are obligated to implement policies that do not infringe upon citizens' freedoms.

FIGURE 7-1 The U.S. Constitution promotes both liberty and security concerns. *Source: © National Archives*

Some scholars and practitioners see little or no conflict between these important priorities. For instance, K.A. Taipale states:

Within the public discourse, concerns about domestic security and civil liberties are often asserted as competing and potentially incompatible policy interests requiring the achievement of some tolerable state of balance. Implicit in this notion of balance is the smuggled assumption of a dichotomous rivalry in which security and liberty are traded one for another in a zero-sum political game. But the notion is misleading, for there is no fulcrum—as is implicitly in the metaphor of a balance—at which the correct amount of security and liberty can be achieved. Rather, security and liberty are dual obligations of civil society, and each must be maximized. (2006, p. 1009)

Others, however, are quick to point out that security and liberty are conflicting goals rather than values that can be achieved in a harmonious fashion. Justin Hood observes that "our history has shown us that insecurity threatens liberty. Yet if our liberties are curtailed, we lose the values that we are struggling to defend" (Murphy 2006, p. 1047). Jonathan White asks a question that may be difficult to answer: "Is there a point where civil liberties can be curtailed in the name of public safety?" (White 2002, p. 275). These issues and inquiries illustrate there is an uneasy relationship between security and liberty (Gaughan 2015). Unfortunately, "terrorism exploits this tension" (Badey 2008, p. 123).

Self-Check

1. War and terrorism are alike in almost every way. True or false?
2. The Geneva Convention is a law that prohibits terrorism specifically. True or false?
3. The Constitution:
    a. Ignores security
    b. Supports liberty only
    c. Sees security and liberty as important priorities
    d. Denounces liberty
4. As a type of conflict, what are the unique attributes of terrorism?

7.2 Security and Liberty

As noted, there are diverse perspectives on security and liberty. One school of thought asserts that security is an extremely important goal (or perhaps the most important objective) for any government. People holding this view acknowledge that national protection has become even more vital after the attacks against the United States in 2001. For instance, a spokeswoman for the U.S. Department of Justice stated that "Sept. 11th has forced the entire government to change the way we do business. ... Our No. 1 priority right now is to prevent any further terrorist attacks" (in Griset and Mahan 2003, p. 285).

Evidence regarding the scope of the threat indicates that terrorism has become more severe over time. Rosenzweig's research in 2006 revealed:

- 70,000 terrorists were trained in Afghanistan before the United States deposed the Taliban.
- The State Department has compiled a list of at least 100,000 known terrorists around the world.
- Jemaah Islamiyah, a terrorist organization in Indonesia, has 3,000 members.
- As many as 5,000 Al-Qaeda operatives could be present in the United States.
- "Virtually every terrorism expert in and out of government believes that there is a significant risk of another attack" (2006, p. 1022).

Of course, there is also the possibility today of other types of terrorist attacks. Right-wing extremists have been recently regarded to pose grave concerns to the well-being of our nation (Jones, Doxsee, and Harrington 2020). There are also growing anti-Semitic sentiments being shared within the United States due to the Israeli response to the Hamas attack. This is a concern that may result in domestic acts of violence and terrorism.

Furthermore, views favoring enhanced security recognize the negative impacts of terrorism. It seems as if more and more people are willing to engage in terrorism. Groups like ISIS have been increasingly passionate about their violent goals. Their desire has been to kill as many people as they could and in the most gruesome ways imaginable. Because of the possibility of acquiring weapons of mass destruction, terrorists may develop the ability to kill hundreds, thousands, or even millions of people today. If it is possible to conduct these types of attacks, they will certainly affect the economy and disrupt daily life in the process as well.

In The Real World

The Legality of the Imprisonment of Ibrahim al-Qosi

Ibrahim al-Qosi was a jihadist leader who served as a cook, accountant, and chauffeur to Osama bin Laden. He was captured by U.S. troops in Afghanistan in December 2001. Shortly thereafter, al-Qosi was detained at the Guantanamo Bay prison for more than 10 years—most of this time without any formal charges. The United States returned al-Qosi to Sudan in 2012 following his guilty plea to war crimes in 2010.

The hope was that al-Qosi would mend his ways and avoid a return to terrorist activity. Unfortunately, al-Qosi reemerged in a video released by the Guardians of Sharia group in December 2015. The propaganda film covered his biography and showed al-Qosi and other commanders of Al-Qaeda in the Arabian Peninsula encouraging jihadist attacks against the United States and other Western countries. As a high-profile leader, al-Qosi's recidivism sparked a lot of controversy, and some U.S. leaders questioned whether he should have been freed in the first place. This case illustrates the potential danger of releasing terrorists since there is a possibility of them re-engaging in violence. For example, at least 235 enemy combatants engaged in terrorism after being released from the Guantanamo Bay prison. However, it is necessary to point out that the actual rate of recidivism is difficult to calculate with certainty or even lower than anticipated (Hodwitz 2019). Therefore, is it right to detain terrorists without formal charges? The al-Qosi case illustrates the tension between security and legal rights.

Along these lines, another assertion is that it may be impossible to avoid taking drastic counterterrorism measures. One intelligence chief for the Department of Homeland Security previously believed "things are changing, and this change is happening because [attacks] can be brought to us that we cannot afford to absorb. We can't deal with them, so we're going to … do something ahead of time to preclude them. Is that going to change your lives? It already has" (Murphy 2006, p. 1047). Under this perspective, combating terrorism is seen as a crucial moral responsibility. It is perhaps a goal that is even more important than other government priorities or the protection of civil liberties. In other words, it is better to forego a few rights, the argument goes, instead of suffering the negative consequences of terrorism. This argument is more likely to be supported if the reduction of liberties occurs outside of the United States and if the associated actions are effective (Garcia 2016).

Others disagree vehemently with this point of view. For instance, it is reported that Ben Franklin once stated that those who give up liberty for security deserve neither (see Figure 7-2). His assertion is that the loss of rights was unwarranted and could actually result in insecurity. Franklin assumed that stable governments can only be established and maintained when citizens' liberty is guaranteed. Under this perspective, our rights are as valuable—or even more important—than security itself.

FIGURE 7-2   Ben Franklin declared that those who give up liberty for security deserve neither. *Source: Oil on canvas painting by Joseph Siffred Duplessis (1725–1802).*

People who adopt this latter viewpoint feel that the government has gone too far in its attempt to ensure security against terrorism (Demmer 2004, p. 149). Some even assert that having fewer rights in the pursuit of increased security will aggravate animosity toward the government and could ironically produce more terrorist attacks (Dragu 2016). For instance:

- "Critics of the government's ... policy have long argued that lawmakers overrate terrorist threats to achieve their political goals, with the result that civil liberties are sacrificed" (Griset and Mahan 2003, p. 282). This has increased frustration with relevant political decisions in some cases.

- Others assert that profiling is an intelligence measure that is inherently discriminatory. "**Profiling** is defined as the practice of law enforcement officials (including security personnel) using race, ethnicity, religion, or national origin as the decisive factors in targeting an individual for suspicion of a wrongdoing" (Starks 2008, p. 241). While profiling was very effective in helping to find out who the Unabomber was, people believe this type of law enforcement activity may infringe on rights and therefore create animosity with government policies.

- Many people today are concerned about the collection of **metadata**. Metadata include information gathered from communication through electronic devices. Besides containing the content of such messages, metadata include phone numbers, e-mail addresses, the device used, and the electronic service provider. Other examples of metadata are automated license plate recognition systems and digital photography in public places. People worry that this private information is being collected about them and given to others without permission.

As can be seen, many people assert that the government has gone or is going too far in the war against terrorism. The American Civil Liberties Union (ACLU) states, "If we give up our freedoms, the terrorists win" (Murphy 2006, p. 1047). Perhaps for this reason, a movement has emerged that might be called "anti-antiterrorism." The argument is that the steps being taken domestically to prevent potential terrorist attacks are too intrusive and threaten civil liberties (Rosenzweig 2006, p. 1013). Many concerned citizens and groups assert that such measures must be avoided at all costs.

The argument in favor of rights has clearly gained a great deal of attention since the attacks on 9/11. However, Paul Rosenzweig, an adjunct professor at George Mason University, lists several reasons why this viewpoint cannot be blamed solely on a reaction to the policies that were passed after the attacks in New York, Washington, D.C., and Pennsylvania. He states that the liberty view has become more popular for six reasons:

1. **A more activist court.** The judicial system in the last 40 years has tended to overturn executive branch actions and congressional decisions.

2. **A more partisan Congress.** Congressional investigative authority has expanded oversight of the president's power.

CAREER OPPORTUNITY | Lawyer

Government responses to threatened or actual terrorist attacks and other criminal activity must be based on governing principles and regulatory requirements as specified in the Constitution and the legal code. For this reason, those working in homeland security and emergency management may wish to hire and consult lawyers to help determine what course of action should be taken to address the situation at hand while also avoiding potential lawsuits and other negative legal repercussions.

For instance, the Department of Homeland Security has over 3,000 attorneys dedicated to making sure that policies are implemented in a way to protect the rights and liberties of Americans. These lawyers are responsible for decisions pertaining to Immigration and Customs Enforcement, the Federal Emergency Management Agency, and other federal government organizations that get involved in border issues, homeland security, disasters, international criminal organizations, and so on. Lawyers are also hired at the local and state levels.

The general employment outlook for lawyers is positive (about 8% over the next decade), and specialization in the areas of homeland security and emergency management will likely keep or outpace the overall trend. These positions require graduate degrees in law in addition to strong skills in analytical reasoning, written communication, and oral presentation. Such employees often make between $130,000 and $200,000 per year.

For more information, see **https://www.bls.gov/ooh/legal/lawyers.htm**.

3. **Investigative journalism.** Reporters have increasingly focused on activities that some might want to keep out of the public light.

4. **Public interest groups.** Organizations like the ACLU are heavily involved in public information and litigation actions.

5. **Technology.** Computers and the Internet have augmented people's ability to monitor the government.

6. **Greater awareness of civil liberties.** Citizens are more educated about rights than ever before.

Thus, the perspective favoring liberty downplays the focus on security and remains a popular paradigm to this day.

In The Real World

Physical Security Versus Civil Liberties

Paul Rosenzweig, an adjunct professor at the George Mason University School of Law, states that some people believe physical security is becoming more important than civil liberties (2006, p. 1023). He uses a hypothetical case of placing 10 alleged terrorists in prison. In the past, people in the United States generally accepted the rule that it is "better 10 terrorists go free than one innocent be mistakenly punished." Today, because of the horrendous nature of terrorism, more people believe it is better to punish 10 alleged terrorists even if only one of them is actually guilty. This line of reasoning is perhaps accepted because the potential for a terrorist attack increases if the guilty are released back into the general population. Such cases show the difficulty of balancing security and liberty in the United States.

Self-Check

1. There are less than 10,000 terrorists around the world. True or false?

2. Some people believe that security is more important than liberty, but others believe the reverse is also true. True or false?

3. Ben Franklin stated:
   a. Lawmakers exaggerate the threat of terrorism.
   b. If we give up our liberties, the terrorists win.
   c. Combating terrorism is a crucial moral responsibility.
   d. Those who give up liberty for security deserve neither.

4. Explain why the anti-antiterrorism movement has gained momentum in recent years.

7.3 Cases and Considerations

As can be seen, there are strong feelings about both security and liberty. At times, these political values have been in direct conflict with one another. By trying to ensure security, policy makers have implemented decisions that have had dramatic repercussions on liberty. Rosenzweig (2006) and Griset and Mahan (2003) provide several examples from history:

- When the Civil War broke out in the United States, President Abraham Lincoln declared that anyone guilty of disloyal practices would be imprisoned. Those convicted could not seek redress through the **writ of habeas corpus** (i.e., there was no way to protect citizens from unlawful imprisonment).

- During World War I, over 2,000 people were locked up because of their opposition to the global conflict. This eliminated virtually all antiwar sentiment in the United States, but it also jeopardized free speech.

- After Pearl Harbor was attacked and World War II commenced, President Franklin D. Roosevelt signed Executive Order 9066. This allowed him to gather more than 110,000 people of Japanese descent and "relocate" them to detention centers in the Western Pacific region of the United States (see Figure 7-3). This occurred even though these individuals were of American citizenship.

- In 1953, Senator Joseph McCarthy raised fear of Russian infiltration into the United States. Under the banner of the **"Red Scare,"** his permanent investigations subcommittee sent numerous Hollywood directors and actors to prison for alleged involvement in communism.

Such incidents have led Geoffrey Stone to suggest that "in time of war … we respond too harshly in our restriction of civil liberties, and then, later, regret our behavior" (in Rosenzweig 2006, p. 1017).

Liberty may also be limited or curtailed in other ways when terrorism is taken into account. For example, citizens of the United States are guaranteed:

- **Freedom of speech.** They are allowed to express their opinions, even when they criticize the government or others.

FIGURE 7-3 During World War II, over 100,000 people of Japanese descent were relocated to detention centers due to security concerns. *Source: © Library of Congress.*

- **Freedom of religion.** They cannot be denied their right to worship according to the dictates of their own conscience.
- **Right to assemble.** People are permitted to join in politically motivated gatherings.
- **Right to bear arms.** Guns can be purchased and owned without government interference.

While these are all liberties that Americans enjoy and generally want to protect, they do present thorny questions for those engaged in creating homeland security policies. For instance, what if people are inciting terrorist attacks as they talk to others? What should be done with a religious group that advocates violent behavior? What if meetings have the purpose of planning attacks? What if weapons will be used by terrorists to kill others?

The rights upheld by American political values could therefore run counter to efforts to stop terrorism. Again, these issues bring up difficult dilemmas. Should liberties be curtailed to promote security? Or, must liberty be treasured more than security? These are not just historical, hypothetical, or philosophical predicaments. We are facing very difficult choices today regarding security and liberty. A variety of recent examples can be given:

- After 9/11, the federal government immediately began to investigate the status of immigrants and visitors in the United States. Over 1,000 foreigners (mostly from Arab and Muslim nations) were detained, and some of them were charged for involvement in the attacks on America, for their expired visas, or for miscellaneous violations (e.g., traffic tickets). Although this action may have prevented other attacks, some people suggest that racial profiling and subsequent detentions were illegal (Griset and Mahan 2003, p. 285).

- As U.S. forces captured Taliban soldiers and Al-Qaeda members in Afghanistan, they sent many of them to prisons at Guantanamo Bay, Cuba (Mahan and Griset 2013). Many of these individuals were held for indefinite periods without being charged with formal crimes. If evidence warranted further investigation, prisoners were tried in military tribunals that lacked juries and civilian oversight. These efforts have limited the number of "enemy combatants" around the world but seem to overlook or downplay our own laws pertaining to illegal seizures and fair trials.

- The State of Florida declined to issue a driver's license to a woman who did not want to remove the veil over her face when pictures were being taken (Long 2003). The woman sued and claimed religious discrimination. Howard Marks, the woman's attorney, states, "There is no public safety issue here with Sultaana Freeman, none whatsoever. Sometimes when there are fears and prejudices in our country, we react and ... go overboard. We go too far. We infringe on the liberties we are fighting for—the liberties that make us the American Society." Judge Janet C. Thorpe disagrees with the lawsuit. She stated, "If you rule for the plaintiff, someday a man will present himself to a driver license office and demand a license, without a photo, based upon his religious convictions. You may have to give it to him, but we will not know if his objectives are sincere or peaceful, or if they are terrible until it is too late." Both comments are justified—but difficult to reconcile in light of the other.

- A professional emergency management organization in one state wrote Congress about its fear that sensitive public information would get into the hands of terrorists. The letter declared: "Over the past several years, government agencies across the state have been assessing our vulnerability to natural disasters and, more recently, terrorist attacks. Unfortunately, these assessments, if publicly disclosed, can provide a road map for a person or groups intent on destroying government facilities, disrupting government services and, in some cases, causing injury and death. Such assessments are now available upon request under provisions of the Open Government Act. In addition to the assessments, other critical information is available which has the same potential to be used against the interests of governments and their constituents. This includes items such as: detailed engineering drawings of facilities such as dams, power plants and government buildings; detailed response plans and procedures; and network diagrams for information management and telecommunications systems. ... We propose that the Open Government Act be amended to allow officials to deny access to certain, carefully defined classes of information such as those described previously. Our goal is to restrict public access to sensitive emergency plans and information that might be valuable to persons attempting to mount an attack." Such limitations seem perfectly logical in light of the terrorist threat. However, are there potential dangers in amending or rescinding the Open Government Act?

- In 2011, the U.S. military used a drone to kill Anwar al-Awlaki. Al-Awlaki was a terrorist propagandist, but he was also an American citizen engaged in Jihad abroad. Lawyers for the Obama administration argued that the Authorization to Use Military Force allowed them to carry out this mission to protect U.S. interests abroad. However, others assert that al-Awlaki's rights were violated. He was not tried in a U.S. court and had no recourse to legal representation. Some may argue that al-Awlaki was punished in a drone strike before even being found guilty of terrorism.

- After killing five Dallas police officers and two civilians on July 7, 2015, during a Black Lives Matter rally, Micah Xavier ran into a parking lot and found himself surrounded. Because Xavier posed a risk to law enforcement officials and others, the police department sent in a remote-controlled robot. After negotiations failed, the police detonated a bomb on the robot and killed Xavier. This was the first time a suspect was killed by a remote-controlled police bomb. The threat was neutralized, but some wonder if Xavier's right to a public trial was neglected.

These types of issues are not limited to the United States. France recently passed a law that requires mosques to respect French values and avoid extreme interpretations of Islam. Israel has implemented many policies that physically separate Palestinians and Jews. Israeli laws also prohibit the commemoration of Palestinian history, mandate the use of Hebrew as the official language, and allow the stripping of citizenship for those who voice opposition to the Jewish state. Some may view these actions as discriminatory and a violation of human rights. For this reason, there has been a resistance movement against these states and policies that some regard to be heavy handed (O'Neill 1978).

In The Real World

Support of Hamas and the Palestinians

After the October 7, 2023, attacks, thousands of people took to the streets in the United States to celebrate the violent actions of Hamas and protest Israel. This was especially prevalent among college professors and students at universities around the nation. For instance, Russell Rickford, a faculty member at Cornell University, stated at a demonstration that he was "exhilarated" by the violent attack on Israel. Students at George Washington University projected the message "Glory to the martyrs" on the outside wall of the Gelman Library. A student group at Harvard issued a statement placing blame for the Hamas attack on Israel. A professor at the University of California Berkeley promised his students extra credit to attend the pro-Palestinian walkout and to call or e-mail local lawmakers about their concerns about the colonial occupation of Gaza. Others advocated a day of resistance and suggested that further violence would be needed to ensure that Palestine would be "free from the rivers to the sea." As a result of these statements and actions, many Jewish students feared for their safety at college campuses around the country. Prominent donors cut ties to Harvard, the University of Pennsylvania, and other institutions around the nation. Many concerned Americans asked that professors be fired and that students retract their anti-Semitic statements. In response, pro-Palestinian groups asserted that their rights to free speech are being jeopardized. This case is a perfect example of the challenges policy makers face when dealing with controversial subjects that relate to terrorism and have heightened emotional attachment.

There are perhaps no simple resolutions to the difficult situations discussed above. For this reason, success is not to be achieved by giving security preference over liberty or by elevating rights concerns over the goal of preventing attacks and protecting communities. Instead, both values should be carefully weighed against the other, and decisions should be somewhat fluid and never set in stone. Paul Rosenzweig affirms: "One possible lesson from history is that we should not be utterly unwilling to adjust our response to liberty and security for the sake of counterterrorism, since we have the capacity to manage that adjustment, and to readjust it as necessary" (Rosenzweig 2006, p. 1020).

Others have also provided recommendations for ensuring liberty even while addressing threats associated with terrorism. David Lowe (2016) believes that reactions to terrorism should be based on proportionality and always be approved by judges. In a statement before Congress, Richard Ben-Veniste and Slade Gorton of the National Commission on Terrorist Attacks Upon the United States stated that:

> *The test is a simple but important one. The burden of proof should be on the proponents of the measure to establish the power or authority being sought would in fact materially enhance national security, and that there will be adequate supervision of the exercise of that power or authority to ensure the protection of civil liberties. If additional powers are granted, there must be adequate guidelines and oversight to properly confine their use. (Murphy 2006, p. 1048)*

In The Real World

FBI Versus Apple

In a technological world, privacy and security are often on two seemingly divergent roads. The privacy versus security argument has been a great source of debate, with the question of whether or not governments have the right to access private, protected information that relates to a criminal case. This became a visible national discussion following the mass shooting in San Bernardino by Syed Rizwan Farook and his wife, Tashfeen Malik, on December 2, 2015.

In the early months of 2016, the Federal Bureau of Investigation (FBI) asked Apple Inc. to assist in unlocking Farook's work iPhone so they could recover valuable information from it. Farook had a four-digit passcode on his iPhone that had to be inputted to access the contents. To unlock it, the FBI wanted Apple to create new software to override the current system that destroys all iPhone data after 10 failed password attempts.

Apple declined to help in this regard, stating that the action would set a lenient privacy precedent and lead to a slippery slope favoring security over confidentiality. Apple and other technology moguls frequently attribute encryption (like the iPhone passcode system) as a way to help secure customers' valuable information and prevent access by uninvited hackers. The FBI, on the other hand, requested Apple's cooperation to aid in their investigation of this important national security matter. As a result, a hearing was scheduled for the end of March 2016. However, prior to the hearing, the FBI found a third-party entity to unlock the phone, so the request to Apple was rescinded.

The case was legally closed, but the ensuing ripples it created in the privacy debate are far from over. Many questions remain. Should there be a balance between privacy and security in the land of the free? Does the average citizen have the right to keep their information private regardless of the impact on society? Should companies like Apple (which build trust with customers by creating products that are meant to ensure personal privacy and security) be forced to violate their own standards for the sake of national security? Should the government have access to private information when it work to protect the safety of others? If so, how should this be facilitated? The case between the FBI and Apple was one of the most high-profile examples of the security–privacy predicament. Cases like these are likely to remain complicated as time goes on.

Laura Murphy, director of the Washington Legislative Office of the ACLU, also suggests we need to apply additional tests (Murphy 2006, pp. 1049–1051):

- Do the costs of liberty outweigh the potential benefits to public safety?
- Will the measure call for, or result in, discrimination based on religion, ethnicity, race, or other group characteristics?
- Is the measure properly tailored to the desired mission, or could it result in unintended and possibly abusive consequences?
- Will the public react negatively to the measure?

If any of these questions are answered in the affirmative, Murphy recommends that the considered action be rejected in favor of less severe alternatives.

Thus, logic and evidence sometimes suggest that security may require some limitations on liberty. The threat of terrorism is significant and should be taken seriously. Nonetheless, liberty should not be fully sacrificed on the security altar.

One of the reasons why we should be extremely careful about this issue is because there is circumstantial evidence that terrorists are likely to emanate from politically repressed societies (Piazza 2015). The United States does not want to create more terrorists in the future as it responds to existing terrorists. If this country denies rights to others, it runs the risk of creating more external enemies in the future. This is nicely summarized in a statement by Lhaj Thami Breze, president of the Union of Islamic Organizations in France. He states, "The majority of Muslims want to practice their religion in peace and in total respect of the laws. But, when you persecute, when you make fun of, when you refuse, when you don't respect beliefs, what is the consequence? The consequence is radicalization" (Associated Press 2004). Along these lines, if domestic rights are severely eroded in the name of homeland security, there is also a chance that citizens in the United States may rise up violently against their own government. Both possibilities should be avoided to the best degree possible.

In The Real World

Treatment of Detainees at Guantanamo Bay

After an investigation of the prison facility at Guantanamo Bay, Cuba, the FBI described the treatment of enemy combatants who were captured in Afghanistan or elsewhere (Ackerman 2008, p. 124). At times, detainees claimed that they had insufficient water or food. Others asserted that they had been chained to the wall and had no choice but to urinate or defecate on themselves. In the investigation, it was also asserted that the temperature of interrogation rooms was set either too cold or too hot. Those captured were exposed to loud music featuring Lil' Kim and Eminem along with repeated videos of a Meow Mix commercial. Each of these incidents was authorized under the Pentagon guidelines. Such cases raise challenging questions for those involved in homeland security. If such techniques could prevent a terrorist attack, would they be justified? Is it ever correct to limit people's rights and liberties? Can a proper balance between security and liberty be found? If so, how? This is a very contentious issue. Some assert that the loss of rights does nothing to prevent terrorism and it may actually promote such acts of violence. Others declare that many lives have been saved by making security the top priority. It is likely that this debate will continue long into the future.

Self-Check

1. There have been no wars that have resulted in the loss of liberties. True or false?
2. Law in the United States protects the right to assemble, but terrorists could use this right to meet to plan attacks. True or false?
3. The search for communists in the United States after World War II was known as:
   a. The writ of habeas corpus
   b. The Red Scare
   c. The Japanese internment
   d. Executive Order 9066
4. What are some tests that could help to protect rights while we fight the war on terrorism?

Summary

As a participant in homeland security, it is necessary that you understand the difference between war and terrorism. In addition, it is imperative that you assess the trade-offs between security and rights. As illustrated in this chapter, you must know why terrorism exploits the tension between these two important priorities. You should be aware of the threats terrorists pose to society and the reasons why freedoms should not be eliminated. Because there are historical and current events that illustrate the delicate balance between security and freedoms, you must educate yourself about the consequences of homeland security policies. Most importantly, you should find ways to protect security while minimizing the loss of liberty. Striving for both objectives is most likely to reduce the possibility of future terrorist attacks. It is key to successful homeland security policies and actions.

Assess Your Understanding

Understand: What Have You Learned?

Go to **www.wiley.com/go/mcentire/homelandsecurity3e** to assess your knowledge of the relationship of terrorism to security and liberty.

Summary Questions

1. Governments take measures to protect themselves from outside aggressors. Otherwise, they may cease to exist. True or false?

2. The president and Congress have no responsibility to protect citizen freedoms and rights. True or false?

3. Security has become a much greater priority since 9/11. True or false?

4. There are no known terrorists in the United States. True or false?

5. Some government leaders feel their most important responsibility is to protect the nation from further attacks. True or false?

6. During the Civil War, President Lincoln sent people to prison if they were involved in disloyal practices. True or false?

7. The right to bear arms implies that people can purchase and own guns without government interference. True or false?

8. U.S. citizens are guaranteed freedom of religion. This will never create a problem for homeland security since people will not use their beliefs to incite terrorist attacks. True or false?

9. What is the name of the international laws that were created to protect soldiers?
   a. The Stafford Agreement
   b. The Hague Convention
   c. The Geneva Convention
   d. The Paris Accords

10. According to the ACLU, if we give up our freedoms:
    a. We are most likely to win the war on terrorism.
    b. The terrorists win.
    c. We will not be able to prevail against terrorism.
    d. The terrorists will not be able to prevail against terrorism.
11. A more activist court implies:
    a. Congressional investigative authority has expanded.
    b. Reporters increasingly focus on rights.
    c. Citizens are more aware of their liberties.
    d. The judicial system has overturned executive decisions.
12. During World War II, President Roosevelt:
    a. Sent 110,000 Japanese Americans to relocation centers.
    b. Sent many Hollywood actors and directors to prison.
    c. Stopped all antiwar sentiment.
    d. Rescinded the writ of habeas corpus.
13. The State of Florida:
    a. Does not want to give sensitive information to the public.
    b. Did not give a license to a woman who would not remove her veil.
    c. Held Arabs and Muslims who were here illegally.
    d. Took the U.S. government to court because it violated the writ of habeas corpus.

Applying This Chapter

1. Why is terrorism different from war?
2. What are some of the reasons that security has become important after 9/11?
3. List three reasons that an appreciation for liberty and rights has increased in recent years.
4. Have there been cases where liberty has been curtailed in the name of security? If so, discuss this case.
5. Discuss how the right to assemble can be abused by terrorists. Does this create a dilemma for policy makers?
6. One professional emergency management organization proposed to limit public access to sensitive government information. Why was this recommendation made?
7. What are five litmus tests that can help us to protect rights as we attempt to augment security?

Be a Homeland Security Professional

The Threat of Terrorism

You have been asked to testify before Congress about the need for new laws to protect our nation. What evidence could you use to describe the threat we are facing?

The Importance of Rights

As a member of a Congressional Committee on Terrorism, you are concerned about rights and liberties. What questions could you ask those who are proposing more stringent security measures?

The Geneva Conventions and Terrorism

Read a book or search the Internet for information about the Geneva Convention. Write a two-page essay describing the relation between the Geneva Convention and terrorism.

Key Terms

Freedom of religion   People cannot be denied their right to worship according to the dictates of their own conscience

Freedom of speech   People are allowed to express their opinions, even when they criticize the government

Geneva Conventions   A set of internationally accepted laws pertaining to the conduct of war

Profiling   Profiling is defined as the practice of law enforcement officials (including security personnel) using race, ethnicity, religion, or national origin as the decisive factor in targeting an individual for suspicion of wrongdoing

Red Scare   Senator McCarthy's fear of communist infiltration into the United States

Right to assemble   People are permitted to join in politically motivated gatherings

Right to bear arms   Guns can be purchased and owned without government interference

Writ of habeas corpus   A law protecting citizens from unlawful imprisonment

References

Ackerman, S. (2008). Island mentality. In: *Homeland Security* (ed. T.J. Badey), 124–128. Dubuque, IA: McGraw-Hill.

Associated Press (2004). Head Scarf Ban Backlash Warning. CNN.com, February 10. http://www.cnn.com/2004/WORLD/europe/02/10/france.headscarves.ap. (Accessed August 17, 2017).

Badey, T.J. (2008). *Homeland Security*. Dubuque, IA: McGraw-Hill.

Demmer, V.L. (2004). Civil liberties and homeland security. In: *Homeland Security* (ed. T.J. Badey), 149–152. Guilford, CT: McGraw-Hill/Dushkin.

Dragu, T. (2016). The moral hazard of terrorism prevention. *Terrorism and Political Violence* 79 (1): 223–236.

Garcia, B.E. (2016). Security versus liberty in the context of counterterrorism: an experimental approach. *Terrorism and Political Violence* 28 (1): 30–38.

Gaughan, A.J. (2015). A delicate balance: liberty and security in the age of terrorism. *Drake Law Review* 63: 1015–1029.

Griset, P.L. and S. Mahan. (2003). *Terrorism in Perspective*. Thousand Oaks, CA: Sage.

Hodwitz, O. (2019). The terrorism recidivism study (TRS) examining recidivism rates for post-9/11 offenders. *Perspectives on Terrorism* 13 (2): 54–64.

Jones, S.G., Doxsee, C., and Harrington, N. (2020). The escalating terrorism problem in the United States. Center for Strategic and International Studies. June 17. https://www.csis.org/analysis/escalating-terrorism-problem-united-states. (Accessed October 12, 2023).

Long, P. (2003). State: ID photo is security issue. *The Miami Herald* (May 30). http://www.miami.com.mld.miamiherald/news/state/5972523.htm. (Accessed December 28, 2005).

Lowe, D. (2016). Surveillance and international terrorism intelligence exchange: balancing the interests of national security and individual liberty. *Terrorism and Political Violence* 28 (4): 653–673.

Mahan, S. and P.L. Griset. (2013). *Terrorism in Perspective*. Thousand Oaks, CA: Sage.

Murphy, L.W. (2006). Principled prudence: civil liberties and the homeland security practitioner. In: *The McGraw-Hill Homeland Security Handbook* (ed. D.G. Kamien), 1045–1062. New York: McGraw-Hill.

O'Neill, B.E. (1978). Towards a typology of political terrorism—the Palestinian resistance movement. *Journal of International Affairs* 32 (1): 17–42.

Piazza, J.A. (2015). Repression and terrorism: a cross-national empirical analysis of types of repression and domestic terrorism. *Terrorism and Political Violence* 29: 102–118.

Rosenzweig, P. (2006). Thinking about civil liberty and terrorism. In: *The McGraw-Hill Homeland Security Handbook* (ed. D.G. Kamien), 1013–1030. New York: McGraw-Hill.

Starks, G.L. (2008). Profiling. In: *Homeland Security Handbook* (ed. J. Pinkowski), 239–250. Boca Raton, FL: CRC Press.

Taipale, K.A. (2006). Introduction to section 12. In: *The McGraw-Hill Homeland Security Handbook* (ed. D.G. Kamien), 1009–1012. New York: McGraw-Hill.

White, J.R. (2002). *Terrorism: An Introduction*. Belmont, CA: Wadsworth.

CHAPTER 8

Preventing Terrorist Attacks

Root Causes, Law, Intelligence, and Counterterrorism

DO YOU ALREADY KNOW?

- The causes of terrorism
- The laws that aim to stop terrorist attacks
- How information is acquired about potential terrorists
- The definition of counterterrorism

For additional questions to assess your current knowledge of measures to prevent terrorist attacks, go to **www.wiley.com/go/mcentire/homelandsecurity3e**

WHAT YOU WILL LEARN

8.1 The difficulty of reducing the causes of terrorism

8.2 Legislation that is passed to prevent terrorist attacks

8.3 The cycle used to acquire information about potential terrorists

8.4 The advantages and disadvantages of counterterrorism activities

WHAT YOU WILL BE ABLE TO DO

- Argue ways to minimize people's desire to attack the United States
- Evaluate policies designed to prevent terrorism
- Plan ways to acquire intelligence about potential attacks
- Critique the need for and impact of preemptive strikes against terrorists

Introduction

As a participant in homeland security, one of your principal objectives is to prevent terrorist attacks against the United States. The central goal is to do all that is possible to avert any event that results in injuries, deaths, damage, and disruption. For this reason, it will be imperative that you take several steps to minimize the probability of terrorism. One such measure is to eliminate—if feasible—grievances that foment terrorist activity. You should also be aware of and advocate for laws that proscribe terrorist actions. Another requirement is to identify potential terrorists and anticipate violent intentions through intelligence-gathering methods. At other times, the military may need to be deployed in counterterrorism operations abroad. Each of these measures may help you to stop terrorist attacks now and in the future.

8.1 Addressing Root Causes

According to some people, the predominant approach to protecting the homeland (i.e., the deployment of soldiers against terrorism in the Middle East and Afghanistan) has been ineffective and even counterproductive (Lustick 2007). The argument is that it wastes tax dollars, results in the death of U.S. military personnel, and intensifies anti-American sentiment. Instead of advocating an aggressive military response to terrorism, some scholars, war protestors, journalists, and politicians around the world assert that the United States should do its best to reverse the root causes of terrorism through other means (Shakilur-Rahman 2007). According to this view, one of the most important—but currently underutilized activities to prevent terrorism—is to eliminate terrorists' motivations to commit such attacks. As noted in Chapter 3, it is believed that terrorism will increase when abject poverty prevails, when political freedoms are limited, when human rights are ignored, and when alternate viewpoints or groups of people are not respected.

For instance, the head of the World Bank stated that "terrorism will not end until poverty is eliminated" (as cited by Francis 2002). This argument asserts that poverty can create animosity on the part of the poor against the rich. It is also suggested that people have an innate desire to participate in and influence political decisions. If individuals and groups cannot shape political decisions, violent activity could result. Others assert that the United States must do more to protect the human rights of everyone around the world—and not just American citizens. Without taking a more universal approach to rights, the argument goes that individuals and nations will have animosity against the United States. In addition, a common opinion in the Middle East is that terrorism would be minimized if the West would show increased consideration of and respect toward Muslim values. There is also a belief that terrorism could be reduced if the United States and Jewish leaders would avoid unfair treatment of Arabs (i.e., it is asserted that the Palestinians are denied freedoms and a political voice in Israel) (Abu-Amara 2002). Furthermore, U.S. foreign policy decisions, in the opinion of President Mohammad Khatami of Iran, may produce opposition against America and generate further terrorist attacks around the world (CNN 2001). Summarizing these perspectives, Jonathan Lash, a reporter for *Science Magazine*, states, "The compound of poverty, powerlessness, lack of opportunity, and injustice is volatile" (2001, p. 1789). It is assumed that these situations must be changed if the number of terrorist attacks is to be diminished in the future.

The idea of addressing such root causes is both controversial and difficult (Franks 2006). For instance, one challenge is that countries in the Middle East may overlook or misunderstand America's intentions to increase political freedoms around the world. Israel has held elections for Arab-Israelis, and the United States has attempted to establish a state in the Middle East to protect the Palestinian people. What is more, the expansion of democracy and human rights to other nations may actually augment the attacks we are trying to prevent because it is viewed as a threat to the theocratic governments that are preferred among certain groups of people. However, violence has continued unabated, and terrorism against Israel and the United States has required additional security measures to be taken against Palestinians and fundamentalist Muslims. The prevalence of attacks has compounded the problem of root causes mentioned earlier.

Furthermore, American corporations may oppose or overlook many economic practices that could not only reduce poverty, but also limit free trade or curtail profits. And it is also true that certain dictators in the world are likewise to blame for the extreme poverty their citizens experience due to their corruption and exploitative policies. There are other root causes that the United States may not be able to control. As an example, children in some countries are often taught to hate Americans and are brought up in a culture of violent behavior (Leiter 2002). It may be virtually impossible to reverse such attitudes when they are deeply ingrained at such a young age. Consequently, there are no easy answers to the challenges terrorism poses.

Regardless of the severity of these dilemmas, it would be advisable that the U.S. government recognize that it must do more to address root causes if homeland security is to be enhanced. Failing to do so will only lead to a more protracted conflict with individual terrorists as well as the groups and nations that support them. Reaching out to moderate Muslims in a partnership may prove useful for the generation of ideas that will help resolve such problems. Diplomatic activity and peaceful actions are therefore the first preference. Of course, it is important to point out that poverty, powerlessness, and discrimination are not just problems that are witnessed at the international level. People in the United States may launch attacks if they are sensitive to these root causes. Thus, the same principles apply internally.

Nonetheless, addressing root causes may not be enough. International or domestic terrorists may attack the United States or its leaders and citizens no matter what this country does. Also, if no steps are taken to prevent immanent terrorist attacks, people will be killed and injured. Therefore, taking a less aggressive and more peaceful approach could jeopardize the rights of the potential victims of terrorist attacks in the United States or elsewhere. This is not acceptable either. Other preventive measures may inevitably be required.

In The Real World

Idealistic Versus Realistic Policies

During the summer 2016 Presidential debates, diplomacy became a heated subject among Democratic and Republican candidates. Most of those from the left asserted the need to engage enemy nations in dialogue and find ways to resolve conflict through diplomacy and peaceful foreign policy. Those on the right declared that negotiations with terrorists and states that support such violence are pointless since they have already stated their desire to attack the United States and kill Americans. Thus, one side saw value in diplomatic endeavors, while the other affirmed no logical reason to negotiate with self-declared enemies. The first approach attempts to address root causes through diplomacy while this practice is discredited by the second perspective. Diplomacy with terrorists or enemy states is a politically contentious issue for sure.

Self-Check

1. Addressing the root causes of terrorism is currently the preferred and most practiced approach to preventing terrorist attacks. True or false?

2. Some people believe that poverty and injustice will augment terrorist attacks against the United States. True or false?

3. Which is not seen as a way to address the root causes of terrorism?

   a. Eliminating the indoctrination of children to hate the United States

   b. Improving U.S. foreign policy

   c. Disregard the rights of Palestinians

   d. Respecting different cultures around the world

4. Will poverty reduction, human rights, and the expansion of democracy prevent all terrorist attacks? Why or why not?

8.2 Policy and Legislation

Another measure that has and should be taken to deal with the threat of terrorism is to create policies and enact laws that make terrorism illegal or criminal. The policies and regulations that outlaw terrorism may emanate from executive orders (e.g., presidential directives), congressional decisions, and state and local codes and ordinances. Although law can be initiated by different branches and at all levels of government, homeland security has been driven mostly by the president, Congress, and federal agencies. Some of these policies and laws were implemented before 9/11. Others were created to promote homeland security after this fateful date.

8.2.1 Laws Prior to 9/11

Many of the laws associated with international conflict evolved from the National Security Act of 1947 and the Goldwater–Nichols Act of 1987. And, even though the United States has often been threatened by terrorism, it only began to address terrorism through policy and legislation in the past few decades. To be sure, the government did prosecute those engaging in terrorism. But it usually did so through other laws that dealt with murder, arson, and property destruction. In the mid-1990s, however, things began to change.

One of the first significant policies to address terrorism directly was Presidential Decision Directive 39 (e.g., the U.S. Policy on Counterterrorism). This policy was issued after major terrorist attacks in Japan and the United States. These two events are particularly noteworthy since they had a dramatic impact on policy.

On March 25, 1995, the apocalyptic cult, Aum Shinrikyo, punctured plastic bags containing liquid sarin on various subways in Tokyo and elsewhere. The attack killed 12 people and sent 5,500 people to hospitals (although only 1,000 had actually sustained injuries). A month later, on April 19, 1995, Timothy McVeigh detonated a bomb in Oklahoma City at the Alfred P. Murrah Federal Building. The blast killed 168 people and injured another 800.

Recognizing the seriousness of these events, President Clinton reiterated in Presidential Decision Directive 39 (PDD-39) that it is the policy of the United States to use all means necessary to defeat terrorism. Clinton vowed that the government would work with friendly nations to pursue terrorists through counterterrorism operations. He also declared that the government would seek to have terrorists who commit acts against our national interests extradited to the United States. President Clinton also assigned terrorism-related roles and responsibilities to many federal organizations and leaders including the Department of State, the Department of Defense, the Department of the Treasury, the Department of Energy, the Department of Transportation, the Attorney General, and the Director of the Central Intelligence Agency (CIA). PDD-39 also separated functions dealing with terrorism into two categories. As noted in Chapter 1, crisis management was a law enforcement responsibility focusing on prevention efforts as well as the apprehension and prosecution of terrorists. Crisis management was to be led by the Federal Bureau of Investigation. Consequence management dealt more with preparedness, response, and recovery activities. The Federal Emergency Management Agency was put in charge of these functions.

In the late 1990s and early part of the new millennium, Congress also began to recognize the growing threat of terrorism against the United States. It passed a few laws that instituted homeland security as a national and foreign policy goal. From this point on, America would be less concerned with enemy states in relative comparison with individual terrorists and terrorist organizations. Understanding this emerging threat, House Resolution 1158 permitted the establishment of the National Homeland Security Agency. This law was passed on March 21, 2001. House Resolution 1292 allowed the president to create a strategy for homeland security. The approach was enacted on March 29, 2001. While each of these laws signified substantial change, it was not until 9/11 that most of the legislation dealing with homeland security was initiated.

8.2.2 Legislation After 9/11

Shortly after the terrorist attacks in New York, Washington, D.C., and Pennsylvania, President Bush issued Executive Order 13228 on October 8, 2001. This law established an Office of Homeland Security in the White House. Tom Ridge was sworn in as the first director of this office. Within a few short weeks, additional homeland security legislation would be passed. These laws would have a profound impact on the United States.

For instance, on October 25, 2001, Congress passed, and the president ratified, the USA PATRIOT Act. The **USA PATRIOT Act** is an acronym for "Uniting and Strengthening America by Providing Appropriate Tools Required to Intercept and Obstruct Terrorism." Its goal was to prevent terrorist attacks and enhance law enforcement ability to investigate and punish offenders. This law relaxed restrictions on sharing information among the CIA and FBI, permitted roving wiretaps and increased surveillance over computer communications, and allowed for the detention of terrorists. The act also mandated new measures to prevent the funding of terrorists, prohibited the harboring of terrorists, and augmented the number of border agents. It also eliminated the statute of limitations for terrorist acts, meaning that a person could be tried for attacks regardless of how long ago they occurred. Another aspect of the PATRIOT Act deals with material support. People can be convicted for providing aid and assistance to terrorists. Although highly controversial because of the perceived infringement of rights and liberties (Etzioni 2004), the PATRIOT Act did enhance the ability of law enforcement to find, apprehend, and prosecute terrorists. This law has since been amended, but it remains in effect today.

A short time later, the **Transportation Security Act** was passed. On November 19, 2001, Congress created this law to protect transportation systems in the United States. From this point on, only passengers could enter airport terminals. The federal government also took over most airport screening operations and purchased new equipment to detect explosives in carry-on and checked baggage. These activities are intended to make it more difficult for terrorists to use airplanes in future attacks.

A number of other important policies and laws were implemented in 2002 and 2003. The first of which was Executive Order 13224. On August 9, 2002, President Bush acknowledged that the government would freeze any assets belonging to terrorists to limit resources that could be used in attacks against the United States. Courts have since upheld this decision, and funds supporting terrorism have been seized in the United States and elsewhere. The Justice Department released a statement noting that "freezing the assets of organizations that bankroll terror is a legitimate and important role in the government's arsenal for fighting the war against terrorism." In the most notable case, the Holy Land Foundation for Relief and Development was indicted for funding terrorism. It is believed that this organization sent over $12 million to Hamas. This court case began in July 2007 and resulted in a hung jury in October. However, at a retrial in 2008, the jury found the defendants guilty of each of the charges against them.

Another very significant law was passed on November 25, 2002. President Bush signed the **Homeland Security Act**, which would become active on January 24, 2003. This law resulted in the most dramatic reformulation of government in 50 years and led to the creation of the Department of Homeland Security (DHS) (see Figure 8-1). The Homeland Security Act outlined the mission of DHS and listed its responsibilities dealing with border security, infrastructure protection, and the like. This law also focused on countermeasures to deal with weapons of mass destruction

FIGURE 8-1   The Seal of the Department of Homeland Security (DHS) reflects its mission. The eagle's claws represent activities during peace and war. The circles surrounding the eagle symbolize DHS's aims to coordinate with other levels of government. *Source:* © *FEMA.*

(WMDs) (see Chapter 14) and explained why the initiation of homeland security would take place. The Homeland Security Act resulted in major challenges for the federal government due to massive organizational realignments, but virtually everyone agreed that this was a first step to dealing with the threat of terrorism.

An additional important piece of legislation to be mentioned here, the **Comprehensive Homeland Security Act**, was passed on January 7, 2003. This legislation resulted in many specific recommendations to implement the responsibilities of the DHS. For instance, this law related to the security of critical infrastructure and railroads. It also implemented more stringent measures pertaining to border control and identified ways to halt the proliferation of WMD. Other portions of the law deal with improving intelligence gathering. The law released funds to local law enforcement agencies to support national homeland security goals.

Further policies and pieces of legislation have been issued or passed over the past several years. In fact, new laws are being approved and implemented on a regular basis. If you are working in homeland security, it will be imperative that you stay on top of new developments. Some laws have been changed or repealed, while others are being created at a frantic pace. This can be overwhelming. Nevertheless, resources are available to help you understand your role. For instance, the National Governors Association created a Governor's Guide to Homeland Security, which can be accessed at **https://www.nga.org/files/live/sites/NGA/files/pdf/1011GOV-GUIDEHS.PDF**.

As can be seen, the government is desperately working to prevent terrorist attacks. Unfortunately, some of these laws have been hastily put together or are controversial for other reasons. For instance, there is a debate about the authorities given in U.S.C. Title 10 and U.S.C. Title 50. Title 10 governs the organization and function of the armed forces (e.g., Army, Navy, Air Force, Marine Corps, Coast Guard, and Reserve). Title 50 determines how the United States declares and conducts war and focuses more on intelligence issues. Some argue that the two laws overlap substantially. However, others like Wall (2011) argue that there is a clear difference between military and intelligence operations. In addition, legislation has not always provided the funding that is necessary to make homeland security effective and allocated resources have at times been misdirected (as will be seen in Chapter 16). It is likely that the challenges confronting legislators will continue into the foreseeable future.

In The Real World

The Comprehensive Homeland Security Act

The Comprehensive Homeland Security Act of 2003 is one of the most important pieces of legislation which guides homeland security activities. This bill, put forward by Senator Thomas Daschle (D-SD), focuses heavily on a variety of measures to prevent terrorism. It established a task force to protect nuclear power plants from terrorist attacks. The goal of this designated group is to assess the vulnerability of these structures and outline actions to secure them against terrorist threats. Besides being concerned about terrorist attacks involving radioactive material, the act mandated a national smallpox vaccination program, promoted ways to foster intelligence dissemination, increased regulations pertaining to the use of hazardous materials, and augmented financial and technical assistance to law enforcement agencies in the United States. This law will have a lasting impact on homeland security for the foreseeable future.

Self-Check

1. Terrorism has always been an important subject of legislation in the United States. True or false?
2. Homeland security was discussed as a policy function before 9/11. True or false?
3. Which terrorist attacks resulted in Presidential Decision Directive 39?
   a. 9/11 and the Oklahoma City bombing
   b. The Tokyo subway attack and Oklahoma City bombing
   c. The Tokyo subway attack and 9/11
   d. The sarin gas attack and 9/11
4. What are some of the features of the Homeland Security Act and Comprehensive Homeland Security Act?

8.3 Intelligence

Because of events like 9/11 and the recent Hamas attack on Israel, there is ample concern about the lack of prior warning. Therefore, one of the best ways to prevent terrorist attacks is to identify potential terrorists in advance and have a sound understanding of their plans to attack the United States or its citizens. This brings up the important function of intelligence.

Intelligence is a word that describes the function of collecting, assessing, and distributing information about an enemy, criminal, or terrorist. Governments have long been concerned about the activities of unfriendly countries. Their desire is to become aware of the threats posed by rival nations so they may take necessary defensive or offensive measures. Law enforcement officials also seek information about citizens or nonresidents who plan to break the laws that bring order to society. In this case, detectives attempt to avert theft, murder, and other crimes or to arrest those who are intent on violating the law. In the context of homeland security, intelligence about terrorist intentions is the most important priority. It is imperative to acquire knowledge about potential terrorist attacks to prevent their negative consequences of death, damage, and disruption.

8.3.1 The Need for Intelligence

As noted, intelligence has been a major feature of America's national security interests since our birth as a nation. Efforts have always been undertaken to comprehend the activities of enemy nations and the threat they pose to the United States. This was especially the case after World War II. During this period, intelligence efforts were bound to a "Cold War" mentality. Although animosity with the former Soviet Union dissipated in the late 1980s, American political and military leaders remained fearful of potential enemy states. In particular, the United States was spending time and energy on understanding rising powers (i.e., China, North Korea, and Iran) during the 1990s. On the domestic front, law enforcement agencies were giving attention to the war on drugs, the sale or use of illegal weapons, and civil rights violations (e.g., hate crimes). It was under this context that foreign terrorists took America by surprise.

For instance, during President Clinton's tenure in office, the United States witnessed several terrorist attacks against American interests. On February 26, 1993, Middle Eastern terrorists bombed one of the towers at the World Trade Center. The car bomb killed 6 people and injured 1,042. Fortunately, it did not bring the building down. High-profile attacks also occurred toward the end of the 1990s. Two U.S. embassies were bombed in the African nations of Tanzania and Kenya on August 7, 1998. The near-simultaneous attacks killed over 200 people. Another attack with explosives took place against the USS Cole (a military vessel) in Yemen port on October 12, 2000. This event killed 17 and injured 39 others. Sadly, none of these were detected and stopped in advance. However, the intelligence community was able to deter a major terrorist incident during this period. The **Bojinka plot**—a planned attack on airliners over the Pacific Ocean—was thwarted in 1995. In spite of this success, it appeared that not enough was being done to anticipate the growing threat of terrorism.

This trend of downplaying the threat of terrorism continued for a time when George W. Bush was elected as president. The CIA and FBI began to collect evidence specifying that Al-Qaeda was intent on attacking the United States. One analyst in Phoenix, Arizona, noted in July 2001 that potential terrorist operatives were taking flight training lessons at U.S. schools. On August 6, 2001, President Bush was given a brief entitled "Bin Laden Determined to Strike in US." It stated that Al-Qaeda was intent on bringing the fight to America and that it was recruiting and operating in places like New York City. In spite of these clues, U.S. officials appeared to be oblivious to the impending attacks on 9/11. After a careful study of what led up to this fateful day, the 9/11 Commission (2004) listed several reasons why the attacks were not fully anticipated. This included an overwhelming number of intelligence reports with insufficient details, an inability to share intelligence across government agencies, and a failure to appreciate the tenacity and creativity of terrorists. Terrorist attacks and government panels revealed that intelligence must be a key feature if homeland security is to be successful (Githens and Hughbank 2010).

8.3.2 The Intelligence Cycle

In an attempt to understand and improve intelligence activities, Patrick Deucy (2006) recommended that those involved in counterterrorism rely on the **intelligence cycle**. This includes a four-step process of gathering, understanding, and synthesizing data and then sharing it with those who will use it (see Figure 8-2):

1. **Intelligence collection**—Activities to gather information about terrorist organizations and their operations and potential attacks. Intelligence collection may be obtained through a variety of means including:

   - **OPINT**—Open-source intelligence is acquired through publicly available materials including academic research, newspaper articles, library books, and so on.

   - **SIGINT**—Interception and interpretation of electronic signal communications such as phone conversations and e-mails.

   - **IMINT**—Geospatial imagery collected by satellites and aircraft.

   - **MASINT**—Measurement and signature intelligence looks for the characteristics of certain types of actions and evidence (e.g., the presence of nuclear material when one is trying to develop, smuggle, or use a nuclear weapon or dirty bomb).

   - **HUMINT**—Intelligence collected by people from people (and can be done overtly or covertly).

FIGURE 8-2   Thousands of employees help to collect, interpret, and disseminate intelligence to the President, Congress, and others who need to know. *Source: © FBI.*

2. **Intelligence analysis**—Efforts to make sense of the voluminous data that are gathered from the field. This is often performed at the headquarters level where individuals and groups process information to determine what is really going on at the field level.

3. **Intelligence production**—The creation of publications, briefings, images, or maps to influence policy or operational decisions. Written reports, oral presentations, and supporting documents are examples of this stage of the intelligence cycle.

4. **Intelligence dissemination**—Sharing information with end users to increase agency awareness, warn the public, and take steps to "preempt, disrupt or defeat terrorism" (Kauppi 2006, p. 423). This may include policy makers, FBI special agents, homeland security personnel, the military, etc.

Without a doubt, additional intelligence will be required when data collection is incomplete. In addition, the analysis phase seeks to "connect the dots," the production phase generates new questions, and the dissemination of results phase will open up new questions about future attacks. It is perhaps necessary to add another step to this cycle and this could be labeled as **intelligence adjustment**. Intelligence adjustment implies that the intelligence cycle must adapt repeatedly based on intelligence shortfalls, insufficient information, changing priorities, new leads, and unfolding questions and concerns. The intelligence cycle therefore requires a great deal of planning, direction, and follow-up (Kauppi 2006, p. 415).

8.3.3   Challenges Facing the Intelligence Community

As can be seen, the intelligence cycle is not a simple linear process, nor is it exempt from significant challenges. At least four major difficulties must be overcome. The first problem relates to the types of intelligence products that are useful for homeland security purposes. During the Cold War, the United States relied heavily upon IMINT and MASINT to detect troop movements and the building of missile sites or bomb bunkers. HUMINT was also utilized during this time with spies and secret agents who understood the politics, language, history, and culture of Russia. The problem with these sources of intelligence is that they are less useful in homeland

In The Real World

DHS Office of Intelligence and Analysis

In 2007, the Department of Homeland Security established the Office of Intelligence and Analysis (I&A). The goal of this organization was to help overcome some of the intelligence failures as outlined in the 9/11 Commission Act of 2007. Specifically, the Office of Intelligence and Analysis was charged with improving the sharing of intelligence among federal, state, tribal, and territorial jurisdictions along with the private sector. The I&A has five mission centers:

1. The Counterterrorism Mission Center collects and shares intelligence about possible terrorist attacks.
2. The Transnational Organized Crime Mission Center focuses on dealing with drug smuggling, human trafficking, and money laundering.
3. The Cyber Mission Center analyzes the threat of cyberattacks, whether they be foreign or domestic.
4. The Counterintelligence Mission Center fosters measures to counter intelligence services by foreign adversaries.
5. The Economic Security Mission Center concentrates on ways to reduce intellectual property theft, supply chain threats, and illicit trade.

The I&A also has other centers and divisions that manage watchlists, monitor unfolding threats 24/7, and share information through the network of fusion centers across the nation. The Office of Intelligence and Analysis is therefore one more organization to help prevent terrorist attacks in the United States.

security. IMINT and MASINT can still be utilized to detect terrorism activities, but they are of reduced value because terrorists are especially adept at operating in stealth.

The HUMINT system built up during the Cold War is also not suited to address today's homeland security needs. Besides needing a substantially larger number of agents, the United States must have operatives who understand Arabic languages, the Islamic religion, and Middle Eastern cultural practices. This is no small feat when one considers the complexity of these subjects. Even as the American intelligence community retools for today's homeland security objectives, it is extremely difficult to penetrate terrorist groups and cells. Terrorists have close-knit relations, and they are not likely to allow significant infiltration into their organizations. The United States will therefore have to work through Arab partners and others around the world who already understand and can assimilate into terrorist goals and activities.

A second problem with the function of intelligence is that it is heavily dependent on technology. Terrorists communicate, plan, and conduct attacks with the use of technology. Cell phones, fax machines, pagers, teleconferencing, e-mails, and the Internet are all used frequently by terrorists. This technology allows terrorists to operate with a degree of impunity. As noted in Chapter 4, FBI director Louis Freeh stated before a U.S. Senate Commission in March 2000 that "uncrackable encryption is allowing terrorists to communicate without fear of outside intrusion." He further stated that because of this technology "they're thwarting the efforts of law enforcement" (in Thetford 2001, p. 252). Consequently, the intelligence community must have adequate equipment and training to allow them to operate in the modern world of technology.

A third problem is that there are many organizations that provide intelligence services. Most people have heard of the well-known **Central Intelligence Agency (CIA)** and **Federal Bureau of Investigation (FBI)**. These organizations generally operate in the international and domestic spheres, respectively. However, people may not have heard about the National Security Agency, the Defense Intelligence Agency, the National Geospatial-Intelligence Agency, and the National Reconnaissance Office, to name a few. The sheer number of federal agencies (which are not mentioned in their totality here) is eclipsed only when we consider other intelligence units around state and local jurisdictions in the United States. Individed states are developing their own terrorism intelligence units, and even the City of New York has its own dedicated agency to assess threats against this large metropolitan area in the United States.

The number and diversity of organizations obscure intelligence ownership and could at times discourage collaboration. As an example, before 9/11, the FBI and CIA did not have open lines of communication. The federal government did not want the CIA to spy on its own citizens, and the FBI and CIA were prohibited by law from sharing information with each other. Fortunately, new laws (such as the Intelligence Reform and Terrorism Prevention Act of 2004) were passed to enhance the distribution of intelligence. **The Terrorist Threat Integration Center (TTIC)**, which joins the efforts of the FBI and CIA, was created in 2003 to improve cooperation in the federal government. The DHS also aims to collect information on attacks planned against the United States. Efforts have similarly been undertaken to augment the sharing of intelligence internationally and to lower levels of government in the United States. For instance, **INTERPOL** is an international police organization that is involved in intelligence gathering and sharing. It has 186 member nations that collect and distribute information about terrorists and various forms of criminal behavior. The United States also works with the North Atlantic Treaty Organization and other friendly nations to discuss terrorist threats. There are also many daily briefs that are distributed to state and local governments from the FBI. Other guidelines are provided to assist state, local, and tribal law enforcement agencies to gather and act on intelligence (Carter 2009).

The various actors involved in this important function bring up a fourth challenge: determining with whom to share intelligence. For instance, while some information is sensitive but unclassified, much of it is classified. **Classified intelligence** is highly sensitive information given only to a very specific and limited number of people to protect sources of acquisition and deny adversaries information that would lead them to alter their communications or operations (Deucy 2006, p. 401). Classified intelligence is thus given only on a need-to-know basis and is strictly guarded. Unfortunately, failing to share information with others could prove dangerous as we found out on 9/11. The CIA and FBI were prevented from sharing information because of their unique missions internationally and domestically. The 9/11 Commission reiterated that it is imperative to find an appropriate balance between safeguarding intelligence and sharing it with those who "need to know."

A final problem concerns gathering intelligence about the funding of terrorism. Since 9/11, the government has been very interested in disrupting terrorist finances. Numerous laws have been passed to make it illegal to support terrorism

In The Real World

Fusion and Intelligence Integration Centers

In light of the information-sharing weaknesses made evident on 9/11, the federal government has implemented additional measures to increase collaboration across intelligence agencies. This includes fusion centers and intelligence integration centers. (see Figure 8-3).

The concept of fusion centers started in the 1990s when Los Angeles County created a Terrorism Early Warning Center. However, after the terrorist attacks on 9/11, both the DHS and the Department of Justice recognized the need to improve the collection and sharing of information about terrorism and crime. For this reason, federal funding was provided to create fusion centers around the nation. These centers are typically located within local and state law enforcement agencies. However, fusion centers vary dramatically in terms of organization, reporting chains, and so on. Sizes range from 4 to over 200 personnel. Some centers stress drugs and organized crime, while others focus more on disasters and emergency management. Nevertheless, fusion centers play an important role in collecting information within their jurisdictions and relaying intelligence products up to the federal government or down to local communities. Guidelines for fusion centers can be accessed at **https://bja.ojp.gov/sites/g/files/xyckuh186/ files/media/document/fusion_center_executive_summary.pdf**.

Other organizations were created to help address intelligence gathering and sharing. On May 1, 2003, the Terrorism Threat Integration Center (TTIC) was formed in the FBI. According to the government, the TTIC is an interagency body intended "to provide a comprehensive, all-source-based picture of potential terrorist threats to U.S. interests." It is composed of personnel from many organizations, including the CIA, the FBI, and the DHS. Each organization will continue to gather data from their respective areas of specialization: the CIA focusing on terrorists at the international level, the FBI focusing on terrorists at the domestic level, and DHS focusing on other ways to reduce vulnerabilities to known threats. The overall objective of the TTIC is to increase information sharing and avert any disconnects relating to intelligence collection and processing. In other words, the FBI, CIA, and DHS can help one another "fill in the blanks" when intelligence is lacking and find ways to "connect the dots" where it is present. A similar organization helps to inform the president regarding intelligence and counterterrorism planning. The National Counterterrorism Center was established by Presidential Executive Order 13354 in August 2004. It produces the president's Daily Brief and the National Terrorism Bulletin.

through monetary means. A challenge arises due to the fact that terrorists are very careful about how they channel money to each other. **Money laundering** is the process of hiding where money is coming from and what it is being used for. Terrorists are adept at collecting and transferring money through legal and illegal means.

As an example, terrorists may obtain funds from legitimate charities and businesses as well as fraudulent companies and organizations. These finances may then be sent through a variety of accounts in banks around the world. This covert movement of money creates significant obstacles for intelligence officers who are trying to cut off the financial lifeblood of terrorist organizations. For this reason, financial institutions are now required to report to the government wire transfers and other bank transactions that involve a significant amount of money (usually anything over $10,000).

8.3.4 Successes

In spite of notable difficulties, the intelligence community has been successful on various occasions at uncovering terrorist conspiracies in the United States and around the world. For instance, on August 10, 2006, 24 suspects were arrested in the United Kingdom. These individuals intended to board 10 jets from Britain to California, New York, and Washington, D.C. They wanted to smuggle on board gel explosives in sports drink bottles and detonate them with the use of an electronic signal from an iPod or cell phone. Fortunately, officials in the United States and the United Kingdom worked closely to investigate and thwart the plot.

On May 8, 2007, six men were arrested for planning to purchase automatic weapons and attack the Fort Dix Army Base in New Jersey. The men talked to an employee at a store and wanted him to transfer video footage of them firing weapons to a DVD. The thoughtful citizen promptly notified the FBI, and the attack was foiled. In another case, four men were intent on blowing up terminal buildings, fuel pipelines, and fuel tanks at JFK Airport in New York. This plot was prevented on June 2, 2007, and may have been only days away from implementation.

Another positive example was seen on January 3, 2020. Intelligence agencies discovered the location of an Iranian man named Qasem Soleimani, a major general in the Islamic Revolutionary Guard Corps. The U.S. military then sent a drone to the Baghdad International Airport in Iraq and killed him and a few of his associates with a missile strike. As noted in Chapter 6, some scholars, UN officials, and media representatives asserted that this assassination occurred against existing international law. However, President Trump replied that Soleimani killed and wounded hundreds of American soldiers who were operating in the Middle East when he provided improvised explosive devices to terrorists in the area. This operation finally accomplished a goal that was discussed by U.S. intelligence and military officials as early as 2007.

Such cases indicate the importance of working with other nations, the need for citizen involvement, and the benefit of anticipatory action. While these and other accomplishments are praiseworthy, the intelligence community must not be complacent. Terrorist leaders have said they are sending suicide bombers to the United States, and many fear it is only a matter of time before these and other attacks occur in the future. It is a common adage that terrorists only have to be successful once, while intelligence officials have to be right 100% of the time. This brings up the notion of counter-terrorism operations.

In The Real World

Gathering Intelligence

Information is vital for those involved in gathering intelligence. To keep track of potential terrorists, the government is turning to corporations for assistance. Companies such as ChoicePoint and LexisNexis have worked closely with homeland security and intelligence officials. These businesses obtain and retain personal data about people including information about homes, relatives, and criminal records. Software has also been developed to find links in the data that would be indicative of terrorist activity. While some people like former Attorney General John Ashcroft argued that these new tools are vital for homeland security. Others in the FBI feared that there is not sufficient oversight or restrictions pertaining to corporations collecting information and then sharing it with the government (O'Harrow 2008). Intelligence operations can be important activities in homeland security, but could also be regarded as controversial measures depending on your point of view.

Self-Check

1. There were no clues that the attacks would take place on 9/11. True or false?
2. IMINT is intelligence collected by electronic communication. True or false?
3. Which part of the intelligence cycle is concerned with the creation of briefings and maps?
   a. Intelligence collection
   b. Intelligence analysis
   c. Intelligence production
   d. Intelligence feedback loop
4. What are some of the challenges facing the intelligence community?

8.4 Counterterrorism

One of the ways the United States deals with potential threats to our nation is to engage in counterterrorism operations (Jones 2014). **Counterterrorism** is the active pursuit of known terrorists that includes preemptive military strikes or the involvement of law enforcement officials. It is different from the detention and interrogation of terrorists by law enforcement, military custody, and CIA control. Counterterrorism relies heavily on intelligence but these operations take place prior to attacks occurring or even while they are taking place. The goal of such activities is to neutralize terrorist threats before they materialize or rescue hostages and minimize the loss of life afterward. There are numerous counterterrorism organizations in the United

---

CAREER OPPORTUNITY | Military Personnel

The military plays significant roles in intelligence-gathering and counterterrorism operations. In fact, many of the intelligence operations regarding terrorism reside in the Department of Defense and the armed forces (comprised of the Air Force, Army, Coast Guard, Navy, Marine Corps, and Space Force) are key participants in a variety of homeland security and emergency management functions.

Military personnel may be classified as reserve or active duty, and they may be designated as enlisted or officer in terms of rank. Employees of the armed forces may work domestically or internationally. Those serving in the military could be responsible for a whole host of things ranging from administrative affairs, combat, and construction to human resources, protective service and logistics, and transportation.

The growth of jobs in the military is expected to remain solid in the years to come due to ongoing and expanding threats from traditional enemies, terrorist organizations, and possibly even international criminal actors such as the Mexican drug cartels.

Pay for military personnel may start off at a lower scale than the private sector, but salaries rise with experience. The positions also come with extraordinary benefits (e.g., housing, housing allowance, education) as well as the possibility of retirement after 20 years.

For further information, see **https://www.bls.gov/ooh/military/military-careers.htm** or **https://www.dfas.mil/militarymembers/payentitlements/Pay-Tables/**.

States and other nations. Some of their operations illustrate the importance of adequate training. Other experiences provide lessons about failed operations or missions that have met their objectives.

The United States has several organizations that are involved in counterterrorism operations. This may include covert operatives from the CIA or undercover law enforcement officials from the FBI. The CIA infiltrates enemy organizations and operates on foreign soil. Its main purpose today is to gather intelligence about terrorist activities internationally. However, the CIA may also actively pursue foreign individuals in the United States if it is believed they pose a threat to the country. The FBI is also interested in intelligence issues, and it does collaborate with other law enforcement agencies abroad. In fact, the FBI has offices in at least 40 nations around the world and it desires to protect U.S. interests abroad such as an embassy. However, and in contrast to the CIA, the FBI is generally interested in preventing terrorist attacks at home. FBI special agents seek to arrest individuals who threaten or plan attacks in the United States. At times, efforts for apprehension may turn violent if the alleged criminals or terrorists resist.

The military is also heavily involved in counterterrorism operations, and the armed forces have operated under mandates such as the 2001 Authorization for Use of Military Force, which helped the United States find the perpetrators of 9/11 and combat groups like the Taliban who harbored these perpetrators. When authorized, the Special Operations Command plays an important role in coordinating counterterrorism organizations. It oversees special reconnaissance, direct action, unconventional warfare, and counterterrorism activities. It coordinates closely with special forces of the U.S. military.

In The Real World

Counterterrorism in Afghanistan

Shortly after 9/11, intelligence agencies confirmed that Osama bin Laden and Al-Qaeda were responsible for the attacks against the United States. It was also a well-known fact that the Taliban was providing these terrorists with sanctuary in Afghanistan. On January 22, 2002, special U.S. forces flew into the mountains of Kandahar to stop the instruction of terrorists at a training camp (Zaroya 2003). Before arriving at the compound, they launched a grenade to catch the terrorists off guard. They then entered the facility and began shooting at those offering resistance. While searching through the premises, one U.S. soldier was hit over the shoulder with a large stick. He then began to fight the terrorists by hand. The soldier broke his collarbone in the altercation but prevailed in the close-quarters combat. The training, tactics, night vision goggles, and other equipment were believed to have been some of the reasons for the success of this counterterrorism operation. Allies in Afghanistan who opposed the Taliban were also very helpful in this surprise attack.

Delta Force, or First Special Forces Operational Detachment-Delta, is one of the well-known special force units of the U.S. Army. It was formed in 1977 under Colonel Charles Beckwith and is modeled after the British SAS. Although the Pentagon does not publicly report on its specific activities, Delta Force is believed to be involved with reconnaissance, rescuing hostages, and fighting against enemy forces. Delta Force is made up of about 1,000 soldiers who have come from the Green Berets and the 75th Ranger Regiment. Candidates seeking to qualify for Delta Force membership

FIGURE 8-3 The role of counterterrorism centers is to thwart attacks before they happen.
Source: © *FBI.*

are considered on an invitation-only policy. They must have been in the military for at least 4½ years and then must undergo extensive physical and psychological tests. Delta Force members look like ordinary citizens (they can let their hair grow) and often have no markings on their uniforms. If accepted, they will be given a M1191 .45 caliber pistol. They are expert marksmen with M14, M21, and M25 rifles. Delta Force personnel train with many other special forces around the world, and they have been deployed in numerous counterterrorism operations.

Counterterrorism is also carried out by other military forces. Some of the more well-known teams include the Special Warfare Units (U.S. Navy), the Special Operations Wing (U.S. Air Force), and the Anti-Terrorism Battalion (U.S. Marine Corps).

In The Real World

Finding bin Laden Through Operation Neptune Spear

The most notable counterterrorism operation in U.S. history was the assassination of Osama bin Laden on May 2, 2011. Because the United States was increasingly interested in finding the whereabouts of bin Laden, the CIA established the Bin Laden Issue (Alec) Station from 1996 to 2005. Although initially unsuccessful in their efforts to find bin Laden, the CIA started to track a courier who was connected to this infamous terrorist mastermind. Intelligence agencies began to closely monitor a compound in Pakistan. Their leads indicated that bin Laden was likely living in the residence with his wives and children. In the middle of the night, SEAL Team Six flew secretly into Pakistan in helicopters, landed in the compound, and sealed off the streets in the area. Team members then proceeded to clear rooms in the building. They took out bin Laden's security forces and gathered women and children in rooms until the conflict was over. When the Navy Seals found bin Laden, they shot and killed him. Before they left, the Navy Seals gathered bin Laden's documents and computers to acquire additional intelligence about his operations and key partnerships. Although one of the helicopters was damaged during the landing and could not return to base, the event was regarded to be a huge success.

These teams are sent around the world to eliminate terrorist organizations and prevent attacks. Counterterrorism forces may work with soldiers in foreign nations, as is the case in the Philippines. Others have been deployed to Afghanistan and Iraq to capture Al-Qaeda members or other terrorist insurgents. In several cases, U.S. Special Forces have been highly effective at disrupting terrorists and keeping them on the run. States and local governments are now developing counterterrorism squads. They work closely with the FBI on Joint Terrorism Task Forces.

8.4.1 Risky Operations

Counterterrorism operations can be very risky and may involve a significant loss of life. Two of the most problematic counterterrorism operations in history are Operation Eagle Claw and the Moscow theater crisis. Operation Eagle Claw occurred on April 24, 1980, after 53 hostages were taken hostage in the U.S. Embassy in Tehran, Iran. In an attempt to resolve the situation, two actions were required. First, a staging base would have to be established in Iran. Second, military units would need to approach the embassy, rescue the hostages, and extract them out of Iran. Unfortunately, things went horribly wrong with this operation from the start. Two helicopters got lost in a sandstorm and another experienced a mechanical problem. A bus of civilian Iranians also came too close to the staging area and had to be captured to limit detection and notification of the operations. Because of these challenges, President Carter called off the rescue attempt. To make matters worse, a helicopter crashed into a C-130 transport plane when the mission was aborted, and eight airmen and marines were killed. The fiasco underscored the need for increased counterterrorism funding, better equipment, and improved training and communications.

Another counterterrorism operation also had deadly results (see Figure 8-4). On October 23, 2002, approximately 40 terrorists entered a Broadway-style theater in Russia with explosives and took about 850 hostages. The Islamic terrorists demanded the withdrawal of Russian military forces from Chechnya. The Kremlin would not give in to such demands, and the situation deteriorated when a girl was shot by the terrorists.

FIGURE 8-4 Russian Special Forces were able to kill terrorists at a Russian theater. Unfortunately, many citizens died in the process. *Source: © Getty Images News/Getty Images.*

After a two-and-a-half-day siege, Russian leaders decided to spray fentanyl (an anesthetic) into the building through the ventilation system, which caused everyone to fall asleep. OSNAZ (Russian Special Forces) then entered the building and killed about 30 of the terrorists with gunfire. The hostages were then taken out of the building and laid on the sidewalk. Unfortunately, many of the citizens suffered adverse reactions due to the chemicals. Complicating the matter, there were insufficient ambulances on the scene, and military leaders did not want to release information about the chemical agent to doctors. It is estimated that about 128 hostages died as a result.

8.4.2 Learning from Other Nations

The United States has worked diligently to develop one of the most sophisticated counterterrorism forces in the world. It has no doubt gleaned additional knowledge from countries that have a long history of responding to violent terrorist organizations. According to Cindy Combs (2000, pp. 169–176), Germany, Britain, and Israel are notable examples. A successful counterterrorism operation can also be witnessed from activities in Peru in 1996.

The **GSG9** is a German counterterrorism organization, whose name means "Border Guards, Group 9." The GSG9, which is now part of the Federal Police, was created after the terrorist attacks during the Olympics in Munich in 1972 (see Figure 8-5). As noted in a prior chapter, terrorists entered the Olympic Village where Israeli personnel were staying. The extremists killed some athletes and took other citizens as hostages. The response by the German government was less than desirable because it resulted in the death of both law enforcement officials and civilians. Wishing to minimize such problems in the future, this organization developed specialized training in counterterrorism operations. Members of GSG9 are extremely proficient in weapons use and are equipped with BMWs and BO 105 helicopters. In 1977, the GSG9 was able to enter a hijacked plane (Lufthansa Boeing 737) in Somalia and rescue 82 passengers. Four terrorists were killed in the altercation. The GSG9 has been involved in several operations against the Red Army Faction, a left-wing terrorist organization in Germany. Today they take measures against radical Islamic terrorists.

FIGURE 8-5   Counterterrorism organizations like Grenzschutzgruppe 9, the tactical unit of the German Federal Police, participate in training exercises.

Because Britain has experienced many terrorist attacks from the Irish Republican Army, it also has an acclaimed counterterrorism organization—the **Special Air Service (SAS)**. The SAS has developed a great deal of expertise as it has been in existence for over three-quarters of a century. Those wishing to become members of this group must serve in the military for a minimum of three years. They must then pass rigorous tests of strength and endurance. According to Combs,

> *Out of every 100 men who apply, only about 19 will meet the physical and mental requirements. The initial tests include a series of treks across the Welsh hills, carrying weighted packs. The final trek covers 37 miles, carrying a 55-pound pack, over some of the toughest country in the Brecon Beacons. It must be covered in 20 hours, and it is literally a killer course. Men have died trying to complete it (Combs, 2000, p. 173).*

If accepted, SAS recruits will subsequently be trained in hand-to-hand combat, water warfare, and emergency medicine. They are also experienced with the use of pistols, machine guns, and enemy weapons.

The British SAS carried out a successful counterterrorism operation in 1980. During the first few days in May, members of the Democratic Revolutionary Movement for the Liberation of Arabistan took over the Iranian embassy in downtown London. They demanded the freedom of their associates who were imprisoned and also wanted to gain possession of oil fields in Iran.

After a hostage was killed, officials in the United Kingdom decided to implement Operation Nimrod. To move SAS forces into place without detection, the government ordered that all airplanes flying into London approach the airport much lower than normal. Leaders also had a British utility company work in the area to create as much noise as possible. When the timing was right, the SAS set off a charge in the building and cut the power to the facility. SAS members then swung down from ropes and burst into the windows (after placing explosives on them first). As they entered the building, they set off stun grenades. Within a short period of time, they killed five terrorists and captured the last one remaining. Only one hostage was killed in the incident, and the rest were safely escorted from the building.

Of all of the counterterrorism organizations, few are as well-known as the **Sayeret Matkal** in Israel. Sayeret Matkal is a strike team devoted to finding terrorists before they attack Israeli interests. They were also established after the Munich massacre and shortly thereafter killed the mastermind of this attack (Ali Hassan Salameh) with a car bomb in Beirut. Because of the long-standing threat of terrorism against this country, this group of Special Forces in Israel is carefully selected. On a biannual basis, Camp Gibush is held to determine who can join the organization as new recruits. If applicants pass the constant and close monitoring of military officials, doctors, and psychologists, they then must train for an additional 20 months. This training is rigorous and includes, among other things, basic and advanced infantry training, parachuting, and other classes on reconnaissance, martial arts, and counterterrorism operations.

The Sayeret Matkal illustrated its capabilities when a Belgian aircraft was hijacked from Vienna to Tel Aviv on May 8, 1972. While the airplane was on the tarmac in Israel, 16 commandos (who were dressed as airplane technicians) told the terrorists the plane needed servicing. The terrorists let the commandos in disguise approach the plane, and the rescue operation commenced. Within 10 minutes, two terrorists were killed and others were taken captive. Three civilians were injured, and one died as a result. Benjamin Netanyahu (an Israeli prime minister) was a participant in Sayeret Matkal at the time.

Peru has also established special military forces to deal with the Shining Path and other left-wing and drug-related terrorist organizations. While Peru's special units are not as renowned as those in Germany, Britain, or Israel, they did participate in a successful counterterrorism operation. On December 17, 1996, 14 members of the Tupac Amaru terrorist organization stormed a Japanese embassy. They took at least 72 people hostage, including the brother of President Alberto Fujimori. The Tupac Amaru wanted 400 people released from prison, but the government would not acquiesce to these demands. Instead, counterterrorism forces smuggled listening devices into the embassy through humanitarian workers and closely monitored the activities of the terrorists. They noticed that the terrorists would play soccer every day at 3:00 p.m. and decided this would be the best time to conduct the raid. For the next several days, the Special Forces implemented Operation Chavin de Huantar (a name referring to a pre-Inca site with secret passages). While loud music was playing nearby, six tunnels were dug into the compound from nearby buildings. On the designated day (April 22, 1997), several explosive charges were detonated. Special Forces then emerged from the ground, entered the embassy, and began to exchange gunfire with the terrorists. During the conflict, hostages took cover (since they were previously notified to do so through an intelligence agent who posed as a doctor). When all was said and done, each of the 14 members of the Tupac Amaru were killed. Sadly, one hostage and one soldier were killed. Nine others were also injured. Nevertheless, the operation was hailed as a national success, even though international human rights monitors believe some of the terrorists were summarily executed.

In The Real World

Airstrike on Senior Hamas Commander

On October 26, 2023, Israeli Defense Force fighter jets conducted an air strike that killed Hassan Al-Abdullah, a leader of the Hamas terrorist organization. The Israeli Security Authority obtained intelligence about this official who was responsible for the Northern Khan Yunis Rockets Array. Precise air strikes were undertaken to ensure that this individual could never attack the state of Israel again. Over 250 similar strikes were ordered on Hamas targets in a 24-hour period. The goal of these military operations was not only to enact retaliation but to prevent further Hamas attacks in the future. Similar Israeli counterterrorism operations occur to this day against Hamas and other known terrorist organizations.

8.4.3   Controversy Regarding Counterterrorism

In spite of the successes and perhaps as a result of notable failures, counterterrorism is frequently controversial. Those supporting counterterrorism argue that there are many in and outside of the United States that wish to do us harm. They assert that the government has a moral responsibility to protect its citizens and prevent attacks. Others decry counterterrorism and suggest that it is not a reliable activity. Some protestors are more critical and assert that counterterrorism is an immoral undertaking. These individuals and groups fear intelligence operations and oppose the preemptive and deadly use of force.

The debate about counterterrorism operations is especially evident in discussions about the use of unmanned aerial vehicles—also known as drones (Mahan and Griset 2013). Drones are popular among politicians and military leaders because they reduce the cost of conflict and minimize risk to personnel. Drones can also be effective weapons. From 2004 to 2011, U.S. drones killed between 1,300 and 2,100 people. For these reasons, it is asserted that drones are "cheap, safe and precise." However, others assert that drones violate the airspace of other nations as well as the rights of individuals. Opponents also declare that drone strikes may have unintended consequences (e.g., the death of innocent civilians). These individuals consequently see drones as immoral weapons.

While the opponents to counterterrorism and drones seem to overlook the fact that enemies wish to do us harm, they note that counterterrorism may aggravate hostilities and breed additional violence in the future (Kamau 2021). For these reasons, counterterrorism will likely remain a contentious and divisive policy option going forward.

Self-Check

1. Preemptive military strikes are known as counterterrorism operations. True or false?
2. The FBI does not have offices in foreign countries. True or false?
3. Which nation has special forces that rescued hostages in the Iranian embassy?
   a. The United States
   b. Britain
   c. Peru
   d. Germany
4. Are there any drawbacks to counterterrorism operations?

Summary

Preventing terrorist attacks is one of the main pillars of our national homeland security strategy. If you are to succeed in stopping attacks before they occur, it is vital that you consider efforts that will reduce the root causes of terrorism. However, it is also important to understand the limitations of focusing solely on efforts to reduce poverty and protect human rights. You should therefore promote laws that prohibit terrorism and punish those who support such violent behavior. Relying on human and other sources of intelligence may help you to apprehend terrorists before they strike. Of course, this requires that the collected information is accurate and shared with those who need to know. At times, confronting terrorists may also require the involvement of the military in counterterrorism operations. There are countless measures you can take to prevent acts of terrorism against American citizens and others around the world.

Assess Your Understanding

Understand: What Have You Learned?

Go to **www.wiley.com/go/mcentire/homelandsecurity3e** to assess your knowledge of measures to prevent terrorist attacks.

Summary Questions

1. Not everyone agrees with the military response to the threat of terrorism. True or false?
2. The United States cannot be held responsible for each of the causes of terrorism around the world. True or false?
3. Policy and legislation may emanate from presidential directives but not congressional decisions. True or false?
4. The USA PATRIOT Act attempts to prevent terrorist attacks by strengthening law enforcement abilities. True or false?
5. The 9/11 terrorist attacks took the United States by surprise because it was operating under a Cold War mentality. True or false?
6. HUMINT includes intelligence gathered from academic research and books. True or false?
7. Counterterrorism can be defined as preemptive military strikes against terrorists. True or false?
8. The SAS is a counterterrorism force in Israel. True or false?
9. Which of the following may be considered a root cause of terrorism but cannot always be controlled by the United States?
   a. The treatment of Muslims around the world
   b. The upbringing of children in the Middle East in a culture of hate and violence
   c. American foreign policy
   d. Respect for culture and religion
10. Which is not a feature of the Comprehensive Homeland Security Act?
    a. Creation of the Department of Homeland Security
    b. Protection of nuclear power plants
    c. Increased border control
    d. Regulations pertaining to rail security
11. PDD-39 reaffirmed:
    a. The importance of border control
    b. The need for improved intelligence
    c. The value of rights and liberties
    d. The policy of the United States to defeat terrorism
12. Which component of the intelligence cycle deals with sharing information with policy makers?
    a. Intelligence collection
    b. Intelligence analysis
    c. Intelligence production
    d. Intelligence dissemination
13. Intelligence given to specific and limited individuals is known as:
    a. HUMINT
    b. Classified intelligence
    c. Categorized intelligence
    d. IMINT

14. Which counterterrorism force is the most well known around the world?

   a. SAS

   b. GSG9

   c. Sayeret Matkal

   d. Shining Path

Applying This Chapter

1. While watching the news on TV one night, you hear a scholar suggest that America must address the root causes of terrorism. What does this mean, and will "addressing root causes" eliminate all terrorist attacks?

2. If you were working in homeland security and someone suggested that laws will not help to reduce terrorism, what would you say?

3. A coworker who just started working with the FBI asserts that the United States should spend more money on SIGINT. Will this type of intelligence provide the most accurate view of terrorist operations? Is there another type that is preferable?

4. As a concerned citizen, you frequently watch the presidential debates broadcast on national television. On one occasion, there was a heated discussion about the merits of counterterrorism operations. What are the pros and cons of these types of military actions?

Be a Homeland Security Professional

Homeland Security Laws

Discuss laws relating to homeland security with another student. Discuss why they are necessary and provide some examples.

Assignment: Understanding the Intelligence Cycle

Write a paper about the intelligence cycle. Be sure to mention the steps that must be taken to gather intelligence about potential terrorists.

A New Recruit

You have just joined the military, and you desire to participate in counterterrorism operations. What kind of training might you undertake in the future?

Key Terms

Bojinka plot   A planned attack on airliners over the Pacific Ocean

Central Intelligence Agency (CIA)   The federal agency in charge of international intelligence operations

Classified intelligence   Information given only to a very specific and limited number of people to protect sources of acquisition and deny adversaries information that would lead them to alter their communications or operations

Comprehensive Homeland Security Act A law passed in 2003 containing new regulations for critical infrastructure security, railroad security, and more stringent measures related to border control and WMDs

Counterterrorism The active pursuit of known terrorists that includes preemptive military strikes

Federal Bureau of Investigation (FBI) The federal agency in charge of domestic intelligence operations

GSG9 A German counterterrorism organization, whose name means "Border Guards, Group 9"

Homeland Security Act A law passed in 2002 that mandated the creation of the Department of Homeland Security

HUMINT Intelligence collected by people from people (and can be done overtly or covertly)

IMINT Geospatial imagery collected by satellites and aircraft

Intelligence The function of collecting, assessing, and distributing information about an enemy, criminal, or terrorist

Intelligence adjustment Adaptation of the intelligence cycle is required when the collection is incomplete, the analysis seeks to "connect the dots," production generates new questions, and dissemination results in the anticipation of future concerns

Intelligence analysis Efforts to make sense of the voluminous data that are gathered from the field

Intelligence collection Activities to gather information about terrorist organizations and their operations and potential attacks

Intelligence cycle A four-step process of gathering, understanding and synthesizing data, and then sharing it with those who will use it

Intelligence dissemination Sharing information with end users (e.g., policy makers, FBI special agents, and homeland security personnel)

Intelligence production The creation of written reports, briefings, images, or maps to influence operational decisions

INTERPOL An international police organization that is involved in intelligence

MASINT Measurement and signature intelligence that looks for the characteristics of certain types of actions (e.g., the presence of nuclear material when one is trying to develop a nuclear weapon)

Money laundering The process of hiding where money is coming from and what it is being used for

OPINT Open-source intelligence acquired through publicly available materials including academic research, newspaper articles, library books, and so on

Sayeret Matkal A strike team devoted to finding terrorists before they attack Israeli interests

SIGINT Interception and interpretation of electronic communications such as phone conversations and e-mails

Special Air Service (SAS) A British counterterrorism organization

The Terrorist Threat Integration Center (TTIC) A government organization that attempts to improve coordination between the FBI and the CIA

Transportation Security Act A law designed to protect transportation systems in the United States

USA PATRIOT Act The name of a homeland security law that stands for "Uniting and Strengthening America by Providing Appropriate Tools Required to Intercept and Obstruct Terrorism." This law aims to prevent terrorist attacks and enhance law enforcement's ability to investigate and punish offenders

References

9/11 Commission. (2004). *The 9/11 Commission Report: Final Report of the National Commission on Terrorist Attacks Upon the United States*. New York: W.W. Norton and Company, Inc.

Abu-Amara, S. (2002). Terrorism is a result of Israel's denying a people its rights. *Dallas Morning News*. Viewpoints (Monday, July 22), p. 9A.

Carter, D.L. (2009). *Law Enforcement Intelligence: A Guide for State, Local, and Tribal Law Enforcement Agencies*. Washington, DC: US Department of Justice.

CNN (2001). Iranian president: "root of terrorism" must be addressed. *CNN* (November 12). http://archives.cnn.com/2001/WORLD/meast/11/12/khatami.interview/index.html. (Accessed April 19, 2017).

Combs, C.C. (2000). *Terrorism in the Twenty-First Century*. Upper Saddle River, NJ: Prentice Hall.

Deucy, C.P. (2006). Intelligence and information sharing in counterterrorism. In: *The McGraw-Hill Homeland Security Handbook* (ed. D.G. Kamien), 391–412. New York: McGraw-Hill.

Etzioni, A. (2004). *How Patriotic Is the Patriot Act? Freedom Versus Security in the Age of Terrorism*. Philadelphia, PA: Routledge.

Francis, D.R. (2002). Poverty and low education don't cause terrorism. *The NBER Digest* (September).

Franks, J. (2006). *Rethinking the Roots of Terrorism*. New York: Palgrave Macmillan.

Githens, D. and R. Hughbank. (2010). Intelligence and its role in protecting against terrorism. *Journal of Strategic Security* 3 (1): 31–38.

Jones, K. (2014). *U.S. Counterterrorism Programs in East and Northwest Africa*. New York: Nova Science Publishers, Inc.

Kamau, J.W. (2021). Is counter-terrorism counterproductive? A case study of Kenya's response to terrorism, 1998–2020. *South African Journal of International Affairs* 28 (2): 203–231.

Kauppi, M.V. (2006). Counterterrorism analysis and homeland security. In: *The McGraw-Hill Homeland Security Handbook* (ed. D.G.Kamien), 413–430. New York: McGraw-Hill.

Lash, J. (2001). Dealing with the Tinder as well as the Flint. *Science* 294 (5548): 1789.

Leiter, K. (2002). Palestinians must remake their society before peace can come. *Dallas Morning News*. Viewpoints (July 22), p. 9A.

Lustick, Ian. (2007). The war on terror: when the response is the catastrophe. Paper presented at FEMA Higher Education Conference, Emmitsburg, MD (June 5).

Mahan, S. and P.L. Griset. (2013). *Terrorism in Perspective*. Thousand Oaks, CA: Sage.

O'Harrow, R. (2008). Mining personal data: one company keeps tabs on the public in the name of post-9/11 security. In: *Annual Editions Homeland Security* (ed. T.J.Badey), 141–143. Dubuque, IA: McGraw Hill.

Russell, H., Forest, J., and Moore, J. (2006). *Homeland Security and Terrorism: Readings and Interpretations*. McGraw-Hill Homeland Security Series, 1e. New York: McGraw-Hill.

Shakil-ur-Rahman, M. (2007). Root causes of terrorism must be addressed: PM. *The International News* (Monday, August 13). http://www.thenews.com.pk/top_story_detail. asp?Id=5452. (Accessed January 30, 2008).

Thetford, R.T. (2001). The challenge of cyberterrorism. In: *Terrorism: Defensive Strategies for Individuals, Companies and Governments* (ed. L.J. Hogan), 238–257. Frederick, MD: Amlex, Inc.

Wall, A. (2011). Demystifying the title 10—title 50 debate: distinguishing military operations, intelligence activities & covert action. *Harvard National Security Journal* 3: http://harvardnsj.org/wp-content/uploads/2012/01/Vol-3-Wall.pdf.

Zaroya, G. (2003). Inches divide life, death in the Afghan darkness. *USA Today* (October 19), p. 4A.

CHAPTER 9

Securing the Nation
Border Control and Sector Safety

DO YOU ALREADY KNOW?

- Methods to enhance border control
- Measures to limit attacks against air transportation
- Steps to secure railroads and rail yards
- Efforts to avert terrorism against sea ports
- Actions to reduce threats facing chemical facilities

For additional questions to assess your current knowledge of measures secure the nation, go to **www.wiley.com/go/mcentire/homelandsecurity3e**

WHAT YOU WILL LEARN

9.1 Agencies related to border control

9.2 The mandate of the Transportation Security Administration

9.3 Why VIPR teams patrol rail transportation systems

9.4 The role of the private sector in port security

9.5 How chemical facilities can safeguard themselves against attacks

WHAT YOU WILL BE ABLE TO DO

- Critique policies relating to border security
- Identify ways to secure air transportation from terrorism
- Anticipate alternative measures to reduce attacks on railroad systems
- Implement procedures to secure seaports and maritime trade
- Assess the effectiveness of chemical facility safety standards

Introduction

A major priority to minimize the probability and consequences of terrorism is to secure the nation against possible attacks. If you are working in the homeland security profession, you must do all you can to stop the infiltration of terrorists into the United States. This suggests that the border needs to be carefully controlled and that immigration processes need to be strengthened. Policies must also be implemented to limit the risks associated with air, rail, and sea transportation. This is imperative since terrorists often attack such transportation systems. In addition, it will be imperative that you enhance chemical facilities through the enforcement of widely accepted security and safety standards. Each of the measures discussed in this chapter may help to halt terrorist attacks now and in the future.

9.1 Border Control

Since its very foundation, the United States has been established on the principle of immigration and the Statue of Liberty is a symbol of a welcoming attitude toward those who arrived in this country. The migration of foreigners into provided many significant benefits over time. For instance, immigration brought diversity as well as a notable aspiration to succeed, new knowledge, unique skills, cutting-edge innovation, and rapid economic growth to America. Thus, there have been many advantages to immigration over time. However, some scholars and practitioners suggest that legal migration may diffuse terrorism around the world (Bove and Bomelt 2016). That is to say, the movement of people to other countries allows the ideological forces of terrorism along with logistical support to spread elsewhere too.

Of even more concern is illegal immigration, which may be a conduit by which terrorists enter other nations to launch attacks. Of course, it is true that the people who come to the United States illegally often want to seek freedom and economic prosperity. Nevertheless, a former Secretary of the Department of Homeland Security pointed out that the illegal entry of foreigners is "a national security vulnerability that must be addressed" (Chertoff 2009, p. 41). The assertion is that, if anyone can enter the United States, terrorists can as well. Conversely, if the government can control the flow of people and goods into this nation, then there will obviously be less likelihood that enemies from abroad will be able to implement terrorist acts against us. Therefore, reducing or eliminating the infiltration of terrorists and weapons of mass destruction (WMDs) into the United States is a very important way to prevent possible terrorist attacks. This brings up the important subject of borders and border control.

9.1.1 What Is the Border?

The **border** is the territorial boundary of any nation along with its various points of entry. Jack Riley, an associate director of RAND Infrastructure, Safety, and Environment, stated that people and goods can enter the United States from one of four locations (Riley 2006, p. 587). This includes airports, seaports, guarded land points, and unguarded land borders and shorelines.

Airports are the transportation hubs where foreigners fly into the United States and are allowed or denied entry based on valid passports and visas for business, education, or vacation purposes. Seaports are also transportation networks that make up portions of the border. They facilitate the movement of people and likewise play a role in commerce due to the trade they facilitate. Guarded land points include border stations where people, vehicles, and trucks are processed for entry. Unguarded land borders are those locations between the United States and its neighboring countries (i.e., Canada to the north and Mexico to the south). These border locations may be protected at times by fences and walls but are generally unsecured in most cases. There are also coastlines to the northeast, east, west, and south of the United States, which are considered a part of national borders. The waterways near the Great Lakes, the Atlantic Ocean, the Pacific Ocean, and the Gulf of Mexico are often open to human movement and maritime shipping with limited or at least minimal oversight. Unfortunately, the sheer size and openness of our nation is an invitation to those who wish to do us harm.

9.1.2 Our Porous Border

Statistics regarding the size and fluid nature of the border are impressive or overwhelming, depending on your point of view. For instance, there are more than 100 international airports in the United States (Riley 2006, p. 587). America also has 90,000 miles of coastline (Hoffman 2006, p. 143) and "more than 1000 harbor channels and 25 000 miles of inland, intra-coastal, and coastal waterways, serving over 300 ports, with more than 3700 passenger terminals" (Benztzel 2006, p. 631). These numbers do not include, of course, the hundreds of other foreign airports and seaports around the world that are connected to counterparts in this country. What is more, there are also 9,000 miles of land borders (Hoffman 2006, p. 143). The scope of the border is therefore awe-inspiring.

The extent of the border is only dwarfed by the dizzying pace of activity that takes place around it. In a single year, "half a billion people, 125 million cars, 12 million trucks, and 33 million overseas shipments—including nearly 6 million shipping containers, 800 million planes, 2 million railcars, and over 200,000 ships" enter U.S. borders (Hoffman 2006, p. 143).

For instance, there are thousands or even hundreds of thousands of employees that enter the country on a daily basis to work in the agricultural sector or in factories called maquiladoras along the U.S.–Mexican border. Another interesting fact about the border is that international commerce has increased substantially over the past few decades due to improvements in transportation and the high demand for foreign goods and services. Corporations prefer the ability to send things quickly from other countries to the United States. In fact, "many manufacturers and retailers now use the 'just-in-time strategy' to reduce the costs of carrying and storing inventory" (Riley 2006, p. 588). This commercial activity across national borders has boosted global trade in dramatic ways.

There are many other reasons why the border is so fluid. Millions of people travel to the United States to conduct business operations or to vacation in one of the many tourist locations throughout the nation. The United States is one of the most significant economic hubs in the world, so there are many negotiations and transactions occurring at any given point in time. In addition, people love to visit major attractions in Alaska, Arizona, California, Colorado, Florida, Hawaii, New York, and Utah among other states.

However, it is also true that a sizable and even overwhelming number of people enter the country illegally. Hundreds of thousands and even millions of people cross the border without permission each year (Knickerbocker 2006). While the number of illegal immigrants declined during President Trump's tenure, the trend reversed recently and in dramatic fashion. For example, in 2022, there were an estimated 2.2 million people who crossed the border illegally (PBS 2023). Since President Biden took office, it is estimated that that there have been more than 8 million illegal border crossings (Blankley 2023) and perhaps even more. This may be a direct result of Biden's campaign promises during the 2020 election cycle. He stated that he wanted to welcome immigrants to the United States and implied that they would not be sent back if they came. Tim Murtaugh of the Heritage Foundation (2021) asserts that the democratic nominee also pledged "an array of enticements for people to ... enter this country [such as] ... amnesty for those already here, taxpayer-provided health care, work permits, support for sanctuary cities, and a cessation of deportations."

Millions of individuals and families therefore come to the United States to improve the quality of their lives. Immigrants have traditionally been comprised of Mexican nationals and others from Latin America. In addition, hundreds of thousands of Venezuelans have arrived in recent years to escape the repressive, corrupt, and inept authoritarian regime in that country. There is an increasing number of immigrants from African and Asian nations who also seek to improve their financial well-being.

There are diverse sentiments about this type of immigration and the extent of illegal crossings. Politicians on the left and right have historically preferred an open border at times, even if they will not admit it publicly. Some politicians may want to extend America's freedoms and prosperity to anyone wishing to obtain these coveted benefits. Democrats, in particular, often welcome immigrants to the United States because these lower-income individuals may eventually result in increased votes for the candidates representing their political party. In contrast, conservatives have frequently desired a constant supply of cheap labor to keep corporate profits high, although they have much less of an appetite for illegal immigrants today than in the past.

Citizens and local/state governments also have divergent opinions about illegal immigration. Some Americans feel bad about the harsh political, economic, social, and other conditions facing people around the world and want to help these people overcome their challenges by encouraging and facilitating their arrival to the United States. In addition, certain political jurisdictions like Chicago, New York, and San Francisco have advertised their sanctuary status (implying that migrants are welcome will not be reported to federal border officials). However, citizens and jurisdictions along the southern border are extremely frustrated by the millions of illegal immigrants making their way to the United States. They have increasingly voiced their concerns about the open border and frequently share their anger about being overwhelmed by the significant needs of these immigrants (e.g., food, housing, and other social services). Due to this opposition, the Biden administration started to transport illegal immigrants by bus and plane to cities throughout our country. Republican governors like Greg Abbott (Texas) and Ron DeSantis (Floria) also chartered places and buses to send illegal immigrants to Chicago, New York, and even the exclusive Martha's Vineyard community. Their goal was to illustrate how just a few hundred immigrants could easily outstrip available resources. This concern appears to be justified. Some communities have converted sports parks to encampments for the immigrants. They have shut down schools meant for the children of taxpayers to make way for immigrants. Others are spending enormous portions of the budget to shelter immigrants in hotels. In fact, some estimate that the United States is spending

In The Real World

Immigration Debate

During the 2016 Presidential Election, Donald J. Trump, the Republican candidate, repeatedly announced his closed-door policy toward certain groups of Muslims who wished to enter the United States. He also advocated for deporting all illegal immigrants from the United States but eased up on this perspective to some extent by clarifying his goals for immigration would focus on those who pose a danger.

Following the San Bernardino, California, shooting attack in December 2015, Trump suggested the nation should again enforce a ban on Muslim immigrants. He subsequently altered his stance slightly, choosing to focus on preventing immigrants from entering the United States if they were from countries that had any ties to terrorists. This sparked an intense debate on Trump's policies, even for an already highly controversial candidate.

Notably, there are pros and cons to having such a restricted approach to immigration as Trump proposed. He and his supporters believed that the thorough monitoring and screening of immigrants would help identify those who have links to terrorists or those who condone jihadist ideals. This would then allow border officials to prevent prospective enemies from infiltrating the United States. Trump also vowed to build a wall along the entire U.S.–Mexico border as part of his plan, and he made some progress toward this goal. Stricter enforcement of the border through tightened immigration standards and a wall would logically make our nation safer and decrease our vulnerability to terrorist attacks on U.S. soil. In addition, a more regimented border could mean increased attention to citizens' needs rather than spending tax dollars on illegal immigrants' welfare or education. Employment opportunities and government spending could be focused more on helping legal citizens provide for themselves and help care for their families.

However, these stronger immigration policies are a costly financial endeavor. Tightening immigration regulations and increasing border security as Trump advocated would require the allocation of significant financial and human resources. Furthermore, a closed-door immigration policy might result in economic losses since millions of jobs are filled by immigrants. Also, those who favor open immigration standards criticized Trump since they believed the United States should be open to all people who wish to take advantage of the wonderful opportunities available in this nation. Others likewise asserted the country should remain a melting pot of varying cultures, ideas, ethnicities, religions, and races. Lastly, it was stated that having more stringent regulations could potentially cause discrimination against immigrants who are already legally and lawfully residing in the United States. It is evident that there are a multitude of advantages and disadvantages to Trump's immigration propositions, depending on your political viewpoints.

In recent years, the debate about the border has intensified due to President Biden's reversal of Trump's policies. The arrival of millions of immigrants since 2021 has placed a significant burden on local and state governments throughout the nation and a majority of citizens and politicians now have serious reservations about illegal immigration. But, besides the financial and other sociocultural strains, there are increased concerns about the possibility that terrorists may enter the United States illegally. An October 20, 2023, memo issued by the Border and Customs Protection (CBP) stated that individuals "inspired by, or reacting to, the current Israel-Hamas may attempt to travel ... from the area of hostilities in the Middle East via circuitous transit across the Southwest border" (Shaw and Melugin 2023). The FBI Director has also expressed concerns about increased risks of terrorism owing to Biden's immigration policies. This could pose a serious security threat to the citizens, infrastructure, and government of the United States. Clearly, the topic of illegal immigration will continue to be discussed and disputed as time goes on.

hundreds of billions of dollars each year to care for those who have arrived in this country. As a result, and ironically, even the leaders of many sanctuary cities (including Chicago's Lori Lightfoot and New York's Eric Adams) have begun to oppose the massive influx of immigrants due to the enormous financial burden this is placing on municipal purse strings. They and many others have publicly criticized the Biden administration for its open border policies.

Finally, the challenge and expense of securing our borders is monumental. In 2006, airport officials estimated that less than 10% of cargo was physically inspected, and most airports lacked "the equipment to conduct inspections, especially of large containers" (Riley 2006, p. 593). While the inspection of airport cargo has improved over the past decade, only about 25% of truck cargo and 90% of rail cargo are inspected before or when they come into the nation. One of the problems is that there is a shortage of personnel to quickly process the vast amount of goods that come to the United States on a daily basis. This is to say nothing about the number of personnel needed to prevent and deter illegal immigration or process legal visitors. Protecting the border therefore requires serious investment of resources on an ongoing basis.

Regardless of the cause for the movement in and around our borders and the expense to protect them, citizens and experts alike recognize the potential danger to homeland security. If the border cannot be adequately controlled or regulated, terrorists may be able to enter the United States along with weapons that could be used in terrorist attacks. This difficulty of ensuring only friendly immigrants arrive in the United States has already been exploited by those who wished to do us harm on 9/11. Frank Hoffman stated that bin Laden "struck at America's Achilles' heel—its porous borders, transportation networks, and vulnerable economic portals" (Hoffman 2006, p. 142). He also asserted that "our most daunting domestic challenge is posed by the open nature of America's society: borders" (Hoffman 2006, p. 143).

9.1.3 Participants Involved in Border Control

There is no single entity involved in border control. Instead, there are a variety of government agencies interested in the securing of our border (see Figure 9-1). This includes Customs and Border Protection, Immigration and Customs Enforcement, and the Coast Guard. Corporations and international organizations also play a unique role in securing our border.

Customs and Border Protection (CBP) is the largest organization within the Department of Homeland Security that enforces immigration law at and between ports of entry. The 60,000 employees working for CBP process legal immigration into the United States and attempt to deter illegal immigration as well as the smuggling of money, drugs, and other materials that support terrorism.

Immigration and Customs Enforcement (ICE) is another organization within the Department of Homeland Security that enforces immigration and customs laws in the internal portions of the United States (i.e., away from the border). It was created in March 2003 by combining the Immigration and Naturalization Service and the U.S. Customs Service. In an attempt to protect the homeland, ICE looks for fraudulent passports, investigates employers that hire illegal aliens, and stops the transport of weapons and sensitive military technologies. It works alongside other government entities including the Small Business Administration and the U.S. Citizenship and Immigration Services that enforce E-Verify (an internet-based data system that relays information about eligibility to work in the United States).

FIGURE 9-1 There are many government employees who work to secure U.S. borders.
Source: © U.S. Department of Homeland Security.

CAREER OPPORTUNITY | Border Agents and Transportation Security Screeners

To protect the border and secure our transportation systems, the federal government employs over 100,000 workers in a variety of agencies including CBP, the Transportation and Security Administration, and the U.S. Coast Guard.

The employees in these agencies deter illegal entry into the United States, screen visitors, and passengers and investigate and prosecute the smuggling of people, weapons, and illegal drugs. Consequently, border patrol agents and transportation screeners perform law enforcement functions and must pass physical and background checks to ensure they can perform the job and ensure that there are no insider threats.

The growth rate in this area varies by position and agency, but most of these positions will expand by 3% over the next 10 years. Border patrol agents make about $45,000 per year, while transportation screeners make on average about $60,000 annually. Salaries for those in the Coast Guard vary the most from about $23,000 to $86,000.

For more information, see Adapted from **https://www.bls.gov/oes/current/oes339093.htm** or **https://www.tsa.gov/about/jobs-at-tsa** or **https://www.gocoastguard.com/careers/enlisted**.

The **U.S. Coast Guard (USCG)** is an organization of the Department of Homeland Security. It employs over 43,000 people. In terms of the border, the USCG interdicts illegal aliens and seizes illegal contraband. One of its goals is to prevent terrorists from entering U.S. waters or gaining access to the shores of U.S. lands. Because the Coast Guard is a military branch, its employees receive the same pay and benefits as other active and reserve personnel in the branches of the armed forces.

Besides these agencies in the Department of Homeland Security, others play a role in border control. For instance, "private companies such as airlines, truckers, container shippers, and manufacturers—as well as companies whose employees travel over these borders—are stakeholders" (Riley 2006, p. 588). Observant employees can do much to enhance security at the border. Corporate involvement is also vital since any attack along the border could have drastic economic consequences for their bottom line.

9.1.4 Measures to Secure Borders

Those charged with border control are implementing a number of programs to prevent terrorist attacks against the United States. The most well-known program is U.S.-VISIT. The government and even citizens have undertaken other measures to protect the integrity of the border.

The **United States Visitor and Immigrant Status Indicator Technology (U.S.-VISIT)** is a computer database created and maintained by the Department of Homeland Security. It contains information about passengers who wish to travel to the United States. Those seeking visas will have their biometric data recorded (e.g., digital finger scans and photographs). When the traveler arrives in the United States, his or her biometric data will then be compared with the information stored on the system and checked against known criminals and terrorists around the world. According to Riley (2006, p. 592), "the system was first tested in Atlanta in late 2003 and became operational at all 115 international airports on January 5, 2004. Simultaneously, it was introduced at 14 major seaports served by cruise liners." At least 90 million people have had their biometric information recorded by U.S.-VISIT since its inception. However, citizens of 27 allied countries have been excused from participating in this system. This is a major weakness in that some of the terrorists on 9/11 came from these exempted nations.

The government is also looking into or taking other measures to secure our borders and prevent further terrorist attacks. Some officials in the government assert that passports need to be changed continually to prevent tampering and falsification. Smart lanes and electronic passes (much like toll tags) are being developed and used to speed up the processing of trucks carrying goods into the United States. Unmanned aerial drones and stationary cameras are being used along the borders to detect locations where people are entering the United States illegally. To increase the effectiveness of border agents, the government has opened up a border patrol academy in Artesia, New Mexico. At this school, recruits are taught Spanish, learn about immigration law, practice using weapons, and acquire or enhance other law enforcement techniques and skills. Patrol of the border has also had to adapt to changing circumstances, which has led to the discovery of tunnels and the capture of illegal aliens. The goal is to make it more difficult for illegal aliens and terrorists to enter the United States.

Finally, citizens have become more involved in measures to enhance border security. After feeling that the government was not doing enough to protect the United States against the entry of illegal aliens and potential terrorists, many living on or near the border are protesting current government policies. In addition, a group of concerned individuals in Arizona joined together in 2004 to form the **Minutemen Project**. People who join the Minuteman Civil Defense Corps lobby local, state, and federal officials to do more about border security. They also organize shifts when

In The Real World

Preventing Terrorist Entry into the United States

On January 29, 2017, an executive order was signed by President Trump to allow for the review of standards to prevent the entry of foreign terrorists and criminals into the United States. Trump argued that the most generous immigration system has been "repeatedly exploited by terrorists and other malicious actors who seek to do us harm." To reduce national security risks that were identified under the Obama administration, the president wanted to halt those traveling on passports from Iraq, Syria, Sudan, Iran, Somalia, Libya, and Yemen. He asserted that Congress provided the president under section 212(f) of the Immigration and Nationality Act the ability to prevent entry of individuals who are deemed to pose a threat to national interest. The order also provided the government with the "additional resources, tools and personnel to carry out the critical work of securing our borders, enforcing the immigration laws of our nation, and ensuring that individuals who pose a threat to national security or public safety cannot enter or remain in our country." Although prior presidents have implemented similar measures, the order was met with fierce opposition from some in the public and others in the media. U.S. District Judge Ann Donnelly issued a verdict to bar the travel ban after the American Civil Liberties Union filed a petition in court. This action was later sustained when the Ninth Circuit of Appeals ruled 3–0 to maintain the freeze on the immigration order. In response, Trump later tweeted "SEE YOU IN COURT, THE SECURITY OF OUR NATION IS AT STAKE!" The Trump administration also rewrote the order to facilitate implementation as soon as possible. This, again, was challenged in court. However, in June 2017, portions of the travel ban were approved by the Supreme Court. Some declare that these types of measures may be needed, even if they are not always popular. Over 260 individuals on the terrorist watch list have been stopped at the border between January 1, 2022, through September 30, 2023 (Ainsley 2023). Such actions have logically helped to prevent terrorist attacks. Nevertheless, immigration policies are likely to remain a contentious issue into the future.

---

volunteers can patrol the border and notify the government when people attempt to cross the border illegally. While the group is having an impact on politicians' views and may possibly deter those wishing to enter the United States illegally, some fear that this group may aggravate tensions along the border and be unprepared if any type of conflict erupts. Regardless, hundreds of illegal immigrants have been detained as a result of the efforts of the Minutemen.

In The Real World

Ahmed Ressam

On December 14, 1999, Ahmed Ressam attempted to enter the United States to conduct a terrorist attack known as the Millennial Plot. Ressam hid explosives in his car and was attempting to travel from Canada to California to blow up the Los Angeles International Airport on New Year's Eve. When questioned by a border agent near Seattle, Washington, Ressam became anxious and was evasive. He tried to flee the area and was shortly arrested for his activity along with a colleague from Brooklyn, New York. This case illustrated the importance of observant border patrol agents.

Self-Check

1. Airports are not considered part of the territorial boundary of the United States. True or false?
2. It is relatively easy for the United States to oversee the people and goods that enter our country. True or false?
3. Which agency is responsible for investigating companies that hire illegal aliens?
   a. The Transportation Security Administration
   b. ICE
   c. The U.S. Coast Guard
   d. The United Nations International Maritime Organization
4. What are a few of the measures taken by the Department of Homeland Security to prevent illegal immigration?

9.2 Protecting Air Transportation

As noted in earlier chapters, terrorists have repeatedly targeted ground transportation systems like buses and subways. Airplanes and airports are also favorite targets of terrorists as illustrated throughout many historical examples. In addition, terrorists may launch attacks on railroads and seaports. Chemical facilities are also likely to be targeted in the future. Each of these will be discussed in turn.

The aviation industry comprises a significant portion of the transportation sector across the globe and is consequently a major focus for terrorists. In 2015, the U.S. Department of Transportation calculated that 685 million people flew on domestic airliners. Today, an average of 1.73 million people fly in the United States each day. The vast number of people who use air transportation coupled with airliners' high-speed capabilities and endless destination possibilities makes the aviation industry an extremely vulnerable and enticing form of transportation to exploit by terrorists.

As mentioned in Chapter 1, terrorists infamously turned airplanes into WMDs in the September 11, 2001 (9/11), hijacking attacks. But there have been many terrorist attacks involving aircraft before and after this fateful date. Terrorists have hijacked many flights throughout the history of aviation, and they bombed PAN AM 103 on December 21, 1988. Terrorists also tried to use explosive devices to cause harm to aircraft, such as the sports bottle plot in 2006 and the Underwear Bomber in 2009. On July 17, 2014, Malaysia Airlines Flight 17 was brought down over Ukraine using surface-to-air missiles. During that same year, the threat of laser illumination (when a laser is shined into the cockpit of an aircraft) was reported 3,894 times. In addition, ISIS is suspected of bombing a Russian Metrojet Flight 9268 on October 31, 2015. This event—probably resulted from a bomb placed in a soda can—that killed 224 people was the deadliest air disaster in Russian history.

These examples represent just a handful of the attacks involving or targeting the aviation industry. More will certainly follow. During a one-week period in July 2017, the Transportation Security Administration confiscated 96 handguns while completing passenger screening in the United States alone. Recent intelligence reports indicate that "terrorists continue to target commercial aviation and are aggressively pursuing innovative methods to undertake their attacks, to include smuggling devices in various consumer items" (Jansen 2017, p. 1A). In consequence, this threat prompted an international ban on personal electronics (excluding cell phones) on nine airlines from 10 countries from the Middle East and Northwest Africa. If the affected airlines do not comply, they will not be allowed to fly to the United Kingdom or the United States.

In The Real World

The 2016 Terrorist Attack at the Brussels Airport

On March 22, 2016, two suicide bombers entered the Brussels Airport with large suitcases carrying explosives. After yelling something in Arabic, the bombs were detonated near two check-in counters. This attack, along with another concurrent bombing of a train department at the Maelbeek/Maalbeek metro station, killed 32 people and injured another 300 individuals.

After investigating the attack, it was revealed that Ibrahim El Bakraoul and Najim Laachraoui conducted the coordinated attacks with the help of a few other individuals. These terrorists were members of a cell related to the Islamic State. It is believed that they acted in response to Belgium's involvement in the war in Iraq and the country's counterterrorism operations in France and Belgium from 2014 to 2016.

This attack revealed that aircraft are not the only targets of terrorists. Individuals who adhere to violent ideologies will launch attacks at airports as well. Countless attacks have occurred at these transportation hubs throughout history, including a terrorist attack involving a rifle at the LAX Airport on November 1, 2013.

With all the potential threats to aviation, the government has instituted agencies and programs to help identify and prevent such tragedies from occurring. These include the Transportation Security Administration, the Federal Air Marshal Service, the Computer Assisted Passenger Prescreening System II, and Secure Flight. Airports Council International has also been established to increase the security of airports and the world.

The **Transportation Security Administration (TSA)** is a federal agency under the Department of Homeland Security (see Figure 9-2). It was created after the 9/11 attacks to protect our transportation systems from acts of terrorism. It includes over 47,000 security officers and inspectors who share a unified goal of preventing terrorist attacks on airplanes, subways, and rail systems. The TSA screens passengers, checks luggage and cargo for bombs, deters hijackings and other attacks on transportation systems, and works with local law enforcement agencies to protect transportation systems. The TSA is most visible in airport terminals, but the employees of this organization perform many other activities behind the scenes (e.g., audits of security systems on the airfield or related hangars). This organization is responsible for following security guidelines outlined in the **Transportation Security Act**—the law designed to protect transportation systems in the United States.

The **Federal Air Marshal Service** is another important organization of about 3,000 marshals who act under the TSA. This Air Marshal Service was initially created in 1961 to create and maintain confidence in the civil aviation system when many hijackings started to occur. Today, the Federal Air Marshal Service is most interested in identifying and halting terrorist attacks against U.S. aircraft. Air marshals are armed security officers trained in aircraft-specific engagement tactics. They often blend in with general passengers aboard the aircraft, concealing their status as law enforcement officials. After the 9/11 attacks, the Federal Air Marshal Service grew

FIGURE 9-2   TSA agents have the goal of preventing terrorist attacks against aircraft. *Source:* © *U.S. Department of Homeland Security.*

substantially. Certain "gateway airports" require that an armed security officer is on board at least 48 flights per day. Even though only about 5% or less of flights have an Air Marshal onboard, this is a vital program to protect our aircraft, crew, and passengers.

Another important program for the safety of aviation was the **Computer Assisted Passenger Prescreening System II (CAPPS II). CAPS II** was a database that ensured that passengers are screened by every airliner and airport that operates within the United States. This system was implemented after 9/11 and was maintained by the TSA. CAPS II collected information about passengers (including date of birth, street address, and telephone numbers). Once obtained, these data were then compared with terrorist or most wanted lists as well as federal and state warrants for arrest. A risk score would then be calculated and printed on the boarding pass so airport screeners could be notified. Because of potential rights violations, watchdog groups and the General Accounting Office became very critical of CAPPS II. In the summer of 2004, this program was suspended. However, a new program known as Secure Flight was implemented in its place.

Secure Flight is the more recently instituted advanced passenger screening program that is administered by the TSA. Secure Flight checks passenger information with watch lists maintained by the federal government, such as the No-Fly List. Therefore, Secure Flight functions similarly to the CAPPS II passenger screening. However, a notable difference in Secure Flight compared with programs like CAPPS II is that the TSA is now solely responsible for matching passengers to the watch list. Prior to Secure Flight, airlines had the responsibility for this type of passenger screening.

In addition to U.S. initiatives mentioned earlier, the Airports Council International helps to set global standards regarding the safety and security of airports. This organization has taken a more active role since 9/11. Airports Council International shares information and best practices for training with partner nations from around the globe. The efforts of the Airports Council International are extremely important since the safety of any given aircraft is determined by the security of airports in other parts of the world. Any weak link in the international transportation system could impact the safety of others in faraway nations.

In summary, it is apparent that the aviation industry will likely remain a major target for terrorist activity. Government agencies and international partners must therefore take further precautions and continually advance and improve their safety programs to meet the escalating threat of violence.

In The Real World

Airport Security

U.S. airports are among the safest in the world. Each year, airport screeners stop the smuggling of hundreds of knives and guns on planes. However, there is concern about the 900,000 people who work at airports and have access to most locations in terminals and on the tarmac. While the vast majority of these employees are honest and hardworking, some may have ulterior motives. The bombing of a Russian airliner on October 31, 2015, was the result of an insider who planted explosives on the plane. Metrojet Flight 9268 broke apart in midair and crashed over the Sinai Peninsula, killing all 224 people on board. A similar activity against U.S. aircraft is not out of the realm of possibility. The good news is that employees are being vetted by the TSA. Other airport personnel are also screened by those who are working at airports and in the aviation industry.

Self-Check

1. Terrorists desire to bomb aircraft to disrupt the transportation system. True or false?
2. The TSA was created before 9/11. True or false?
3. What program puts armed security personnel on aircraft?
   a. Air Marshals
   b. TSA Secure Flight
   c. AIR SWAT
   d. VIPR Teams
4. Explain why Secure Flight replaced CAPPS II.

9.3 Rail Transportation Security

Railroads are another important part of the U.S. transportation network and have a major impact on the economy. As an example, in 2023, the major class 1 railroads employed over 177,000 people and created around $90 billion in revenue. In addition, these railroad companies move over 1.6 billion tons of freight. At one point, this represented 41% of all the freight ton-miles carried in the United States (Goel, Hartong, and Wijesekera 2008).

FIGURE 9-3 Trains may be the target of terrorism due to the confinement of a large number of people in a small area. *Source: © U.S. Department of Homeland Security.*

With this information in mind, it is apparent why terrorists would potentially target rail transportation networks (see Figure 9-3). Terrorists could cause significant economic harm by damaging railway infrastructure, which would prevent the flow of people and goods. Trains can be derailed by removing the spikes that hold the rail and crossties together or by throwing a switch that causes a fast-moving train to topple on a curved track. Tunnels, railways, or bridges can be destroyed with explosives, and train engines can also be incapacitated through other nefarious means. All of these tactics and many more could halt or have a negative impact on rail transportation (Goel, Hartong, and Wijesekera 2008).

Railways are also vulnerable because they transport millions of people for both short and long distances. Passenger railroads are generally classified into two different types of rail transportation based on the distance the train travels. A commuter railroad travels short distances within metropolitan areas. In contrast, an intercity passenger train travels long distances. In the United States, most passenger rail transportation is carried out by commuter railroads, with over one million riders per day (American Public Transportation Association). However, Amtrak, the only domestic intercity railroad, reported a 51% increase in their passenger travel from 2001 to 2013. In 2013, 31.6 million passengers rode Amtrak (Imam 2014). Today, about 87,000 passengers continue to ride Amtrack trains on a daily basis.

Terrorists are therefore likely to target passenger cars or train stations. Such an attack would be extremely deadly because these cars and locations have large numbers of people in confined spaces. For instance, approximately 1.4 million people ride one of 32 commuter trains each day (Goel, Hartong, and Wijesekera 2008). In fact, the MTA Long Island Railroad in New York alone has an average ridership of 337,800 passengers per day (American Public Transportation Association 2014). For this reason, security at train stations is generally less stringent than at airports. The goal is to move people to their destinations as fast as possible, and security can get in the way of that objective. These lower standards and the lack of sufficient monitoring make it easier for terrorists to launch an attack and then blend in with the ensuing rush of passengers to escape.

In The Real World

On July 7, 2005, Islamic terrorists launched attacks on the London transportation system. The perpetrators detonated bombs on subways and a double-decker bus. The coordinated blasts killed 52 people and injured over 700 more. The events illustrated that transportation systems are extremely difficult to protect and defend because of the freedom of movement in and around them. Subways are and busses are accessible to terrorists and prone to attacks because of the presence of a large number of potential victims.

Attacks on trains and subways have occurred around the world, including in countries like Japan, Spain, and India. The attacks by Aum Shinrikyo on March 20, 1995, illustrate the vulnerabilities of subway systems. Members of this terrorist organization entered trains and punctured bags containing liquid Sarin. The attack occurred during rush hour. It immediately killed over 10 people and causes other health problems for more than 1,000 others. On March 11, 2004, simultaneous bombings of the Cercanias commuter train system occurred in Madrid. This attack killed 193 people and injured over 2,000. On July 11, 2007, India experienced an attack on a train in Diwana. This event resulted in at least 70 deaths and scores of other injuries (see Figure 9-4).

Russia also experienced several train and subway attacks in 2003 and 2004. In March 2010, two suicide bombers detonated explosives on two trains in Moscow in near-simultaneous attacks. These terrorists killed 40 people and injured another 100. On April 3, 2017, a blast in a subway station in St. Petersburg killed at least 10 people and injured 50 more. These types of events definitely underscore the vulnerability of the rail transportation system.

FIGURE 9-4 Subways are vulnerable to attacks because they are accessible to terrorists and contain a large number of potential victims. *Source: © Sean Randall/E+ /Getty Images.*

Unfortunately, railroads could be targeted by terrorists due to the cargo they carry. As an example, the **Strategic Rail Corridor Network (STRACNET)** is a network of railroads that transport Department of Defense munitions and other materials, including hazardous items; 21.6 million ton-miles of especially dangerous items, known as toxic-by-inhalation materials, are transported along this network per year. These "gasses or liquids that are known or presumed on the basis of tests to be so toxic to humans as to pose a health hazard in the event of release during transportation" (Goel, Hartong, and Wijesekera 2008).

Tank cars containing toxic-by-inhalation materials can be compromised as well. This could occur through gun fire, explosives, or derailments. The consequences of a tank car rupturing near a major city would be catastrophic to those in the vicinity. For instance, in 2005, a train with a tank car containing chlorine gas was derailed and damaged near Graniteville, South Carolina. This accident released 60 tons of gas, which killed nine people, injured 554 people, and required 5,400 people to be evacuated from nearby homes. Indirect damages of this accident were more than $40 million (Goel, Hartong, and Wijesekera 2008). If terrorists were able to damage a tank car containing hazardous materials in a major city, the impact could be even more disastrous.

In The Real World

The Ohio Train Derailment

While not caused by a terrorist attack, the derailment of a train on February 3, 2023, illustrates the danger of hazardous material releases resulting from these types of transportation systems. A Norfolk Southern freight train with a faulty brake system came off the tracks in East Palestine, Ohio. Twenty of the thirty-eight cars were carrying chemicals, ranging from vinyl chloride and butoxyethanol to isobutylene and benzene. Some of these cars caught on fire due to the accident, and emergency crews determined that a controlled burn of the remaining cars would be the safest way to deal with this unfolding disaster. The accident and controversial decision required the evacuation of people living within a 1-mile radius of the derailment site. The derailment and subsequent fires caused serious health and environmental concerns and resulted in several hearings and lawsuits. The incident presents a dire picture of what might happen if a train is derailed or the cars are intentionally breached by terrorists.

Because of these obvious threats to security, the U.S. government and the railroad industry have implemented their own security measures to keep rail transportation as safe as possible. The railroad industry created the **Surface Transportation and Public Transportation Information Sharing and Analysis Center (ST-PT ISAC)** at the request of the U.S. Department of Transportation. This entity interfaces with government leaders, intelligence agencies, law enforcement personnel, and computer emergency response teams to spread top-secret threat information among railroad operators. The railroad industry also coordinates and regulates when and where certain things are transported, such as hazardous materials and munitions.

In addition, the government employs 100 rail security officers from the TSA and 450 rail safety inspectors from the Federal Railroad Administration. These individuals monitor rail activities for potential threats and inspect the rail infrastructure, engines, and tank cars against expected safety standards. Companies such as the Volpe National Transportation Center, Dow Chemical, Union Pacific, and the Union Tank Car Company also collaborate with the federal government to ensure the safety and integrity of tank cars (Goel, Hartong, and Wijesekera 2008).

Many other government organizations and policies are related to railroad security. The Department of Homeland Security obviously has some oversight of rail security. It is concerned about the possibility of attacks against the railroad industry. The Department of Homeland Security therefore issues directives that require protective security measures to be taken by passenger rail operators. However, it is the National Transportation Safety Board that has the specific task of investigating the safety of tank cars to determine which are suitable for transporting hazardous materials (Goel, Hartong, and Wijesekera 2008). The Association of American Railroads likewise tracks and inspects rail shipments to increase safety.

There are a number of laws and regulations to enhance rail security and safety at rail stations/yards. The Material Transportation Act provides the legal basis for assessing the costs and benefits associated with transporting hazardous materials. New laws are being created each year to address the threat of terrorism against the rail system. This is important because the threat of terrorism is dynamic and not static.

As can be seen, railroads are a significant part of the U.S. transportation network, moving either large loads of materials (including potentially toxic substances) or a high volume of passengers to destinations all over the nation on a daily basis. This makes rail transportation a prime target for terrorist attacks. As such, the national government has created agencies and instituted policies to better monitor and track railway activity. The nation and its partners in the railroad industry will continue to find new ways to better secure this form of transportation for the future.

In The Real World

VIPR Teams

The Visible Intermodal Prevention and Response Team (VIPR) is a program in the TSA. It falls under the TSA's Office of Law Enforcement/Federal Air Marshal Service. VIPR teams search and detain risky travelers at railroad and bus stations, on ferries, in tunnels, and at weigh stations and rest areas. They also have special detection devices to identify the presence of chemical, biological, and radioactive materials present on railcars, ferries, and subways or at ports, truck weigh stations, and rest areas. The goal of these teams is to increase security in various modes of transportation. They also develop plans to improve interagency responses to terrorist attacks and other emergencies.

Self-Check

1. Railroads do not ship hazardous materials. True or false?
2. A train could be derailed by removing spikes. True or false?
3. Evacuation of a city due to a chemical release occurred in what state?
   a. South Carolina
   b. Florida
   c. Oregon
   d. Kansas
4. Why is security more problematic at a train station in comparison with the airport?

9.4 Protection of Sea Ports and Maritime Transportation

Like air and rail transportation, seaports and shipping have an incredible economic impact on the United States. According to the American Association of Port Authorities (2018), seaports account for 26% of the U.S. economy, with about $5.4 trillion in total economic activity. In addition, seaports support nearly 31 million jobs and provide more than $378 billion in federal, state, and local taxes in 2018. Without seaports and the vital services they provide, our nation can be severely impacted. This was witnessed to some extent during the COVID-19 pandemic when the ports of Los Angeles and Long Beach were backed up starting in June 2020. At one point, more than 109 ships were waiting to offload their containers onto the shores of Southern California. This had serious implications for the global supply chain (e.g., shipping, logistics, and commerce) as well as inflation for the entire country.

As can be seen, seaports are essential for the United States. Seaports also help to distribute raw materials and finished products around the world. They also act as important hubs for the mass transportation of people. However, the very same characteristics that make them so valuable (e.g., their relative degree of access) also contribute to their inherent vulnerabilities. It is only logical to presume that terrorists will strike U.S. seaports because they are so crucial to our well-being. Sadly, terrorist organizations have made it clear that they wish to harm the United States by spreading fear, causing economic damage, and inflicting mass casualties to accomplish their ideological agendas. Blowing up a seaport and launching other types of attacks against these facilities could, in the minds of terrorists, help them reach their goals.

Terrorists may also attack maritime transportation. For example, terrorists could possibly attack cruise ships to create mass fatalities, subsequently impacting the tourism industry. Such an attack would also generate mass amounts of publicity and would propagate terrorists' political messages and instill fear in others. Not only are ships potential targets for terrorist activity, but smaller sea-going vessels can also be used to facilitate attacks that are intended for other targets. On October 12, 2000, two suicide bombers pulled alongside the U.S.S. Cole and detonated explosives in their small boat. The incident killed 17 sailors and injured 40 others. It is possible that terrorists could also use shipping containers to smuggle WMDs, explosives, or other arms into the United States to inflict damage on inland targets. Ships can be turned into WMDs that impact docks, and terrorists may illegally cross the border through ships to mount attacks elsewhere in our country.

Additionally, terrorists are beginning to finance their violent operations through piracy. These modern-day pirates attempt to gain control over a vessel through the use of fast boats, ladders, and small arms. Most of these piracy cases occur in international waters where there is minimal law enforcement, such as the Gulf of Aden, the Red Sea, or near Indonesia, the Malacca Straits, Malaysia, Singapore Straits, and the South China Sea. The payoff from taking over a ship and its cargo ranges from $8 to $200 million per vessel. If pirates are not given a ransom, they can easily change the identity of the vessel by obtaining a new registry. They may also paint their ship in a remote dock and change flags. Regardless of the intent, modern-day piracy is extremely dangerous, as was depicted in the 2013 film *Captain Phillips*. Because of this threat, the U.S. military is working hard to counter such activities around the world. The Navy has even commissioned special ships to find pirates and prevent their activity in oceans and waterways around the world. Navy seals were successful in killing the pirates who took over MV Maersk Alabama and rescuing Captain Richard Phillips. The Navy is currently involved in addressing a variety of attacks on commercial vessels near the Yemeni coast (on the southern end of the Red Sea).

Because of the various threats posed to seaports and sea transportation, measures must be taken to ensure the security of people and goods on the water and in ports. The USCG, the Container Security Initiative, and the Customs-Trade Partnership Against Terrorism are some of the main organizations and programs that protection sea transportation. The United Nations International Maritime Organization (UNIMO) also works to prevent piracy and improve safety and security of maritime shipping.

As mentioned earlier, the USCG is an organization of the Department of Homeland Security. It enforces maritime law, manages vessel traffic, responds to maritime pollution, and participates in search-and-rescue operations (due to severe weather and sinking or incapacitated boats). The USCG is a vital asset in homeland security efforts to protect the safety of maritime activities (see Figure 9-5). It works to prevent, plan for, and react to potential and actual terrorist attacks in its domain of responsibility.

FIGURE 9-5 The Coast Guard has specially trained teams to deter attacks that could create significant financial losses and disrupt the economy. *Source: © United States Coast Guard.*

The **Container Security Initiative (CSI)** was one of the first measures taken by the government to protect maritime trade and ports against terrorism. Launched in 2004, this program has four components (Hoffman 2006, p. 145). First, shipments are to be placed in tamper-proof containers to limit the possibility that something can be smuggled into the United States. Second, computers are used to identify potential high-risk containers—those thought to be containing questionable goods, such as WMDs. Third, high-risk containers are prescreened to limit the chances that they arrive at the intended location. Finally, high-tech detection equipment is used to identify WMDs before or during shipment or after arrival. To accomplish this goal, U.S. Customs officials are being deployed abroad to push the borders out and speed up the process. Cargo manifests also have to be given to U.S. officials 24 hours before containers are loaded onto ships. Hoffman notes that 18 of the top 20 seaports initially agreed to the CSI. These ports represented about 70% of all containers shipped to the United States (Hoffman 2006, p. 145). Today, more than 61 ports are operating under the CSI in North America, Europe, Asia, Africa, the Middle East, and Latin and Central America (CBP 2023). These ports prescreen 80% of the cargo that is imported into the United States. Other ports need to incorporate CSI principles and practices as time goes by.

The **Customs-Trade Partnership Against Terrorism (C-TPAT)** is an agreement between the public and private sectors to protect international commerce from terrorist attacks. As of 2006, it included at least 1,600 importers, manufacturers, carriers (air, rail, and sea), brokers, port authorities, and others involved in international trade. The organizations participating in C-TPAT voluntarily adhere to high standards pertaining to physical security, access controls, personnel security, training and education, manifest procedures, and other requirements relating to trade. Once certified by the government, these businesses are promised fewer and quicker inspections as long as they do not let their support of the agreement lapse. C-TPAT is a novel way to support efforts to prevent terrorist attacks through international trade.

Finally, the **United Nations International Maritime Organization (UNIMO)** is another organization that is involved in the protection of sea transportation. It has created rules on piracy as well as standard security procedures to prevent the shipment of dangerous goods such as WMDs. The UNIMO also establishes rules regarding access

In The Real World

Terrorism at Sea

On February 27, 2004, an eight-pound trinitrotoluene (TNT) bomb exploded after being placed in a television set, which was brought onto the SuperFerry 14 vessel. The bomb was detonated about an hour after the ship departed a port in Manilla, Philippines. The explosion caused serious damage to the ship and also ignited a fire on board. In response to the bombing, the captain issued an abandon ship order at 1:30 a.m. Those who were not killed in the blast jumped overboard and waited for the arrival of rescue boats. While most of the 744 passengers and 155 crew were spared, the attack killed 116 people. Not only was this the Philippines' deadliest attack, but it was the world's worst terrorist event at sea. After the investigation, it was determined that a man named Redondo Cain Dellos placed the explosive on his bunk when he entered the ship. He then quickly got off the vessel before it disembarked. It is assumed that the bombing occurred because the company that owned the ship refused to give protection money to Abu Sayyaf.

to sensitive locations and broader security measures at seaports. It is a vital partner in preventing maritime terrorism.

As can be seen, the desire to protect seaports and ships from attacks and to prevent terrorists from utilizing sea transportation to further their violent agendas remains a vital concern for homeland security. The hope is that government organizations and programs will continually innovate and improve the safety of people and goods on or near the water.

Self-Check

1. Ports will never be attacked because they are heavily guarded. True or false?
2. The U.S. Coast Guard is a military organization. True or false?
3. Piracy is most likely to occur where?
   a. In international waters
   b. In the United States
   c. In Russian waters
   d. Away from the Middle East
4. Explain the benefits of the CSI.

9.5 Protection of Petrochemical Facilities

As noted through this book, terrorists have made it clear that they wish to cause harm to the United States and its citizens. They can do that by entering our border and damaging the transportation systems described above. But they can also attack other vulnerable locations in the United States, which will result in mass casualties. Facilities producing and containing petrochemicals are one of these high-value targets. These plants contain hazardous materials that are known to be extremely dangerous to human health.

And these chemical facilities are ever present in the United States. In fact, the Environmental Protection Agency identified 1,500 facilities that contain dangerous levels of these chemicals, and many of these facilities are located in highly populated areas (Kaplan 2006). The states of California, Florida, Illinois, Michigan, New Jersey, New York, North Carolina, Ohio, Pennsylvania, and Texas are particularly susceptible due to the number of plants in their jurisdictions.

Terrorists may attempt to enter these facilities to steal hazardous materials to create WMDs for future terrorist attacks. This is very troublesome because of the vulnerable nature of these locations. Studies reveal that some chemical facility sites lack gates, guards, and other protections to prevent or minimize theft and illegal acquisition. As an example, in 2014, ISIS easily took over the chemistry lab at the University of Mosul to test and build deadlier bombs. Therefore, chemical facilities can be overtaken to allow terrorists to manufacture explosive devices. Alternatively, chemical facilities may also be the target of the attack themselves. The threat against chemical facilities and refineries is not hypothetical. On June 26, 2015, Yassin Salhi drove a van with gas canisters into the chemical plant in France, which was owned by Air Products & Chemicals. Although the van exploded, it fortunately did not impact the rest of the targeted facility.

In The Real World

The Beirut Disaster

On August 4, 2020, a massive explosion occurred at the Port of Beirut in Lebanon. After a Moldovan ship was unable to deliver its cargo to Mozambique due to a variety of legal, financial, and operational challenges, it sat motionless in the Port of Beirut. A judge ordered that the ammonium nitrate it was carrying should be stored in Warehouse 12 at the port until the shipping dispute and logistical challenges could be resolved. This chemical was therefore retained at the location for 6 years. Unfortunately, a fire broke out in the area. Platoon 5 was dispatched to the warehouse to extinguish the blaze. The ammonium nitrate exploded in a giant mushroom cloud blast measuring 3.3 on the Richter scale, or the equivalent of 0.5 to 1 kiloton of TNT. The blast killed over 200 people and injured 7,000 others. It left over 300,000 people homeless and caused more than \$15 billion in property damage. Although the Beirut explosion was caused by a series of human mistakes and not a terrorist attack, it revealed the extreme dangers associated with the presence of large quantities of hazardous materials. This event, along with many others throughout history (e.g., Texas City, Bhopal, West Fertilizer Plant), illustrates why extreme caution and security are needed at petrochemical facilities in our nation.

To deal with the threats and vulnerabilities relating to petrochemical plants, Congress initiated the **Chemical Facility Anti-Terrorism Standards** (CFATS) in 2007 (see Figure 9-6). CFATS is a program that identifies risk at chemical facilities (as well as at power plants, refineries, and universities) and regulates standards to ensure sufficient security measures are in place. The law was reauthorized in 2014 under the Protecting and Securing Chemical Facilities from Terrorist Attacks Act. It is a broad regulation that relates to perimeter restrictions, personnel background checks, key or electronic access to sensitive locations, activities related to operations (e.g., shipping and storage), and various threats including theft, sabotage, and cyber-attacks. CFATS also includes periodic inspections to ensure compliance with these safety and security regulations.

FIGURE 9-6   Numerous efforts are being undertaken to prevent terrorist attacks against chemical facilities. *Source: © U.S. Department of Homeland Security.*

While this law and program has received some criticism in the past (Sadiq 2014), it is one of many important measures to protect society from terrorist attacks against chemical facilities and refineries. If you are to work in homeland security, you should be aware of this important regulation. It is one of many measures to protect people from possible terrorist attacks.

In The Real World

Chemical Plots and Facility Security

In 2014, Tom Coburn (a Republican senator from Oklahoma) released a report from the Senate Committee on Homeland Security and Governmental Affairs. In the report, it was revealed that right-wing terrorists planned attacks against facilities in Texas and California in 1997 and 2000, respectively. Law enforcement officials became aware of the plans and were able to arrest the perpetrators involved. Unfortunately, intelligence reveals that Islamic terrorists have also indicated their desire to attack these types of facilities. Due to this threat, the government is doing what it can to fortify chemical facilities. But the efforts may be insufficient. After spending $595 million, only 39 of the 4,011 facilities were initially inspected. Since this time, over 15,000 facilities have been visited to verify compliance. The partnership between the government and industry is therefore an essential way to reduce the vulnerability of chemical facilities going forward. It is one of many efforts to protect the nation from future terrorist attacks.

Self-Check

1. The states of California, Texas, and Ohio may be particularly prone to terrorist attacks due to the chemical plants. True or false?
2. Facilities that contain chemicals may provide terrorists material to launch attacks, or they may be the target of the attack. True or false?
3. CFATS is a federal program that:
   a. Protects seaports
   b. Establishes safety and security standards at chemical facilities
   c. Prevents attacks on airplanes
   d. Corrects unsafe practices on U.S. railways
4. What are some measures to protect chemical facilities?

Summary

As the notion of homeland security implies, numerous efforts must be taken to enhance the safety of our nation. If you are working in this field, one of your priorities will be to limit the entry of terrorists into the United States whether through legal or illegal means. Improved immigration standards, border walls, and other measures may prevent this from happening. You will also need to understand various forms of transportation security and the methods to halt attacks in the air, on rails, and at sea.

These modes of transportation are clearly vulnerable and require numerous proactive measures to ensure their protection. Another important activity is to find ways to prevent or inhibit terrorist attacks that could impact chemical facilities. The dangers of failing to protect these locations could be consequential. For these reasons, border security, transportation security, and chemical facility security are vital to your efforts to minimize the probability and impact of terrorist attacks.

Assess Your Understanding

Understand: What Have You Learned?

Go to **www.wiley.com/go/mcentire/homelandsecurity3e** to assess your knowledge of measures to secure the nation against attacks.

Summary Questions

1. There are many reasons why protecting the border is a monumental task. True or false?
2. Customs and Border Protection (CBP) and Immigration and Customs Enforcement (ICE) both function at the border and fulfill the same role in protecting our country. True or false?
3. The TSA is an agency under the Department of Homeland Security. True or false?
4. VIPR teams have the intent of protecting trains and train stations. True or false?
5. The Customs-Trade Partnership Against Terrorism (C-TPAT) is a program to limit terrorist attacks against aircraft. True or false?
6. The Coast Guard is a military organization that is located in the Department of Homeland Security. True or false?
7. Terrorists target transportation locations because they are heavily populated. True or false?
8. CAPPS II replaced the Secure Flight program. True or false?
9. Which program protects ports and maritime trade?
   a. United Nations Port and Maritime Organization
   b. Container Safety System
   c. Customs and Container Partnership
   d. Customs-Trade Partnership Against Terrorism
10. What is the name of the computer database that is used to screen passengers?
    a. U.S. ASSIST
    b. U.S. PREVENT
    c. U.S. CARGO
    d. U.S.-VISIT
11. The rail network used to transport DOD munitions is known as:
    a. Rail Secure
    b. STRACNET
    c. HazRail
    d. Rail Prevent

12. Which program is meant to protect cargo at sea?
    a. The Cargo Protection Program
    b. The Container Safety Program
    c. The CSI
    d. The Drum and Car Security Program
13. What is the government's effort to protect plants that use hazardous materials in their manufacturing process?
    a. Hazardous Materials Protection Process
    b. Chemical Facility Anti-Terrorism Standards
    c. Chemical Facility Prevention Program
    d. Manufacturing Counter-Terrorism Association

Applying This Chapter

1. You have been hired to advise the president about the threat of terrorists infiltrating the United States through the border. What are the various ways terrorists could enter the United States illegally?
2. What is the U.S. Air Marshal program? How does it relate to terrorism and transportation security?
3. You are a member of the VIPR team? What are you looking for, and how can you help secure the U.S. rail system?
4. Your boss has asked you to explain the Container Security Initiative (CSI) to his peers from a regional trade office. What would you say? Is this initiative important? Why?
5. As an inspector for the Chemical Facility Anti-Terrorism Standards, you need to convince the manager of ACME Manufacturing that he or she needs to implement and continually enforce additional safety measures in his/her company. What would you say?

Be a Homeland Security Professional

Enhancing Border Control

You have been assigned to represent the Department of Homeland Security at a press conference on the border. Why should you be aware of the political issues surrounding border control?

Protecting Transportation Modes

You have been given the assignment to brief the Secretary of the Department of Homeland Security on threats to transportation systems. What information would you relay about terrorist preferences for these targets and the potential impact of these types of attacks?

Security Maritime Trade and Seaports

Your boss at International Maritime Trade, Inc., has hired you to protect their facilities and operations. What threats exist against sea transportation? Who could help you to enhance the security of your company?

Hired as a Security Specialist

You have just been hired as a security specialist for a major industrial firm. You have been tasked with the responsibility of ensuring the safety and security of employees. What measures can you take to reduce the possibility of an attack against your facility?

Key Terms

Border   The territorial boundary of any nation along with its various points of entry

Chemical Facility Anti-Terrorism Standards (CFATS)   A program that identifies risk at chemical facilities and regulates standards to ensure sufficient security measures are in place

Computer Assisted Passenger Prescreening System II (CAPPS II)   A former computer program utilized by the TSA to screen passengers against lists of known terrorists and others with criminal records

Container Security Initiative (CSI)   One of the first measures taken by the government to protect maritime trade and ports against terrorism

Customs and Border Protection (CBP)   The largest organization within the Department of Homeland Security that enforces immigration law at and between ports of entry

Customs-Trade Partnership Against Terrorism (C-TPAT)   An agreement between the public and private sectors to protect international commerce from terrorist attacks

Federal Air Marshal Service   Another important organization of a few thousand marshals who act under the TSA

Immigration and Customs Enforcement (ICE)   An organization within the Department of Homeland Security that enforces immigration and customs laws in the internal portions of the United States (i.e., away from the border)

Minutemen Project   Activities to promote border security carried out by a group of volunteers that founded the Minuteman Civil Defense Corps in Arizona

Secure Flight   The proposed program to replace the Computer Assisted Passenger Prescreening System II

Strategic Rail Corridor Network (STRACNET)   A network of railroads that transport Department of Defense munitions and other materials, including hazardous items

Surface Transportation and Public Transportation Information Sharing and Analysis Center (ST-PT ISAC)   Interfaces with government leaders, intelligence agencies, law enforcement personnel, and computer emergency response teams to spread top-secret threat information among railroad operators

Transportation Security Act   A law designed to protect transportation systems in the United States

Transportation Security Administration (TSA)   A federal agency under the Department of Homeland Security created to protect our transportation systems from terrorist attacks

United Nations International Maritime Organization   A UN organization that establishes security rules regarding access and security at ports

U.S. Coast Guard (USCG)   A military branch within the Department of Homeland Security that is in charge of maritime law, environmental protection of waterways, search-and-rescue operations at sea, and interdiction of illegal aliens and contraband

United States Visitor and Immigrant Status Indicator Technology (U.S.-VISIT)   A computer database used to screen passengers who wish to travel to the United States

References

Ainsley, J. (2023). Number of people on terrorist watchlist stopped at southern U.S. border has risen. NBS News. September 14. https://www.nbcnews.com/politics/national-security/number-people-terror-watchlist-stopped-mexico-us-border-risen-rcna105095. (Accessed October 24, 2023).

American Association of Port Authorities (2018). Exports, Jobs & Economic Growth. No date. https://www.aapa-ports.org/advocating/content.aspx?ItemNumber=21150#:~:text=Seaport%20cargo%20activity%20in%202018,personal%20income%20and%20local%20consumption. (Accessed October 24, 2023).

American Public Transportation Association. (2014). Public Transportation Ridership Report Fourth Quarter & End-of-Year 2014. https://www.apta.com/resources/statistics/Documents/Ridership/2014-q4-ridershipAPTA.pdf.

Benztzel, C. (2006). Port and maritime security. In: *The McGraw-Hill Homeland Security Handbook* (ed. D.G. Kamien), 631–648. New York: McGraw-Hill.

Blankey, B. (2023). Illegal aliens since 2021 total more than individual populations of 38 states; illegal crossings top 8 million since Biden took office. Highland County Press. June 25. https://highlandcountypress.com/illegal-aliens-2021-total-more-individual-populations-38-states-illegal-crossings-top-8-million. (Accessed October 24, 2023).

Bove, V. and T. Bomelt. (2016). Does immigration induce terrorism? *Journal of Politics* 78 (2): 572–588.

CBP. (2023). CSI: Container Security Initiative. https://www.cbp.gov/border-security/ports-entry/cargo-security/csi/csi-brief. (Accessed October 24, 2024).

Chertoff, M. (2009). *Homeland Security*. Philadelphia, PA: University of Pennsylvania Press.

Goel, R., Hartong, M., and Wijesekera, D. (2008). Security and the US rail infrastructure. *International Journal of Critical Infrastructure Protection* 1: 15–28.

Hoffman, F. (2006). Border security: closing the ingenuity gap. In: *Homeland Security and Terrorism: Readings and Interpretations* (ed. R. Howard, J. Forest, and J. Moore), 142–166. New York: McGraw-Hill.

Imam, J. (2014). The surprising comeback of train travel. CNN.com. October 18. http://www.cnn.com/2014/10/15/travel/irpt-train. (Accessed July 7, 2017).

Jansen, B. (2017). U.K. joins U.S. in electronics ban on flights. *USA Today* (March 22), pp. 1A, 5A.

Kaplan, E. (2006). *Targets for terrorists: chemical facilities*. Council on Foreign Relations. www.cfr.org/united-states/targets-terrorists-chemical-facilties/p12207.

Knickerbocker, B. (2006). Illegal immigrants in the US: how many are there? *Christian Science Monitor* (May 16). https://www.csmonitor.com/2006/0516/p01s02-ussc.html.

Murtaugh, T. (2021). Border crisis is not accident. It's Biden making good on campaign promises. July 30. https://www.heritage.org/immigration/commentary/border-crisis-no-accident-its-biden-making-good-campaign-promises, https://www.heritage.org/immigration/commentary/border-crisis-no-accident-its-biden-making-good-campaign-promises. (Accessed October 24, 2023).

PBS. (2023). What's behind the influx of migrants crossing the U.S. southern border? September 21. https://www.pbs.org/newshour/politics/whats-behind-the-influx-of-migrants-crossing-the-u-s-southern-border. (Accessed October 24, 2023).

Riley, K.J. (2006). Border control. In: *The McGraw-Hill Homeland Security Handbook* (ed. D.G. Kamien), 587–612. New York: McGraw-Hill.

Sadiq, A.-A. (2014). Chemical sector security: risk, vulnerabilities and chemical industry representatives' perspectives on CFATS. *Risk, Hazards and Crisis in Public Policy* 4 (3): 164–178.

Shaw, A. and B. Melugin. (2023). CBP memo sounds the alarm on Hamas, Hezbollah fighters potentially using southern border to enter the US. Fox News. October 23. https://www.foxnews.com/politics/cbp-memo-sounds-alarm-hamas-hezbollah-fighters-potentially-using-southern-border-enter-us. (Accessed October 24, 2023).

CHAPTER 10

Protecting Against Potential Attacks

Threat Assessment, Mitigation, and Other Measures

DO YOU ALREADY KNOW?

- How to assess the threat of terrorism

- Definitions of structural and nonstructural mitigation

- Ways to increase physical security

For additional questions to assess your current knowledge of protecting against potential attacks, go to **www.wiley.com/go/mcentire/homelandsecurity3e**

WHAT YOU WILL LEARN

10.1 Components of threat assessments

10.2 The importance of protecting critical infrastructure

10.3 The need to build structures with terrorism in mind

WHAT YOU WILL BE ABLE TO DO

- Evaluate the threat of terrorism against key assets and soft targets.

- Predict the benefits of structural and nonstructural mitigation

- Implement activities that will enhance physical security

Introduction

To minimize the impact of terrorist attacks, it is imperative that you take steps to protect the nation. If you are to work in homeland security, one of your primary responsibilities is to assess the threat of terrorism. This entails an evaluation of all types of potential targets. But it also includes understanding the intents of terrorist organizations as they relate to vulnerable locations. Threat assessments also require that you work closely with other homeland security partners to determine what might be attacked. This includes government entities at the federal, state, and local levels as well as businesses in the private sector. Once your analysis is complete, you must then take measures to secure critical infrastructure and key assets. In addition, it is imperative that you undertake structural and nonstructural mitigation measures ranging from architectural design and improved construction practices to zoning and set-back requirements among other security measures. The overarching goal of this process as noted in this chapter is to make targets less attractive to terrorists and to diminish consequences should deterrence fail.

10.1 Threat Assessment

Because of the nature of terrorism, it will be impossible to prevent every single terrorist attack. For this reason, it is essential that you consider the benefits of strategies to enhance protection. In homeland security, **protection** is a proactive activity designed to deny the possibility of attacks and defend oneself if they occur anyway. If terrorists are able to enter the United States or plan attacks that we are not aware of, we must find ways to minimize the appeal of anticipated targets and reduce the severity of attacks that do occur.

To accomplish these goals of homeland security, a threat assessment of the possible attacks that could be launched against our nation, states, and communities will be required. According to William Jenkins, Jr., a director of Homeland Security and Justice Issues, such an evaluation is needed to "determine which elements of risk should be addressed in what ways within available resources" (in McElreath 2021, 229). A **threat assessment** is a careful study of the targets that might be appealing to terrorists. It is, in some ways, similar to a hazard and vulnerability analysis in emergency management. A hazard and vulnerability analysis is the estimation of risk—or an evaluation of what may occur along with possible consequences (see McEntire 2022). However, instead of determining the potential impact of natural or technological disasters, a threat assessment focuses specifically on potential terrorist attacks and their probable impacts. It is an educated guess about what terrorists might do along with a determination of the outcome of such actions. In this sense, the analysis is equivalent to FEMA's threat and hazard identification and risk assessment (THIRA) process (see **https://www.fema.gov/sites/default/files/2020-06/fema_national-thira-overview-methodology_2019_0.pdf**).

10.1.1 Critical Infrastructure, Key Assets, and Soft Targets

When conducting a threat assessment (or what is often called a hazard and vulnerability analysis in emergency management), it is important that you consider both critical infrastructure and key assets. **Critical infrastructure** is defined as interdependent networks composed of industrial, utility, transportation, and other distribution systems (Edwards 2014; Pupura 2007, p. 359). According to Nancy Wong, the deputy director of the U.S. Critical Infrastructure Assurance Office, examples of this infrastructure include:

- Information and communication systems (computer networks, line-based phone systems, and cell towers)
- Electrical systems (power plants, step-up and step-down stations, transformers)
- Transportation systems (airports, highways, bridges, seaports)
- Petrochemical systems (oil wells, refineries, storage facilities)
- Water systems (dams, sewage treatment plants, distribution lines) (as cited by McEntire, Robinson, and Weber 2001, p. 5).

In addition to this list of vulnerable locations, we may want to add farms, food processing plants, and food distribution networks. They, too, could be targeted by terrorists and may be deemed as critical infrastructure.

All of these infrastructure systems are regarded to be "critical" because "their incapacity or destruction would have debilitating impact on the defense or economic security of the United States" (Clinton 1996). For instance, the loss of the Internet or phone systems would limit our ability to conduct business and communicate with others. A cyberattack (see Chapter 14) on systems that operate ATMs and credit card swiping devices would severely hamper business activity. Disruption of energy systems would prohibit the heating and cooling of homes in the winter or summer and limit visibility at night. As noted in Chapter 9, the destruction of transportation systems would also have an enormous impact on the movement of goods and services. Likewise, if petrochemical plants were taken out of service, many aspects of our lives, including manufacturing and travel, would be in jeopardy. Terrorist attacks on water systems could result in sickness or even death. An attack on food production and food transportation systems could likewise result in severe hunger and the loss of life. Critical infrastructure must therefore be evaluated as part of a comprehensive threat assessment. This type of analysis is also important since "The United States has not invested enough in the long-term maintenance of its levees, dams and power grids" (Chertoff 2009, p. 89).

In The Real World

The Attractiveness of Certain Targets

On June 2, 2007, Russell Defreitas, a resident of Brooklyn, New York, devised a plot to blow up fuel pipelines leading into JFK International Airport. According to the U.S. Justice Department, Defreitas stated that the attack would provide "more bang for the buck" because of the anticipated fire it would produce as well as resulting economic and psychological consequences. Pipelines such as this one may be attractive targets because it is impossible to protect them completely due to their length and presence. The loss of pipelines could lead to a fuel shortage and rising prices. For this reason, the security of pipelines is an important priority in the National Infrastructure Protection Plan. The Department of Homeland Security, under the direction of Homeland Security Presidential Directive 7, is working with oil and petroleum companies to plan responses to future threats in this area. Many corporations are now taking steps to protect the most vulnerable locations. Improved monitoring and shutdown strategies are possibly making pipelines less appealing targets to terrorists (Kimery 2007).

During your threat assessment, it is also imperative that you examine possible attacks against key assets. **Key assets** include "a variety of unique facilities, sites, and structures that require protection" (White House 2003, p. 71). Examples include banks, financial institutions, major corporations, fire and police stations, hospitals, national monuments, and government property (e.g., the White House, Capitol Hill, the Supreme Court Building, the Governor's Office, and City Hall). These buildings and facilities are vital for the well-being, safety, and operation of government and our way of life. Key assets also have unique symbolic importance for the United States.

Besides considering the critical infrastructure and key assets in your threat assessment, you must recognize that any location of public assembly could be attacked by terrorists (Home Office 2012). For instance, sporting venues, fairgrounds, shopping malls, grocery stores, and restaurants are all likely to be attacked. Schools are also feasible targets as was witnessed in Beslan, Russia, on September 1, 2004. During this attack, Chechen separatists took more than 1,200 adults and children hostage at a local school. A few days later, a gun battle broke out between the separatists and government police forces, and the fighting resulted in the death of 334 people (including 186 children). Locations like schools are likely to be attacked because they are regarded to be **soft targets**, meaning they are open and accessible to the public. Soft targets are also extremely vulnerable to attacks due to the high concentration of people in a single location.

In The Real World

Using Drones to Neutralize Drone Attacks on Key Assets

The October 7, 2023, attack by Hamas began with simple drones that dropped grenades on Israeli machine gun posts and communication towers. This limited the ability of the Israel Defense Force to stop a ground assault or to notify others that an attack was unfolding.

The use of this technology by Hamas brings up an important question for those involved in homeland security in the United States: could terrorists utilize drones to attack certain locations and events that are vulnerable from a security perspective (e.g., a military base, the White House, a presidential inauguration, or the Olympics)? A company named Fortem Technologies in Utah has developed a system that is able to reduce the likelihood of this happening.

Fortem has created TrueView radars that are installed at sensitive and vital facilities or areas. This equipment conducts constant reconnaissance and surveillance to enhance airspace safety and facilitate perimeter security. If an unanticipated drone is detected, the TrueView radar system will automatically dispatch DroneHunter interceptors that will fly to the enemy drone and confiscate the nefarious equipment with a net. The DroneHunter will then carry the enemy drone to a safe location where the attack vehicle can be inspected and investigated, and the enemy drones and weapons can be disabled or destroyed. These counter-unmanned aerial systems (C-UAS) have been purchased and increasingly utilized by the military, government agencies, and commercial customers. They are one novel way to enhance security at places like airports and critical infrastructure or venues hosting sporting events and parades. For further information on this advanced homeland security technology, see **https://fortemtech.com**.

10.1.2 Collaboration with Others to Identify Threats

To accurately assess the threat of attacks against critical infrastructure, key assets, and soft targets, it will be imperative that you work closely with others. Government agencies, department leaders, and members of the business community have special knowledge and expertise to help you determine the likelihood of an attack along with possible consequences (McEntire 2022).

At the local and state levels, there are also many departments and individuals that can help you assess terrorist threats and risks. For instance, the Department of Transportation will have vital information about traffic patterns, key interchanges, and vulnerable bridges. Transportation officials will also possess vital information about subway systems and bus terminals. River authorities can provide you with data about the age and capacity of dams, while utility departments can help you assess the criticality of wastewater systems. Engineering departments will have statistics on building age and occupancy rates and should be consulted as part of comprehensive threat assessments. Planning and development agencies can assist you in acquiring information about zoning requirements, demographic data, and census numbers. Parks and recreation departments can be contacted to gain knowledge about athletic venues, water parks, and large community gatherings. The Chamber of Commerce can likewise assist you in identifying commercial districts and industrial areas that could be targeted. Public health agencies may be able to determine the dangers associated with the use of biological agents.

The private sector could also be a great partner when you conduct your threat assessment. Businesses are heavily involved with critical infrastructure. In fact, it is estimated that approximately 85% of all infrastructure is owned or operated by the private sector (Ewing 2004). Therefore, contacting companies operating within communications, energy production, and other economic sectors can help you understand the threats posed to phone lines, gas systems, and electrical grids. Manufacturers and safety managers at petrochemical plants will also be able to assist you in the threat assessment. They possess knowledge about hazardous materials that could be used in or targeted by a terrorist attack. You can also plan with hospitals to identify the consequences of major public health emergencies associated with terrorist attacks. Insurance companies are also a great resource that will enable you to identify the economic losses associated with potential attack scenarios.

There are many other helpful partners to help you complete your threat assessment. Farmers will have vital information about cattle and agricultural production. Biotechnology research firms may have data about the diseases and medical advances they are studying in your community. Rail companies and trucking firms are good assets to contact if you want to know what is being shipped through your county.

Emergency managers are another wonderful resource for those working in homeland security (McEntire 2022). They are increasingly adept at conducting these types of analyses because they have undertaken many assessments to determine the risks associated with various hazards and vulnerabilities. The key point is to recognize that you can and should tap into the expertise of others by networking and collaborating. Without involving specialists in your threat assessment, you will fail to accurately estimate the probability of an attack in your community and the associated negative potential.

Obviously, the FBI is one of the most important agencies involved in domestic threat assessments (Tromblay 2023). It gathers intelligence about the plans of terrorists, which is then filtered down to state and local law enforcement officials. It also works to prevent attacks on computer systems and the Internet. In addition, FBI special agents can help you identify the threats associated with National Special Security Events (e.g., NCAA Final Four or the Super Bowl) through what is known as a Special Event Assessment Rating (SEAR). A **SEAR** assessment is an evaluation of the public importance of an event along with the potential risk of an attack.

Philip Purpura, the director of the Security Training Institute and Resource Center, has identified several additional partners at the federal level that could help you assess risks (Purpura 2007, p. 361). Such stakeholders, along with their areas of expertise, include:

- **Department of Defense.** Is responsible for the physical security of military installations and the defense industrial base.
- **Department of Energy.** Oversees the safeguarding of power plants, energy infrastructure, and nuclear weapons production facilities.
- **Department of Health and Human Services.** Concentrates on health-care assets.
- **Environmental Protection Agency.** Deals with water and wastewater systems.
- **Department of Agriculture.** Focuses on the need to protect agriculture and food infrastructure.
- **Department of Treasury.** Accountable for banking and financial institutions.
- **Department of the Interior.** Conscientious about the security of national monuments, such as the Statue of Liberty and Mount Rushmore.

Finally, the Department of Homeland Security (DHS) is charged with the protection of numerous sites that could be the target of terrorist attacks. These locations include telecommunication systems, postal and shipping sectors, emergency services, and transportation systems (in conjunction with the Department of Transportation). DHS also works closely with the Nuclear Regulatory Commission to oversee the security of nuclear reactors and nuclear waste sites. It likewise collaborates with the Secret Service to oversee special events and special units within the department to monitor cyber security (see Chapter 14). Another responsibility of DHS is the protection of government facilities. The Federal Protective Service helps DHS to reach this vital goal. For this reason, no threat assessment would be complete without some type of consultation with the DHS. There are also published guidelines to help you assess the risks to and vulnerability of critical infrastructure (DHS 2011, 2013).

10.1.3 Points of Consideration

As you assess threats and associated vulnerabilities, you should take into consideration many important questions, for instance:

- Are there known groups in your jurisdiction that may desire to carry out a terrorist attack?
- What are the possible targets that they may wish to destroy?
- How easy would it be to carry out an attack against critical infrastructure through sabotage?

- What is the likelihood of key assets being attacked with traditional explosives?
- Which soft targets would result in more fatalities due to the concentration of people in the area?
- Which of the aforementioned locations are more likely to be targeted than others?
- Are any measures in place to defend these locations from an attack?
- What are the probable consequences of a terrorist attack at that specific location?
- What types of attacks would have the greatest impact on life, property, the environment, government functions, corporate operations, and general social disruption?

Answering these questions can lead to very formal and complex threat assessments, which could take months and even years to complete. Of course, there is always a difficulty in knowing what you should focus on: probability, consequences, or both likelihood and impact (McEntire 2022) See Figure 10-1. Some attacks may be highly probable but result in limited consequences. As an example, a conventional

In The Real World

Risk-Based Funding Prioritization

In June 2006, the Intelligence and Analysis Office in the DHS released the findings of a study about the risk facing states around the nation. The study examined several factors including population size and density, the importance of critical infrastructure, proximity to the border and ports, the presence of hazardous materials, and iconic value, among other variables. The results provided one view on how to prioritize homeland security funding around the nation. The following is a list of states and territories in rank order:

1. California	18. Virginia	35. Arkansas
2. Texas	19. Kentucky	36. New Mexico
3. New York	20. Minnesota	37. Utah
4. Florida	21. Maryland	38. West Virginia
5. Illinois	22. Missouri	39. Nebraska
6. District of Columbia	23. Tennessee	40. Maine
7. Michigan	24. Kansas	41. Rhode Island
8. Ohio	25. Alabama	42. Wyoming
9. New Jersey	26. Wisconsin	43. North Dakota
10. Pennsylvania	27. Mississippi	44. Hawaii
11. Georgia	28. Colorado	45. Vermont
12. Louisiana	29. Connecticut	46. Idaho
13. Arizona	30. Oregon	47. Delaware
14. Massachusetts	31. South Carolina	48. Montana
15. Washington	32. Oklahoma	49. New Hampshire
16. North Carolina	33. Iowa	50. South Dakota
17. Indiana	34. Nevada	51. Alaska

explosive might fall into this category. Other attacks may be improbable but produce significant impacts. A biological attack is one type of event because it could kill hundreds of thousands of people. It is thus a challenge to know how to weigh each scenario and its resulting severity.

Nevertheless, the process of conducting a threat assessment can be simplified if you imagine yourself in the position of a terrorist. People playing this role are known in war games as the "red team" (while the "blue team" would include those anticipating and responding to the actions of red team). If you were a terrorist, what critical infrastructure, key assets, and soft targets would give you maximum publicity for your cause? Which of these targets would be easiest to infiltrate and operate in? How would you implement the attack to have the greatest potential impact? These thoughts are morbid but are necessary if you are to be successful in assessing threats.

Another method to help you determine threats and vulnerabilities is a risk assessment matrix (see Figure 10-1). David Alexander, a well-known scholar of emergency management, has developed a table to determine the risk of natural and technological disasters. It is based on an examination of probability and severity (Alexander 2002, p. 57). Adapting this table to the threat of terrorism, different types of events (e.g., assassination, bombings, mass shootings, attacks involving weapons of mass destruction) can be categorized in Table 10-1.

Specifically, terrorist attacks placed in low probability and low severity cells are described as having an acceptable level of risk. Attacks listed with a moderate degree of probability and severity are defined as having significant risk. Events described as being frequent and catastrophic are regarded to be of critical risk.

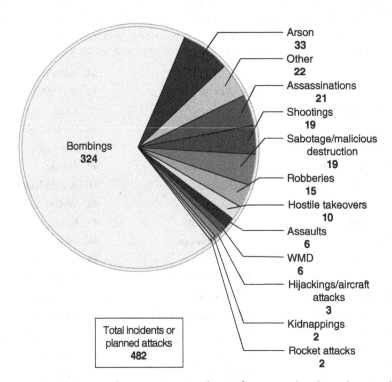

Arson
33
Other
22
Assassinations
21
Shootings
19
Sabotage/malicious
destruction
19
Robberies
15
Hostile takeovers
10
Assaults
6
WMD
6
Hijackings/aircraft
attacks
3
Kidnappings
2
Rocket attacks
2

Bombings
324

Total incidents or
planned attacks
482

FIGURE 10-1 Bombings are the most common threats from terrorists, but other attacks are possible and may have even greater consequences.

TABLE 10-1 Risk assessment matrix

Severity	Probability of occurrence				
	Impossible	Improbable	Occasional	Probable	Frequent
Negligible	Acceptable	Acceptable	Acceptable	Acceptable	Acceptable
Marginal	Acceptable	Acceptable	Acceptable	Acceptable	Significant
Moderate	Acceptable	Acceptable	Acceptable	Significant	Critical
Serious	Acceptable	Acceptable	Significant	Critical	Critical
Catastrophic	Acceptable	Significant	Critical	Critical	Critical

In The Real World

Catastrophic Scenarios

In 2004, U.S. homeland security officials identified 15 possible catastrophic scenarios (see Figure 10-2). It is interesting to note that only two of them are natural disasters, even though they are historically more prevalent than terrorist attacks. However, this list provides several unique possibilities that must be considered by those working in homeland security:

Scenario 1: Nuclear detonation—10-kiloton improvised nuclear weapon

Scenario 2: Biological attack—Aerosol anthrax

Scenario 3: Biological disease outbreak—Pandemic influenza

Scenario 4: Biological attack—Plague

Scenario 5: Chemical attack—Blister agent

Scenario 6: Chemical attack—Toxic industrial chemicals

Scenario 7: Chemical attack—Nerve agent

Scenario 8: Chemical attack—Chlorine tank explosion

Scenario 9: Natural disaster—Major earthquake

Scenario 10: Natural disaster—Major hurricane

Scenario 11: Radiological attack—Radiological dispersal devices

Scenario 12: Explosives attack—Bombing using improvised explosive device

Scenario 13: Biological attack—Food contamination

Scenario 14: Biological attack—Foreign animal disease (foot and mouth disease)

Scenario 15: Cyber attack

As you describe possible terrorist attack scenarios, you should determine your top priorities. The Federal Emergency Management Agency has a course *Building Design for Homeland Security*, that discusses how to accurately assess and minimize risks. This manual can be accessed at **https://www.fema.gov/pdf/plan/prevent/rms/155/e155ig_cvr-agenda.pdf**. It will help you concentrate on attacks that are deemed to be significant or critical. However, the process of determining risks is subjective and cannot be determined with 100% certainty. While risk estimation may include complex mathematical models that incorporate an evaluation of the costs and benefits of specific mitigation measures, the unpredictability and adaptability of terrorists make threat assessments problematic at best. Regardless, doing your best to evaluate risk, even when imperfect, is preferable to doing nothing at all.

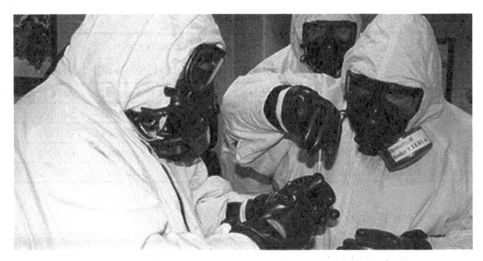

FIGURE 10-2 Chemical attacks are 1 of 15 anticipated attacks against the United States.
Source: © US Department of Homeland Security

In The Real World

Building Security Services & Systems

There is a growing number of companies that provide consulting services regarding physical security—whether it is for terrorism, crime, or many other potential risks. Building Security Services & Systems is one of them. This firm has been working in this area for over three decades. Building Security Services & Systems helps businesses assess threats, review policies, and develop security and contingency plans.

This company has served many different types of clients including real estate firms, hotels, universities, construction enterprises, manufacturers, warehouse operators, and many others. The goal of Building Security Services & Systems is to point out risks and make recommendations to reduce threats and vulnerabilities.

For more information, see **https://www.buildingsecurity.com/consulting/**

Self-Check

1. A threat assessment is a study of what types of attacks could occur and how bad they might be. True or false?

2. Farms and food processing plants are not considered critical infrastructure. True or false?

3. Which of the following is considered a soft target?

   a. A computer network

   b. A petrochemical plant

   c. A school

   d. A sewage treatment plant

4. Why is it important to work with others to assess the threat of terrorism?

10.2   Structural and Nonstructural Mitigation

Besides assessing threats against critical infrastructure, key assets, and soft targets, protecting against terrorist attacks may require two types of mitigation, which are commonly discussed in emergency management (McEntire, Robinson, and Weber 2001). These two measures are known as structural and nonstructural mitigation. **Structural mitigation** involves special construction practices and materials to limit the consequences of terrorist attacks. For instance, installing blast-resistant glass on government buildings is an example of structural mitigation. **Nonstructural mitigation** includes other methods (beyond construction) that may limit the possibility or impact of terrorist attacks. Regulations on building location or the use of security guards are examples of nonstructural mitigation. Both structural and nonstructural approaches are vital to homeland security. They can help to reduce injuries and deaths, limit economic losses, and minimize the disruption associated with terrorism.

In The Real World

> **Four Layers of Defense**
>
> The Federal Emergency Management Agency has described four layers of defense against terrorist attacks. These layers include the following:
>
> **Deter:** Limiting access to a target or countering the weapon or tactic being used. This usually includes perimeter control, fencing, locks, and lighting.
>
> **Deny:** Designing buildings and infrastructure to withstand blasts or minimize the effects of chemical, biological, or radiological weapons.
>
> **Devalue:** Finding ways to minimize the consequence of an attack, thereby discouraging the desire of terrorists to attack.
>
> **Detect:** Gathering intelligence to monitor threats and using security to prevent access to buildings.

10.2.1   Architectural Design and Construction

In the *Reference Manual to Mitigate Potential Terrorist Attacks Against Buildings*, the DHS notes that building design and fabrication plays a large role in the potential impact of terrorist attacks (U.S. Department of Homeland Security 2003). For instance, "U"- or "L"-shaped buildings should be avoided since they channel and exacerbate the force of shock waves. The use of steel reinforcement in concrete buildings as well as the number and redundancy of support columns, may minimize the damage or collapses created by explosive devices. Windows glazed with laminate security film may inhibit or slow down flying shrapnel and other debris. Sprinklers and fire-resistant materials can be installed to prevent arson-induced fires from spreading. Ventilation systems can also be protected in certain ways to discourage or stop hazardous materials releases. For instance, intake vents can be isolated and inaccessible to those without authorization. The great benefit of implementing these types of structural mitigation measures in light of the threat of terrorism is that they

---

CAREER OPPORTUNITY | Site Security Designer/Engineer

The location, layout, and construction of buildings may have an enormous impact on the damage, injuries, and disruption caused by terrorist attacks and other acts of violence. In consequence, there is a need for specially educated architects, engineers, and building designers who are able to identify possible threats and draw up plans for facilities that reduce overall vulnerability.

The typical architect and engineer understands normal set-back requirements and routine building layouts—whether they be in reference to buildings in the public or private sector. However, site security designers go beyond these routine aspects and also look for ways to integrate functionality with additional safety measures such as constructing vehicle barriers/walls/fencing, controlling access to the building in general and sensitive areas in particular, utilizing higher-grade glass panels during construction, and installing cameras, lighting, and motion detectors. In these cases, the structure of the building must align closely with and support the needs of both building occupants and security personnel.

To become an architect, one must complete a bachelor's degree, gain experience through an internship or other type of apprentice program, and pass the Architect Registration Examinations. Additional training for specializations in site security may also be required if one is to design sensitive buildings (e.g., city hall, county courthouse, state regional offices, and federal headquarters).

Architects make about $82,000 per year (median income), and this sector is anticipated to grow by 5% over the coming decade.

For further information, see **https://www.bls.gov/ooh/achritecture-and-engineering/architects.htm**.

---

will often be applicable to other hazards (e.g., earthquakes, hurricanes, tornadoes, and electrical fires). In fact, it is well known in the emergency management community that these hazards are often more likely to occur than terrorist attacks. Terrorism provides yet another reason why architects and construction companies should take structural mitigation seriously.

10.2.2   Zoning and Set-Back Regulations

There are many ways to protect against terrorist attacks (see Table 10-2). At the local level, this may include zoning and set-back requirements as well as unique devices to prevent the movement of people and vehicles (McEntire, Robinson, and Weber 2001). **Zoning** involves regulations that delineate where buildings can be located. Just as you would not want an elementary school located near an industrial plant, you do not want a shopping mall located near a rail yard that transports chemical tankers. In either case, the accidental or intentional release of hazardous materials could injure or kill scores of people or even hundreds or thousands of individuals. Prescribing the best location for businesses, homes, sporting venues, and critical infrastructure could do much to limit the impact of terrorist attacks. Planning with spatial relationships in mind can be extremely beneficial. For instance, if no homes are located in flood plains downstream, terrorists will be less likely to blow up dams and levees. In this case, there is minimal incentive to carry out such an attack because the consequences

TABLE 10-2   Measures to mitigate attacks

1. Locate assets stored on site but outside the building within view of occupied rooms in the facility.

2. Eliminate parking beneath buildings.

3. Minimize exterior signage or other indications of asset locations.

4. Locate trash receptacles as far from the building as possible.

5. Eliminate lines of approach perpendicular to the building.

6. Locate parking to obtain a standoff distance from the building.

7. Illuminate building exteriors or sites where exposed assets are located.

8. Minimize vehicle access points.

9. Eliminate potential hiding places near the building; provide an unobstructed view around the building.

10. Site building within view of other occupied buildings on the site.

11. Maximize distance from the building to the site boundary.

12. Locate the building away from natural or manmade vantage points.

13. Secure access to power/heat plants, gas mains, water supplies, and electrical service.

There are many things that can be done to buildings to minimize the probability of an attack.
Source: © *U.S. Department of Defense.*

will be less severe. The added benefit of such planning is that repetitive flood losses from excessive precipitation will also be minimized as a result. This could also save many lives and countless dollars on the part of taxpayers.

In contrast to zoning, **set-back requirements** describe the proximity of buildings to roads and parking lots. Many, if not most, terrorist attacks involve vehicle-delivered bombs. Creating fewer vehicle access points near a building, avoiding parking lots beneath structures, and moving parking sites away from edifices will dampen the effects of explosive detonations. For instance, if the driveway leading next to the Murrah Federal Building in Oklahoma City did not exist, the impact of Timothy McVeigh's 1995 bomb would have logically been minimized. A bomb half a block away will produce far less damage to a building and its occupants than an explosive device placed within a few yards of the structure. Standoff zones can be a great defense against acts of terrorism.

10.2.3   Other Protective Measures

There are countless other mitigation measures that can be taken to reduce the probability and consequences of terrorist attacks. Purpura (2007) and Kelly (2006) have identified several of them:

- **Bollards,** which are metal or concrete posts installed into the ground or cement, can be used to keep vehicles from entering restricted areas. They make it difficult for cars or trucks to come closer to buildings (see Figure 10-3).

- Trees and vegetation may be planted around buildings to limit the blast from explosives. They serve as a natural barrier during terrorist attacks.

FIGURE 10-3 Bollards like these can prevent vehicle-delivered bombs from reaching their intended targets. *Source: © US Department of Homeland Security*

- Walls, fences, and barbed wire can be installed around critical infrastructure and key assets to maintain a secure perimeter. While these devices can be broken or cut, they are a vital layer of security.

- Proper lighting can create a psychological deterrent for terrorists and enable detection if they attempt to carry out attacks at night. Terrorists might seek a location that has no lighting since this might limit the possibility of apprehension.

- Guards and guard dogs can help patrol facilities and control employee and visitor traffic. They can sense when behavior is suspicious and take appropriate measures.

- Gates and doors should have locks (whether mechanical or electromagnetic). This will prevent unauthorized access to the building or its rooms.

- Metal detectors and X-ray scanners may be used to detect weapons and explosives. Such devices are essential at courthouses and other key assets.

- Background checks can be conducted on prospective employees to determine who should be hired. This will prevent terrorism from internal sources.

- ID badges and access cards allow or restrict movement into sensitive areas. This will help ensure that authorized employees are only permitted to roam freely in certain locations. Note: The United States implemented the REAL ID program in 2005 to improve the security, authentication, and issuance procedures standards for state driver's licenses. All states are to be compliant by the year 2025.

- PIN numbers and biometric identifiers (e.g., fingerprint or retinal scans) are other means to permit or confine the movement of employees. They are useful to gain or restrict access to mission-critical areas.

- Special attention should be given to security in lobbies, loading docks, and mail rooms. These are logical locations where terrorists might attack.

- Cameras, alarms, and intrusion systems (e.g., motion detectors) are valuable tools to monitor physical security. They can help to warn of unusual activity or solve questions of who was responsible for terrorist attacks.

In addition, special efforts need to be undertaken by homeland security officials for events and festivals. Advanced planning—perhaps with the use of the eSAFE system (Hu and Racherla 2008)—could help to protect such vulnerable events. The goal is to consider what could happen and take steps to enhance safety and security. In addition, when key leaders and dignitaries (e.g., CEOs and the president) are to be present at a specific location, physical security becomes even more vital. Areas may need to be "sniffed" or "swept" in advance for bombs, and access to the building or area is carefully controlled. Additional security guards, bodyguards, snipers, and special force teams may need to be deployed to protect such officials. Furthermore, it is imperative to plan for emergency response needs should something unexpected occur. Escape routes and medical care for dignitaries are just of few of the things that need to be considered when national or state leaders are involved in public events and gatherings.

Where possible and when required, it would also be wise to hire someone to be in charge of security measures at any given location. According to the U.S. Government Accountability Office, "having a chief security officer position for physical assets is recognized in the security industry as essential in organization with large numbers of mission-critical facilities" (U.S. Government Accountability Office 2005, pp. 43–44). This person should be familiar with the guidelines provided by the American Society for Industrial Security since it has additional information about physical security. This person might also want to "visit a local airport, jail or courthouse to see the levels of complexity in equipment, operations and management of security systems" (Evans 2001, p. 96). However, the security specialist should not be viewed as the only one in charge of security. Everyone must take more interest in security if terrorist attacks are to be minimized. This includes anyone working in the public and private sectors as well as the public at large. Attacks have been prevented by citizens who saw something out of the ordinary and reported suspicious behavior. The adage "see something, say something" is a good motto for everyone to remember and follow.

In The Real World

Virginia Protects Itself

After 9/11, the state of Virginia felt a need to increase security at water treatment plants, power facilities, and tunnels—all potential targets for terrorist attacks. State officials determined that these locations are extremely vulnerable and that employees could take advantage of this risk-laden situation if they had ill intent. For this reason, it was decided that background checks would be performed on all employees who apply for sensitive jobs. The Cooperative Extension Service will also verify that employees in the agricultural sector are legal residents of the United States. These are only a few of the vital measures that can be taken to mitigate terrorist attacks.

Self-Check

1. ID cards and pin numbers are examples of structural mitigation. True or false?
2. The shape of buildings can reduce or exacerbate the blast waves from explosions. True or false?
3. The proximity of buildings to roads and parking lots pertains to:
    a. Zoning
    b. Set-back requirements
    c. Structural mitigation
    d. Bollards
4. What special measures need to be taken to protect dignitaries?

Summary

Because it may be impossible to prevent terrorism, it will also be necessary to protect locations against potential terrorist attacks. This being the case, you will need to work with others to assess the threats posed to critical infrastructure, key assets, and soft targets. In addition, it is important that you recognize the benefits of mitigation practices as well as the difference between structural and nonstructural mitigation measures. In particular, you should promote the design and construction of buildings that take terrorism into account. You will likewise need to consider zoning and set-back requirements as well as a variety of other measures to augment physical security. While no single measure will be foolproof, a strong and redundant system of self-defense will diminish the possibility and effects of terrorism.

Assess Your Understanding

Understand: What Have You Learned?

Go to **www.wiley.com/go/mcentire/homelandsecurity3e** to assess your knowledge of protecting against potential attacks.

Summary Questions

1. It is possible to prevent all terrorist attacks from occurring. True or false?
2. A threat assessment is an evaluation of anticipated targets of terrorist attacks. True or false?
3. Infrastructure is regarded to be "critical" if its destruction would have a severe impact on the well-being of the United States. True or false?
4. The Department of the Interior can help you assess the threat of terrorism posed to power plants and nuclear facilities. True or false?
5. An important question to consider when assessing threat is: Are there any measures in place to deter terrorism from occurring at that location? True or false?
6. Structural and nonstructural mitigation will never reduce death, economic losses, and disruption caused by terrorist attacks. True or false?
7. Walls and fences may inhibit the ability of terrorists to launch attacks at businesses and manufacturing plants. True or false?
8. It might be wise for any organization deemed to be a likely target for attack to hire a security specialist. True or false?
9. A threat assessment is similar to:
   a. A structural mitigation technique
   b. A hazard and vulnerability analysis
   c. A key asset target identifier
   d. A set-back requirement

10. Which is not considered to be included in critical infrastructure?
    a. Transportation systems
    b. Electrical systems
    c. The White House
    d. Information and communication systems

11. Which of the following is most likely to be requested as a soft target?
    a. Restaurants
    b. Police stations
    c. An oil refinery
    d. The Supreme Court Building

12. What organization could help you assess threats posed against dams?
    a. The FBI
    b. The Department of the Interior
    c. Parks and recreation
    d. River authorities

13. A metal posed placed in the ground to prevent movement of vehicles is best known as:
    a. Structural mitigation
    b. Blast reduction capacitor
    c. Nonstructural mitigation
    d. A bollard

14. Which measure would help to prevent someone from bringing a weapon into a building?
    a. Fences
    b. Background checks
    c. Metal detectors
    d. Cameras

Applying This Chapter

1. After mentioning to your neighbor that you are studying homeland security, he says that there is no point in trying to figure out where terrorists will attack next. He believes they are too unpredictable. What would you say to him about the benefit of threat assessments?

2. As an employee working in homeland security, you have been asked to conduct a threat assessment, looking specifically at the private sector. How could companies help you with your responsibility?

3. As a seasoned emergency manager in your community, you have just received word that the city council has approved your request to relocate the emergency operations center. What structural, zoning, and set-back measures could help protect it against possible terrorist attacks?

4. As the public information officer in your homeland security department, you have been asked to be interviewed about effective security procedures in buildings. What would you say in your remarks?

Be a Homeland Security Professional

Meeting with the FBI

You are an official in the DHS. It is your responsibility to assess potential targets around the nation. The FBI has requested that you meet with them to discuss the consequences of an attack against transportation systems. They fear that a domestic terrorist organization is intent on carrying out such an attack. What would you say about the vulnerability and criticality of transportation systems?

Assignment: Things to Consider When Assessing Threats

Write a two-page paper discussing the concerns you might need to address if you were to complete a threat assessment. Be as thorough as possible.

Hired as a Security Specialist

You have just been hired as a security specialist for a major industrial firm. You have been tasked with the responsibility of ensuring the safety and security of employees. What measures can you take to reduce the possibility of an attack against your facility?

Key Terms

Bollards Metal or concrete posts installed into the earth or cement to keep vehicles from entering restricted areas

Critical infrastructure Interdependent networks composed of industrial, utility, transportation, and other distribution systems

Key assets Facilities, sites, and structures that are believed to require additional protection from terrorist attacks

Nonstructural mitigation Methods beyond construction that may limit the possibility or consequences of terrorist attacks

Protection An attempt to deny attacks and defend oneself from terrorism

SEAR Special Events Assessment Rating is used by the FBI to identify the risk of terrorist attacks against major public gatherings

Set-back requirements Laws that describe the proximity of buildings to roads and parking lots

Soft targets Potential sites of terrorist attacks because they are open and accessible to the public

Structural mitigation Special construction practices and materials to limit the impact of terrorist attacks

Threat assessment A careful study of the targets that might be appealing to terrorists

Zoning Regulations that delineate where buildings can be located

References

Alexander, D. (2002). *Principles of Emergency Planning and Management.* New York: Oxford University Press.

Chertoff, M. (2009). *Homeland Security.* Philadelphia, PA: University of Pennsylvania Press.

Clinton, W.J. (1996). Critical infrastructure protection: executive order 13010. Federal Register 61: 37347–37350.

DHS. (2011). The strategic National risk assessment in support of PPD 8: a comprehensive risk-based approach toward a security and resilient nation. http://dhs.gov/xlibrary/asssets/rma-strategic-national-risk-assessment-ppd8.pdf. (Accessed August 2, 2017).

DHS. (2013). Supplemental tool: executing a critical infrastructure risk management approach. https://www.dhs.gov/sites/default/files/publications/NIPP-2013-Suppment-Executing-A-CI-Risk_Mgmt-Approach-508.pdf. (Accessed August 2, 2017).

Edwards, M. (2014). *Critical Infrastructure Protection. NATO Science for Peace and Security Series.* Amsterdam: IOS Press.

Evans, R.J. (2001). Public works and terrorism. In: *Terrorism: Defensive Strategies for Individuals, Companies and Governments* (ed. L.J.Hogan), 95–100. Washington, DC: Amlex, Inc.

Ewing, L. (2004). The missing link in the partnership. Homeland Security 1.

Home Office. (2012). Crowded Places: The Planning System and Counter-Terrorism. HM Government, Department for Communities and Local Government. January. https://assets.publishing.service.gov.uk/media/5a7df5dded915d74e6223310/Crowded_Places-Planning_System-Jan_2012.pdf. (Accessed October 15, 2023).

Hu, C. and P. Racherla. (2008). eSAFE: the knowledge management system for safe festivals and events. In: *Homeland Security Handbook* (ed. J. Pinkowski), 267–279. Boca Raton, FL: CRC Press.

Kelly, R.J. (2006). Role of corporate security. In: *The McGraw-Hill Homeland Security Handbook* (ed. D.G.Kamien), 745–765. New York: McGraw-Hill.

Kimery, A. (2007). Petrojihad: the next front? *HS Today: Insight and Analysis for Homeland Security Decision Makers* 4 (9): 36–42.

McElreath, D.H., Doss, D.A., Russo, B., Etter, G., Slyke, J.V., Skinner, J., Corey, M., Jensen III, C.J., Wigginton, Jr., M., and Nations, R. (2021). *Introduction to Homeland Security.* Boca Raton, FL: CRC Press.

McEntire, D.A. (2022). *Disaster Response and Recovery: Strategies and Tactics for Resilience.* Hoboken, NJ: Wiley.

McEntire, D.A., Robinson, R.J., and Weber, R.T. (2001). Managing the threat of terrorism. *IQ Rep* 33 (12). Washington, DC: International City/County Management Association.

Purpura, P.P. (2007). *Terrorism and Homeland Security: An Introduction with Applications.* Burlington, MA: Butterworth-Heinemann.

Tromblay, D.E. (2023). Federal bureau of investigation. In: *The Handbook of Homeland Security* (ed. Romaniuk, S.N., Catino, M.S., and C.A. Martin), 71–78. Boca Raton, FL: CRC Press.

U.S. Department of Homeland Security. (2003). *Reference Manual to Mitigate Potential Terrorist Attacks against Buildings.* FEMA: Washington, DC.

U.S. Government Accountability Office. (2005). Homeland security: actions needed to better protect national icons and federal office buildings from terrorism. www.gao.gov/cgi-bin/getrpt?GAO-050681. (Accessed May 8, 2017).

White House. (2003). The national strategy for the physical protection of critical infrastructure and key assets. www.whitehouse.gov. (Accessed October 23, 2017).

CHAPTER 11

Preparing for the Unthinkable

Efforts to Improve Readiness

DO YOU ALREADY KNOW?

- The importance and nature of preparedness
- The role of an advisory council
- The purpose and key elements of emergency operations plans
- The benefit of training and community education

For additional questions to assess your current knowledge of preparedness, go to **www.wiley.com/go/mcentire/homelandsecurity3e**

WHAT YOU WILL LEARN

11.1 How to define preparedness

11.2 Why an advisory council is vital

11.3 The need for budgets and grants

11.4 How to train and educate the community

WHAT YOU WILL BE ABLE TO DO

- Evaluate the different roles of federal, state, tribal and local governments
- Compose a city emergency management ordinance
- Develop an emergency operations plan
- Design and conduct terrorism exercises

Introduction

One of your important roles in homeland security is to anticipate and prepare for possible acts of terrorism. This objective is multifaceted and includes a number of actions that must be undertaken by many professionals and organizations. For this reason, you must understand the concept and elements of preparedness. It is also vital that you are aware of federal and state activities that increase readiness capabilities and steps that must be taken to facilitate tribal and local preparation. In particular, you must establish a preparedness council, designate and equip an emergency operations center, and plan for the impacts of possible terrorist attacks. Training first responders and educating citizens will also improve your ability to react if terrorism occurs. By implementing the measures identified in this chapter, you will enhance your community's capacity to react effectively to the negative effects of terrorist attacks.

11.1 The Importance and Nature of Preparedness

No matter what steps are taken to prevent terrorist attacks or mitigate their adverse impacts, it is always possible that terrorism will occur anyway. A common saying is that homeland security is similar to the game of soccer. Goalies and homeland security professionals must be effective 100% of the time to reach their objectives. In contrast, forwards and terrorists have to be successful only once to accomplish theirs. This reality poses an enormous challenge for those working in this essential field and area of practice. There is simply too much vulnerability to address in comparison to the resources and capacity we have at hand during any given period of time. Richardson (2007, p. 176) puts it this way: "If victory means making the United States invulnerable to terrorist attack, we are never, ever going to be victorious."

For instance, if our intelligence agents and analysts are able to intercept and interpret one communication stream, terrorists will switch to another. If counterterrorism forces close down a training camp in a specific location, it is likely that a new one will emerge elsewhere. If the United States is able to control legal entry into our territory through passports and visas, terrorists will work to find ways to cross the border illegally. If we spend time and energy protecting dams, national monuments, and government buildings, terrorists will simply attack schools, churches, and shopping malls. Prudence therefore dictates that the United States pursues a multifaceted approach to prepare for potential terrorist attacks (Donahue 2014). This brings up the concept of preparedness.

Preparedness is not simple or straightforward. Even though it has been the subject of attention in emergency management for years, there is no agreement on a single definition (Kirschenbaum 2002). Godschalk views preparedness as "actions taken in advance of an emergency to develop operational capabilities to facilitate an effective response" (1991, p. 136). Gillespie and Streeter (1987) suggest that preparedness includes numerous actions taken to improve the safety and effectiveness of a community's response during a disaster. While these views seem to downplay the relation of preparedness to recovery, the Department of Homeland Security (DHS) believes preparedness is related to all aspects of reacting to terrorist attacks.

According to the DHS, preparedness is "the range of deliberate, critical tasks and activities necessary to build, sustain, and improve the operational capability to prevent, protect against, respond to and recover from domestic incidents" (Department of Homeland Security 2004, p. 134). As a reminder from prior chapters, prevention and protection are more concerned about avoidance, deterrence, denial, and mitigation. In contrast, **preparedness** is based on the assumption that terrorist attacks will occur, so this phase of emergency management promotes concerted efforts to improve response and recovery capabilities. Preparedness is thus necessary for more effective post-terrorism operations. It is often driven by policy and other programmatic initiatives at the federal and state levels.

11.1.1 Federal and State Initiatives

Responsibility for national preparedness falls to the federal government in general and the DHS and Federal Emergency Management Agency (FEMA) in particular. However, in an attempt to improve the readiness of the nation for terrorist attacks, the president often issues policies to encourage improved preparedness programs. As an example, on May 4, 2007, President Bush issued Homeland Security Presidential Directive (HSPD) 20. This order establishes continuity of operation requirements for all federal executive departments and agencies. **Continuity of operation** seeks to maintain government functionality after terrorist attacks through the identification of leader succession, alternate work sites, and resumption of operational activities. The idea is that the government must continue to serve the American people if its leaders, departments, and missions are attacked by terrorists. While continuity of operations plans existed among some federal entities before HSPD 20, this directive clarified the purpose of this initiative and mandated further compliance.

Congress also passes many laws to enhance our nation's preparedness for terrorist attacks (see Figure 11-1). Such pieces of legislation started in the 1990s

FIGURE 11-1 Congressional officials pass laws to help the nation prepare for terrorist attacks. *Source: © US Department of Homeland Security.*

and have become more frequent over time. Examples of such laws are numerous and include:

- The National Defense Authorization Law, which was passed after the World Trade Center bombing in 1993. It specified that FEMA and other federal agencies should devote more attention to planning for future terrorist attacks.
- The Antiterrorism and Effective Death Penalty Act of 1996, which required additional training for first responders and provided funds to accomplish this goal.
- The Defense Against Weapons of Mass Destruction Act (also known as the Nunn, Lugar, and Domenici Act), which was also approved in 1996 and mandated further preparedness measures in 120 cities around the United States that were regarded to be vulnerable to terrorist attacks.
- The Public Health Security and Bioterrorism Preparedness and Response Act of 2002, which provided local and state governments $4.6 billion to improve public health readiness around the nation.
- The Comprehensive Homeland Security Act of 2003, which provided specific details about how the DHS must operate and recommends improvement in the area of interoperable communications for first responders.

There are additional laws that deal more specifically with preparedness for all types of disaster scenarios. One important example is the Robert T. Stafford Disaster Relief and Emergency Assistance Act, which was passed on November 23, 1988 (see **https:// www.fema.gov/media-library-data/1519395888776-af5f95a1a9237302af7e 3fd5b0d07d71/StaffordAct.pdf**). This important law describes the federal responsibility to prepare the United States for disasters and outlines various programs to

---

CAREER OPPORTUNITY | Emergency Manager

To deal effectively with terrorist attacks and disasters, governments at the local, tribal, state, and federal levels employ emergency managers.

Emergency managers are generally regarded to be public servants who anticipate all types of hazards and threats and build capabilities relating to mitigation, preparedness, response, and recovery. These individuals spend a great deal of time planning before disasters occur, but they also help coordinate activities when crises take place. Emergency managers are required to have a broad range of knowledge including comprehension of various disciplines including geography, meteorology, sociology, political science, public administration, and even anthropology. These professionals must possess diverse skills including leadership, organizational management, project management, budgeting, and communications. Emergency managers must also demonstrate critical thinking and decision-making skills and be able to work under stressful conditions with many other partners from the public, private, and nonprofit sectors.

Because of anticipated retirements, the growth of emergency management positions will include 900 openings each year. The median pay for emergency managers is about $80,000 with some salaries rising to more than $150,000.

For more information, please see Adapted from **https://www.bls.gov/ooh/ management/emergency-management-directors.htm**.

help state and local governments react when they occur. The Stafford Act recommends that emergency operations plans be developed and reviewed annually. It also discusses federal financial and technological assistance that may be given to state and local governments. Although the Stafford Act has been amended periodically, it continues to serve as the cornerstone of emergency management in the United States.

After numerous failures were witnessed in response to Hurricane Katrina, senators and congressional representatives proposed ways to enhance the ability of the United States to better deal with all types of disasters and not just terrorist attacks. Known as the **Post-Katrina Emergency Management Reform Act**, this law (signed in October 2006) specified ways to avert the slow and disjointed federal response to the catastrophe in New Orleans, Louisiana. For instance, after this chaotic disaster, the close organizational ties that once existed between the director of FEMA and the president were reinstated. Some of the national preparedness programs were returned to FEMA after being integrated elsewhere into the DHS and the FBI. FEMA's budget was also increased since much of its funding was initially diverted to DHS start-up costs. The Reform Act also pointed out ways to enhance the nation's ability to cope with chemical and radiological incidents. These changes were significant and have had a positive impact on preparedness for terrorism and all types of disasters. This law was especially important since terrorists may launch attacks during or after natural and other types of disasters.

In The Real World

Important Legislation and Resources

There are a number of important documents of which you should be aware if you work in homeland security. Some of these may include Homeland Security Policy Directives and Presidential Policy Directives (e.g., HSPD 5 and PPD 8). You should also be aware of the National Strategy Documents, the National Mitigation Framework, the National Response Framework (NRF), and the National Disaster Recovery Framework. There are other laws that are also very important:

Federal Employees' Compensation Act (FECA). Ensures worker's compensation for the family members of federal employees and volunteers of federal agencies who are killed or injured during the course of their duties.

Federal Tort Claims Act (FTCA). Provides immunity to federal government employees from liability in most instances.

National Emergencies Act (NEA). Allows the president to declare a national emergency.

Pandemic and All Hazards Preparedness Act (PAHPA). Addresses organization of public health preparedness and response activities including medical surge capacity and countermeasures for biological threats.

Public Health Service Act (PHSA). Authorizes the Secretary of Health and Human Services to declare a public health emergency.

Public Readiness and Emergency Preparedness Act (PREPA). Protects health workers from liability when responding to public health emergencies.

Volunteer Protection Act (VPA). Ensures immunity from ordinary negligence to volunteers of nonprofit organizations.

National laws do not always solve every problem facing those involved in emergency management. In fact, some people suggest that it is the federal laws themselves that are actually the problem. Regardless of the validity or extent of this argument, organizational and operational changes are required for success and sometimes even the best of intentions fall short at times. One of the strategies developed after 9/11 was the **National Incident Management System (NIMS)**. NIMS is defined as a comprehensive national approach to incident management (see **https://www.fema.gov/nims-doctrine-supporting-guides-tools**). It was initiated after the police and fire units in New York City could not or would not operate jointly when the World Trade Center was attacked. NIMS consequently specified the types of procedures and structures that would improve interoperable communications and collaboration among responding organizations. It also gives guidelines for resource management and promotes compatible technologies. NIMS helps to standardize expectations in disasters and has the goal of improving communication ability in disasters. Thus, NIMS helps first-response organizations coordinate after terrorist attacks and major disasters. Nevertheless, NIMS might focus excessively on technological solutions to communication problems, and it may prove too rigid in the dynamic conditions of disasters. Furthermore, it may not always be applicable because of the politics of response operations, and it may have trouble relating to mitigation and recovery activities (Buck et al. 2006). Therefore, NIMS has been questioned at times by emergency management scholars (see Jensen 2008, 2009, 2011).

Similar charges have been made against the DHS's now defunct National Response Plan. In December 2004, DHS created this planning document for the United States. The **National Response Plan (NRP)** described national procedures for responding to all types of hazards with a multidisciplinary perspective (see **https://www.dhs.gov/xlibrary/assets/NRP_Brochure.pdf**). It listed 15 vital post-disaster functions such as transportation, mass care, and search and rescue and divided those responsibilities among primary and support agencies. The goal of the NRP was to define what federal agencies and all other actors are to do when terrorism or disasters occur. While the intentions were laudable, critics argue that the plan focused too heavily on terrorism and that it was far too complicated (Tierney 2006). The fact that the Federal Response Plan (the existing plan at the time) was not simply revised led to controversy. And the inability of FEMA to significantly influence the content of the NRP proved to be inexcusable. Many of the individuals and agencies that reviewed the initial drafts of the NRP found it complex, confusing, and unwieldy. This is because the relationship between the NRP and NIMS was unclear. The complexity of the NRP was also faulted, in part, for the poor response to Hurricane Katrina. Moreover, those involved in emergency management did not understand the plan and did not receive adequate training on it. As a result, the NRP was rescinded a short time later. Its successor is the National Response Framework.

The **National Response Framework (NRF)** is a document that describes the principles, roles, and structures of response and recovery operations. It replaced the NRP and was written for FEMA, partner agencies, and all elected and appointed leaders at the federal, state, tribal and local levels of government. The NRF is based on five key principles:

- **Engaged partnerships**. Leaders at all levels must communicate and support one another in times of crisis.
- **Tiered response**. Incidents are managed at the lowest level of government and supported by others when needed.

FIGURE 11-2   The National Response Framework is a plan that describes the principles, roles, and structures of response and operations. *Source: © US Department of Homeland Security.*

- **Flexible operational capabilities**. Management activities will change to meet the size, scope, and complexity of events.
- **Unified command**. A clear understanding of the roles of others is required as is collaboration.
- **Readiness to act**. Individuals, families, community organizations, and all levels of government must prepare to deal with incidents of national significance.

Unlike the NRP, the NRF did a better job of clarifying roles and responsibilities, describing what must be done to improve response operations, and explaining how concepts and structures are applied to incident management objectives. Regional Advisory Boards provided feedback on the NRF and other preparedness initiatives. Since the NRF included comments from the emergency management and homeland security communities, it is an important and more comprehensive plan at the national level.

In addition to the federal government, states are heavily involved in preparedness activities. State homeland security departments have been established around the nation. Focus groups in these political jurisdictions meet to consider important policy decisions. Laws are passed to better protect infrastructure and key assets. Tax revenues are invested to improve security, law enforcement, and emergency management functions.

States are also working together to improve post-disaster operations. In 1996, Congress approved the multistate initiative titled the **Emergency Management Assistance Compact (EMAC)**. EMAC is an agreement among states to render assistance to one another in time of need. The goal is to establish guidelines for the sharing of material resources and to resolve in advance legal questions about personnel (e.g., who will pay overtime or death benefits). EMAC has been activated in several

In The Real World

Tribal Governments

Besides federal, state, and local governments, there are many other important participants in national preparedness. One that is often overlooked is tribal government. A **tribal government** is one of 574 federally recognized Indian tribes that have relative autonomy in the United States. These jurisdictions have a unique relationship with federal and state governments. On the one hand, tribal governments reside within the territorial boundary of the United States and individual state jurisdictions. On the other hand, tribal governments are independent and often have full responsibility for governance on their designated lands. In the past, states were often seen as the intermediary responsible for emergency management activities on Indian reservations. That is to say, FEMA would work through states to help tribal governments prepare for disasters and receive federal funding. Today, tribal governments work directly with FEMA and do not necessarily need to coordinate with states to obtain federal resources or implement preparedness programs. If you are working within homeland security, it is imperative that you understand tribal governments. Tribal governments could be the target of terrorist attacks, and reservations are frequently impacted by various hazards including extreme winter weather, atmospheric storms, flooding, drought, and fires. Tribal governments should be viewed as equal and valued partners who may or may not wish to be involved in federal, state, and local emergency management activities. Efforts should be made to understand both the needs and contributions of tribal governments in homeland security.

disasters and is administered by the National Emergency Management Association. The **National Emergency Management Association** is a professional association that is composed mainly of officials from state emergency management agencies. It was created in 1974 when the state directors of emergency management desired to discuss common concerns around the nation. This association has been a key organization for the development of the EMAC (an interstate mutual aid agreement) and the **Emergency Management Accreditation Program** (a standard-based assessment and certification initiative for local and state emergency management agencies).

As has been noted, the federal government is a key asset in emergency management activities throughout the United States. Having federal laws, policies, and plans facilitates a more consistent national approach to emergency management (McEntire 2022). The federal government also provides financial resources that can have a significant impact on national preparedness. However, it is imperative that state governments are prepared to respond as well since they are likely to be on their own for at least 72 hours after a disaster. As an example, it took nearly a week before sufficient help arrived after Hurricane Katrina struck Louisiana. The same principle applies to local governments. Cities and counties should take increased responsibility to be ready for any type of event, including terrorist attacks. Local governments will be impacted most by terrorism and could be on their own for hours or days before outside relief assistance arrives. For this reason, it is imperative that you promote local government preparedness at the county and municipal levels.

In The Real World

The Emergency Management Accreditation Program

After a conference presentation on the importance of standards in emergency management in 1997, the National Emergency Management Association began to discuss the
need to establish recommendations for emergency management programs around the
United States. In time, the Emergency Management Accreditation Program (EMAP) was
developed and implemented. EMAP is a voluntary accreditation initiative that attempts
to improve emergency management capabilities around the nation. It is based on the
National Fire Protection Association's Standard on Disaster/Emergency Management
and Business Continuity Programs that encourages norms regarding laws and authorities, program management, mutual aid, training, and many other preparedness activities.
Emergency management officials who desire accreditation for their programs must write a
self-assessment document and invite a team of independent assessors to review their community's efforts to be prepared. Depending on the findings of the review committee, the
EMAP commission may accredit state and local programs. Accreditation implies that the
jurisdiction is in compliance with widely accepted preparedness guidelines. An added benefit is that external reviews will point out weaknesses that can be corrected in the future.

Self-Check

1. Preparedness deals more with improving response and recovery capabilities rather than
   avoidance and deterrence. True or false?
2. There is complete consensus on the benefits of federal policies and legislation, such as the
   National Incident Management System. True or false?
3. The successor of the NRP is the:
   a. Emergency Management Assistance Compact
   b. Comprehensive Homeland Security Act of 2003
   c. National Response Framework
   d. Post-Katrina Emergency Management Reform Act
4. Why is it essential that local governments take responsibility to prepare for acts of terrorism?

11.2 Foundations of Local Preparedness

To foster preparedness for terrorist attacks, it will be imperative that you work closely
with key stakeholders, establish legal guidelines, and obtain financial resources. You
will therefore need to organize a preparedness committee, draft ordinances, and seek
funding through budget or grants. When this has been accomplished, you should
also consider if your community needs an emergency operations center.

11.2.1 Preparedness Councils

Because homeland security is both complex and interdisciplinary, it will be necessary for you to create a preparedness council. A **preparedness council** is a group
of individuals who come together in an advisory role to share recommendations for

local government policy and assist emergency managers with program administration. Such councils may be composed of representatives from law enforcement/fusion centers, public health, public works, hospitals, key businesses, the American Red Cross, faith-based organizations, and volunteer groups, among others. The goal of these people is to ensure that many diverse viewpoints and areas of expertise are incorporated into community preparedness activities. Although incorporating sufficient members is necessary to get the work done, it is recommended that you avoid having excessive numbers because that may prove overwhelming and unmanageable.

Local Emergency Planning Committees (LEPCs) are one of the most common types of preparedness councils. LEPCs were promoted in the 1980s to help communities prepare for hazardous materials releases (Lindell 1994). During this decade, there was a desire on the part of government leaders to avoid some of the mistakes witnessed in disasters associated with industrial accidents. For instance, doctors had major problems responding to the toxic gas release from the Union Carbide pesticide plant in Bhopal, India, in 1984. The failure to quickly identify the released chemical and appropriately treat victims resulted in nearly 3,000 deaths. LEPCs were accordingly seen as a way to identify hazardous materials risks and help jurisdictions anticipate how best to deal with possible fires and exposure to deadly toxins.

Of course, it should be noted that there will likely be multiple committees to assist with preparedness at the federal, state, tribal, and local levels. One council might focus on weapons of mass destruction, while another is geared toward natural disasters. Some committees are in charge of public health emergencies, and others deal with security and preparedness at community sporting events. Others focus on warning processes and distinct councils concentrate on damage assessment or debris removal. Thus, in reality, there will likely be multiple preparedness committees or subcommittees instead of a single preparedness council. It is also probable that the emergency manager will be the leader of many of these preparedness networks. An **emergency manager** is a local government official in charge of disaster mitigation, preparedness, response, and recovery. Emergency managers have been leading preparedness councils for decades and should be recognized by homeland security officials as experts in the disaster profession.

In The Real World

Local Emergency Planning Committees

Well-known disaster scholar Michael Lindell has conducted several important studies on Local Emergency Planning Committees (LEPCs). His research illustrates that LEPCs are composed of representatives from fire departments, environmental protection agencies, hospitals, and corporations from the petrochemical industry (Lindell 1994). Lindell notes that LEPCs are beneficial in that they promote collaboration among organizations involved in preparedness. LEPCs are also advantageous in that they help to acquire additional funding, foster risk assessment, and rely on the expertise of highly committed members. Each jurisdiction should ensure that they have developed an advisory committee that focuses on terrorism and other types of disasters.

11.2.2 Codes and Ordinances

Passing county and municipal codes and ordinances pertaining to the preparedness aspect of homeland security and emergency management will also be necessary. A **code** or **ordinance** is an authoritative order issued by a government. In terms of homeland security or emergency management, a code or ordinance will justify the need for community preparedness for terrorism and other types of disasters. It will specify the creation of an office or department to deal with these threats and permit the appointment of an official to help the community build capabilities for such events. The duties of this employee will be outlined in the ordinance along with powers that may be granted to him or her in times of crisis. In some cases, mutual aid will be addressed by the decrees of county and municipal governments. **Mutual aid** is a collaborative agreement between jurisdictions when external help is warranted. Such promises are made to assist one another when internal resources prove insufficient. Other issues, including penalties for failure to comply with local laws, may be discussed in ordinances.

Sample codes and ordinances can be obtained from other jurisdictions, and states may have recommended templates for those working in homeland security and emergency management. The input of political leaders (e.g., the mayor, city manager, and city council) and members of the preparedness council will be helpful in the creation of codes and ordinances. Drafts must be approved by the relevant public attorney since these are legal documents. The important point to remember is that codes and ordinances dictate what should be done (and not undertaken) when preparing for terrorism and other types of disasters.

11.2.3 Budgets and Grants

Besides ordinances, every organization will need resources if it is to survive and accomplish its mission. This is also the case with homeland security and emergency management offices. Without people and monetary support, it will be impossible to prepare a county or city for terrorist attacks. Resources can be acquired from local budgets as well as state and federal grants.

Each year, you will be asked to submit a proposed budget to the mayor, city council, or county leadership. This budget may include money needed to fund your position and those of your coworkers and staff. Estimated costs for buildings, utilities, office supplies, computers, phone lines, and other expected expenses will need to be outlined in your recommendation. The city manager, county commissioner, and budget office can help you understand the rules that must be followed in your jurisdiction.

Once your budget draft is completed, you will probably need to present your proposal to city or county leaders. Concise communication, with supporting evidence and documentation, will help you to get as much money for your preparedness program as possible. Because you will be competing for funds with many other departments, you will probably not get everything you desire. However, persuasive argument, accountability for existing funds, and visible activity and achievements in your program will enable you to increase your budget over time.

Since local monetary resources are limited, it may be advisable for you to seek grants from federal or state governments (McEntire 2009). **Grants** are funds given to local or county governments to support or enhance homeland security and emergency management programs. These grants may provide monies for personnel costs and material resources. There are numerous types of grants that can be obtained by local jurisdictions. For instance:

- Emergency Management Performance Grants help fund emergency manager's positions and their general office expenses.
- Assistance to Firefighters Grants provide money to purchase equipment to fight fires.
- Public Safety Interoperable Communications Grants are awarded to acquire and utilize improved communication equipment.
- The Homeland Security Grant Program gives financial assistance to help local jurisdictions prepare to deal with terrorism involving weapons of mass destruction.
- The Infrastructure Protection Program shares money to improve security at seaports, rail stations, bus terminals, and other transportation hubs and networks.
- The Citizen Corps Support Program provides money to help train civilian teams to prepare for terrorist attacks and other disasters.
- The Law Enforcement Terrorism Prevention Program gives resources to police departments to gather intelligence and share information about possible terrorists and criminals.
- The Metropolitan Medical Response System Program funds public health organizations to respond effectively to bioterrorism attacks and cope with mass fatality incidents.

There are also grants for hospital and school planning, urban search-and-rescue teams, and chemical stockpile emergency preparedness, among others.

In The Real World

Urban Area Security Initiative

The most recognized government grant program in homeland security is the Urban Area Security Initiative (UASI). The purpose of UASI is to help local governments build capabilities to deal with terrorist attacks and catastrophic disasters. Funds are directed toward large and dense metropolitan areas that are considered to be vulnerable targets of future terrorist attacks. Those applying for the grants must take a regional approach toward planning. In other words, cities will only be eligible for funds if they collaborate with nearby jurisdictions. UASI grants can be used to purchase equipment, train responders, and conduct exercises. Monies are also given to help prevent and protect against the use of weapons of mass destruction. The UASI program is multidisciplinary and supports numerous organizations working in preparedness, warnings, public health, search and rescue, triage, mass care, firefighting, fatality management, and so on. UASI is funded by the federal government and managed by the state.

These and other grants are awarded on a competitive basis. That is to say, there are a fixed number of grants with limited funding, so they will only be given to select jurisdictions that put together the best applications. As a result, it is imperative that you carefully read the grant instructions and ensure that you have met all requirements in the application documents. Clear writing is key, and it is wise to have the budget or grant office help you prepare your application to ensure compliance with key expectations associated with the grant.

Another important consideration is to illustrate your ability to manage the grant. In some cases, your city will need to fund a portion of the program you are proposing and match federal government monies. If you are awarded the grant, you will also need to work closely with your grant team to accomplish all of the goals that were identified in the proposal. Documentation of expenses must be meticulous and reported to the proper authorities on a periodic basis (depending on the type of grant). Failure to follow up on required paperwork and use funds as outlined could result in termination of the grant or even imprisonment of those failing to meet expectations or misusing government funds. It is therefore wise to be aware of the rules and requirements for grants. Guidelines are available at **https://www.fema. gov/grants/preparedness/manual**.

11.2.4 Emergency Operations Centers

Another fundamental measure to prepare your jurisdiction is to establish an **emergency operations center (EOC)** (see Figure 11-3). An EOC is a location from which disaster response and recovery activities can be overseen and managed (Scanlon 1994; Perry 1995). In these EOCs, government and other community leaders will meet to coordinate each of the functions that must be performed after terrorist attacks or other disasters. Participants in the EOC may include members of the preparedness council and others who have specialized knowledge and skills that are vital when there are emergencies and disasters.

No two EOCs are alike. EOCs could be located in their own separate structure or housed within a designated portion of a building that is adjoined to other rooms (e.g., for top officials, media briefings, a break area, and rest rooms). EOCs include tables, computers, whiteboards, and TV monitors. They will also have other technology and office supplies (e.g., weather monitoring stations, communications equipment, scanners/copiers, pencils, and paper).

FIGURE 11-3 EOCs are equipped locations to help you manage response and recovery operations. *Source: © US Department of Homeland Security.*

While not every city has a designated EOC, these vital facilities are becoming the norm for medium- and large-sized jurisdictions in the United Staters. However, it is possible to set up a temporary or mobile EOC when an incident occurs. For instance, a room that is normally used for meetings or training could be converted into an EOC. Some communities have converted recreational vehicles and trailers in to mobile command posts. Although temporary EOCs can be established, designated EOCs have the advantage of being ready at a moment's notice, whereas temporary EOCs take time to be established and activated.

Regardless of whether the EOC is permanent or temporary, you will need to consider several other issues that relate to these facilities. For instance, you will need to develop procedures for activating the EOC. You will want to determine at what point you will open the EOC (e.g., before, during, or after a potential or actual incident) and how you will contact the participants who should report to that location (e.g., via a phone call, text, or e-mail). You will also need to identify what type of organizational structure will be utilized. The EOC could be set up by departments, functions, or any other type of arrangement. You will want to create an EOC that works for the agencies in your jurisdictions. That is to say, the EOC should correspond to how the city operates and how the county or city will respond to a disaster.

Controlling access to EOCs and having a backup facility is a good idea. On 9/11, New York City's EOC was severely disrupted due to the terrorist attacks and later destroyed due to the resulting collapse of the World Trade Center and resulting fires. As a result, the city had to quickly set up a temporary EOC at a pier on the Hudson River. It is also possible that EOCs could be intentionally selected as targets by terrorist organizations to add to the chaos and disruption associated with attacks. For this reason, EOCs should be carefully protected and guarded as a key asset.

Self-Check

1. Local Emergency Planning Committees are a type of preparedness council. True or false?
2. Ordinances justify the need for community preparedness for terrorism but not other types of disasters. True or false?
3. Who in the community is responsible for most daily disaster mitigation, preparedness, response, and recovery activities?
   a. Mayor
   b. Emergency manager
   c. City manager
   d. Preparedness council
4. What is needed to have a well-equipped EOC?

11.3 Planning

One of the central priorities of preparedness is writing an emergency operations plan (Alexander 2017). An **emergency operations plan (EOP)** is a document that describes what may be anticipated in terms of terrorism (or other emergencies and disasters) and how best to react. The plan may note what types of terrorist attacks or hazards may occur in a community. It also provides an educated guess on the types of issues that will arise from these risks and how they will be met. The EOP outlines who will be in charge of specific post-event functions (as will be outlined in Chapters 12 and 13). As a consequence, plans help to foster coordination and speed up response and recovery operations.

In The Real World

Target Capabilities List (TCL)

In light of the threat of terrorism and other catastrophes, President George W. Bush and Congress advocated for the creation of an improved national preparedness system. The vision proposed by the National Preparedness Guidelines is "A nation prepared with coordinated capabilities to prevent, protect against, respond to, and recover from all hazards in a way that balances risk with resources and need." This objective requires a consensus approach among all of the relevant homeland security stakeholders to build capabilities. These capabilities include common capabilities such as planning and communications; prevention capabilities such as intelligence gathering, counterterrorism, and law enforcement; protection capabilities such as food and agricultural safety and epidemiological surveillance; response capabilities such as incident and EOC management, search and rescue, and warning and evacuation; and recovery capabilities such as damage assessment and lifeline (utility) restoration. Everyone involved in homeland security and emergency management is encouraged to work toward the building of capabilities in these and other areas.

Writing an EOP can be a lengthy and technical task. Fortunately, guidance for writing plans can be obtained from homeland security and emergency management personnel in neighboring jurisdictions or from the state. FEMA has also provided details about planning in SLG 101 (see **https://www.fema.gov/pdf/plan/slg101.pdf**). This document describes the importance of plans and value to the community. It also discusses the format and content of plans so you will be able to help your community prepare for terrorism and other disasters.

EOPs should be based on a thorough assessment known as **Threat and Hazard Identification and Risk Assessment (THIRA)**. As mentioned in Chapter 10, this is a comprehensive evaluation that outlines what could happen along with possible consequences (Wickham et al. 2019). The THIRA should serve as the foundation of EOPs and all other actions relating to prevention, mitigation, preparedness, response, and recovery.

EOPs are commonly divided into three sections. The **basic plan** is an overview of the entire document. It describes the general strategy for dealing with response and recovery operations. The **annexes** discuss specific hazards or functions that will need to be addressed if an event takes place. The basic plan and annexes often contain at least six sections:

- **Authority**. The first section of these parts of the plans often mentions the federal, state, and local laws pertaining to homeland security and emergency management. The goals of the document are derived from the Stafford Act, the NRP, state mandates, and local ordinances.

- **Purpose**. This portion of the document covers the objectives of the plan. It may mention, among other things, the need to protect life, reduce property loss, minimize societal disruption, and promote coordination among participating organizations.

- **Situation and assumptions**. The third part of the plan examines the context of terrorism and other hazards in the community. The potential for attacks and disasters is mentioned along with expected impacts. This segment of the plan is vital for the understanding of what response and recovery functions will have to be performed.

- **Concept of operations**. The fourth component of plans typically identifies, in a brief fashion, what organizations are in charge of specific functions. It often notes that departments will respond based on daily activities and areas of expertise.

- **Organization and assignment of responsibilities**. The segment of the plan describes in detail the roles of each responding organization. It is far more explicit than the concept of the operations section.

- **Direction and control**. The final section of the plan is concerned with the management of the entire post-event operation. It gives the highest attention to top officials in the EOC and their decision making and oversight duties.

Finally, the **appendices** contain additional information to support the plan including resource and contact lists, maps, standard operating procedures, and checklists.

When putting together an EOP, several principles should be kept in mind. First, it is vital that you do not plan alone! Be sure to work with others to develop the plan as it will help you generate better ideas on how to respond to and recover from emergencies and disasters. Your preparedness council, city manager, and legal offices can help you develop a useful and logical document. There are also useful guidelines from FEMA (see Comprehensive Preparedness Guide (CPG) 101) (Department of Homeland Security 2010). Second, your plan should be as comprehensive as possible. It should include all types of events (e.g., terrorism and other disasters) and all actors involved in homeland security and emergency management (e.g., those from the public, private, and nonprofit sectors). Third, your plan should be reviewed and updated annually. Failing to do so may result in the plan being outdated, incomplete, or obsolete because many things can change even within a few short months. Fourth, your plan must ensure that the government is, itself, preparing for disaster. Such continuity of operations plans specifies how to keep your county and municipality functioning in times of crisis as well as who may take over leadership in a line of succession. Finally, you must remember that writing the plan is only one aspect of preparedness. Far too many jurisdictions write a plan and subsequently assume they are now ready to react in an effective manner. Known as the **paper plan syndrome**, this attitude implies that having a plan is all you need to do to be adequately prepared (Auf der Heide 1989). Nothing could be further from the truth. Don't be fooled into thinking that your compliance with state and federal planning mandates is sufficient. Dwight D. Eisenhower once stated: "In preparing for battle, I have always found that plans are useless. But planning is indispensable!" Planning is therefore vital. But do not let that overshadow other preparedness activities that will also build community capacity.

In The Real World

> **The Paper Plan Syndrome and Fantasy Documents**
>
> Public officials sometimes comment that they are prepared for any terrorist attack because "we have a plan." Writing an emergency operations plan is necessary since it describes what could happen and how response and recovery functions are to be performed. Planning also helps to clarify roles and therefore increases the effectiveness and efficiency of service delivery. While planning is important, it should not be regarded as a panacea. Planning is a process, not a checklist that is completed once and never revisited. Concentrating on plans ignores the greatest challenge in emergency management—developing capabilities to enhance your ability to react successfully in a crisis situation. Preparedness cannot be pursued without planning, but writing an emergency operations plan does not ensure you are ready for a terrorist attack. It is imperative that those working in homeland security and emergency management do not fall into the paper plan syndrome (Auf der Heide 1989; Dynes 1994; Clarke 1999).

Self-Check

1. Writing an emergency operations plan ensures that a local government is ready for a terrorist attack. True or false?
2. The appendices in an EOP discuss specific hazards or functions that will need to be addressed if an event takes place. True or false?
3. Which of the following is NOT one of the principles for putting together an EOP?
    a. The plan should be as comprehensive as possible.
    b. The plan should be created by one person.
    c. The plan should be reviewed and updated annually.
    d. The plan must be considered as only one aspect of preparedness.
4. What are the usual sections contained within the basic plan and annexes of an EOP?

11.4 Other Measures

Preparedness entails much more than those activities described earlier (McEntire and Myers 2004). Training, exercises, and community education are vital if a jurisdiction is to be ready to deal with terrorism. Training helps police, fire, emergency medical personnel, and others to respond in a safe and effective manner. Exercises identify weaknesses in the plan and illustrate room for improvement. Community education enlists the support of citizens since response and recovery require a joint effort with the populace. Each one of these preparedness activities will be discussed in turn.

11.4.1 Training

According to the well-known disaster sociologist E.L. Quarantelli, training is a vital component of community preparedness (Quarantelli 1984, p. 29). **Training** includes information sharing in a classroom or practical skill development in the field setting to help familiarize people with the plan and response and recovery protocol (see Figure 11-4).

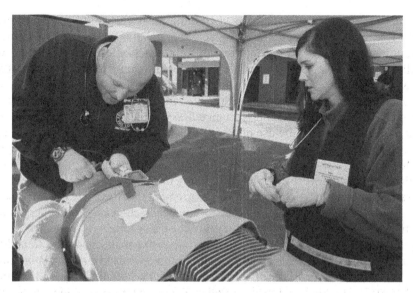

FIGURE 11-4   Training is a great way to educate your public servants or first responders about terrorist attacks. *Source: © US Department of Homeland Security.*

Training helps all of those involved in response and recovery anticipate what could happen and how best to react. It may be focused on **first responders**, such as police, firefighters, and emergency medical technicians, who know how to save lives and communicate one with another. Training can help the directors of public works, the water department and other units know their roles when terrorist attacks occur. Mayors and city managers can also be taught how to seek state and federal assistance through training. Everyone—regardless of position or department—should have some training about terrorism and how best to deal with it.

Training courses are provided by FEMA as well as the FBI, the Department of Defense, the Environmental Protection Agency, the Department of Transportation, and other federal organizations. State agencies including public health, public safety, and emergency management departments likewise provide training on a variety of subjects pertaining to terrorism and on a rotating basis. Emergency management associations and regional councils of government organizations will also be aware of training opportunities for you and your jurisdiction.

In The Real World

The Coming Storm

In October of 2015, the Federal Bureau of Investigation (FBI) released a 40-minute video entitled *The Coming Storm*. Part of the movie is a 13-minute documentary called *Managing the Storm*. This short training film portrays a mass shooting at a college campus and details lessons and important information pertinent to those types of events. It features interviews with both victims and first responders from previous shootings and explains what steps need to be taken immediately following these tragedies. Additionally, this training tool illustrates what FBI resources are available to assist local law enforcement and contains extensive research on the most effective aftermath strategies. The film was distributed nationwide to police departments and other first responders. The hope is that this video will spur future discussions on preparedness and response capabilities in active shooting situations.

One of the most important types of training for terrorism preparedness relates to SWAT teams. **SWAT** stands for Special Weapons and Tactics, and this program and unit has the purpose of developing well-trained police forces. SWAT teams were initially created in the 1960s to handle riots. However, during the 1980s and 1990s, they were used in the war on drugs and in terrorism. In 2005, SWAT teams were deployed more than 50,000 times in the United States. The fact that terrorist attacks are increasing underscores the importance of SWAT teams. For this reason, these teams must be trained in the use of assault rifles, riot control agents, and stun grenades. They must also be well equipped with body armor, ballistic shields, armored vehicles, and night vision devices. Their deployment in many terrorist attacks, mass shootings, and civil disturbances illustrates why they are vital to society.

Another important type of training relates to emergency medical care and responses to weapons of mass destruction. Emergency medical technicians and paramedics are vital in responses to terrorist attacks. They must be trained not only in medical care but ways to protect themselves during violent activities like mass shootings. In addition, other medical personnel—such as doctors and nurses—must be educated on how to deal with victims who have been affected by terrorist attacks (Goralnick, Van Trimpont, and Cali 2017). The Ebola outbreak in Dallas, Texas, illustrated that more needs to be done to decontaminate victims, treat symptoms, and dispose of hazardous material waste (see Chapter 15). There are also major medical concerns about explosives, chemical weapons, and biological agents (as will be noted in Chapter 14).

If you are responsible for training, make sure you recognize that it must be ongoing and continual. Long-time employees will retire. New employees will be hired. People will take on new positions in city government. Each person can forget what they have learned in the past. New policies and procedures will be instituted by federal and state officials. Consequently, a training program will be incomplete and limited in impact if it is not updated and repeated over time.

11.4.2 Exercises

Because disasters and terrorist attacks are sometimes infrequent for any given jurisdiction, experience in dealing with them is generally limited. For this reason, homeland security and emergency management officials should be proactive and take initiative in their efforts to develop and participate in exercises (Roud et al. 2020; Perry 2004; Peterson and Perry 1999). **Exercises** are drills and mock events that test the knowledge and skills of those in charge of or respond to attacks (McEntire 2022). They are the semirealistic methods to evaluate and test the validity of the EOP (Daines 1991). Exercises indicate where planning and training fall short so they can be remedied in the future.

There are three general types of exercises. **Tabletop exercises** are informal discussions about hypothetical scenarios. They occur in an office setting and often involved the key leaders of each department. Tabletop exercises are useful to help decision makers reflect upon how they would respond should a terrorist attack occur. **Functional exercises** are drills that explore one or a few of the annexes in the plan. Such exercises may include a field component as well as equipment and a mild degree of stress. **Full-scale exercises** are major scenarios that test many functions or the entire response system. They often have an EOC and field component and explore the interaction of the broad array of responding agencies. They may include realistic props, "moulaged" victims, and most department leaders from the community.

The City of Denton, Texas, participates in each of these types of exercises. The emergency manager frequently invites his/her preparedness council to review how they would respond to different scenarios. These tabletop exercises are a great way to contemplate what you might do in a difficult situation. At other times, the city has tested its ability to decontaminate victims in a mock exercise of a school bus crashing into a tanker truck carrying hazardous materials. Functional exercises such as this can test any operation that might be needed after a terrorist attack. In a final type of exercise, departments in Denton responded to a mock bombing at a utility company head-quarters. This full-scale exercise involved communications, emergency medical care, law enforcement, and the EOC.

In The Real World

TOPOFF Exercises

In order to prepare for anticipated terrorist attacks, the federal government has instituted an exercise program known as TOPOFF. TOPOFF stands for "top officials" and it therefore has the purpose of assessing how key government leaders and others respond to fictitious terrorist attacks. One TOPOFF exercise, which took place on October 15–19, 2007, focused on intelligence gathering and analysis, victim decontamination, coordination with the military, and the implementation of large-scale recovery activities. Thousands of federal, state, and local participants played roles in this exercise. The lessons learned from such events are vital in that they can facilitate national preparedness for future terrorist attacks. For more information, see **https://www.cnn.com/2013/10/30/us/ operation-topoff-national-level-exercise-fast-facts/index.html**.

In order to develop a solid exercise program, it is first necessary to identify the potential weaknesses in the plan or emergency management system. Once you have determined what areas need improvement, an exercise scenario can then be developed with the assistance of an exercise design team (probably to include the members of your preparedness council). The scenario may include some contextual information about the event as well as specific injects from participants who act out certain roles (e.g., a victim calling 911, a firefighter in the field, or an emergency manager from a neighboring jurisdiction). As the scenario is being developed, you should also schedule the date and location of the exercise and notify each of the participants of these details. Be sure to assign evaluators to determine the success or shortcomings in responding to the mock event.

During the exercise, a controller will make sure the scenario is unfolding at a logical pace. Those responding to the event will then react according to the plan, their training, and/or the specific nature of the scenario. While there are many "right" ways to react to terrorist events, it is important that evaluators look for major errors so they can be corrected in the future. Mistakes may include safety violations, failure to communicate and coordinate with others, and general ineffectiveness in dealing with the problems presented to those involved during the fictitious scenario.

When the exercise is over, evaluators should write up findings so they can be addressed. Many emergency management organizations often fail to follow up on lessons learned, which can make the exercise a useless waste of time, energy, and taxpayer dollars. In addition, sometimes local politicians do not want to hear recommendations for improvement because they worry about how this might impact political support. Regardless, you will also need to submit paperwork to state and federal officials to explain exercise findings and mention your actions to address weaknesses. Failure to do so may result in you losing grants or your good standing among your peers. It would also be wise to fully understand the **Homeland Security Exercise and Evaluation Program (HSEEP)**. This federal program is based on several guiding principles including being: guided by senior leaders, informed by risk, capability-based, objective-driven, progressive in nature, based on common methodology, and supportive of the whole cmmunity (see **https:// www.fema.gov/emergency-managers/national-preparedness/exercises/ hseep**). HSEEP will help you follow up on lessons learned after exercises are undertaken to help you improve planning and build capabilities.

11.4.3 Community Education

Because the government is not the only entity involved in response and recovery activities, it is vital that you educate businesses, churches, volunteer groups, and citizens about terrorism (McEntire 2022). These nonofficial partners can be of great assistance to you if you understand how to harness their potential and channel their efforts in constructive ways (see Figure 11-5). Alternatively, if you ignore or neglect these groups and individuals, your response and recovery operations can be more problematic. For instance, people may visit the scene of attacks out of curiosity and put themselves at risk. They may also donate goods and supplies that really are not needed after most terrorist attacks and disasters. Educating others about terrorism, homeland security, and emergency management is a great way to foster compliance and increase effectiveness.

FIGURE 11-5 The Red Cross can help you educate your community about terrorist attacks and other disasters. *Source: © US Department of Homeland Security*

There are many ways to share information with the public. Community education may include speeches at schools, booths at fairs and other community gatherings,
and the distribution of pamphlets and related information about homeland security
and emergency management. Developing a useful website or social media channels
can also facilitate people's acquisition of information about what they should do to be
prepared for the threat of terrorism. For instance, the DHS has a website (www.ready.
gov) that helps individuals and families develop their own plans, anticipate supplies
that will be needed, and become more self-sufficient until further help arrives. Since
most people rely on computers or cell phones to acquire information today, it is
imperative that your website and social media apps be as user friendly as possible.

In The Real World

Citizen Corps

Citizen Corps is a network of volunteer associations totalling more than 2,000 in the
United States. Its mission is to "harness the power of every individual through education, training, and volunteer service to make communities safer, stronger, and better
prepared to respond to the threats of terrorism, crime, public health issues and disasters
of all kinds." Besides Community Emergency Response Team, Citizen Corps includes
Neighborhood Watch, Volunteers in Police Service, Medical Reserve Corps, and the
Fire Corps. According to the Citizen Corps website (**https://www. ready.gov/citizen-
corps**), Local Citizen Corps Councils will:

- Promote and strengthen the Citizen Corps programs at the community level,
  such as Volunteers in Police Service programs, CERTs, Medical Reserve Corps
  units, and Neighborhood Watch groups.
- Provide opportunities for special skills and interests.
- Develop targeted outreach for the community, including special needs groups.
- Provide opportunities for training in first aid and emergency preparedness.
- Organize special projects and community events.
- Encourage cooperation and collaboration among community leaders.
- Capture smart practices and report accomplishments.
- Create opportunities for all residents to participate.

Unfortunately, many government programs like Citizen Corps could be dismantled
in the future because of funding issues.

Another excellent way to educate your citizens is to develop a **Community Emergency Response Team (CERT)** (Frankie and Simpson 2004). A CERT is a group
of citizens who receive basic training on response operations. They are taught general information about terrorism and disasters and learn how to perform small-scale
firefighting, search-and-rescue, and medical functions. CERT members also gain
knowledge about shutting off gas valves and helping victims with emotional distress.
Establishing and maintaining multiple CERTs in your jurisdiction will therefore
augment preparedness beyond normal levels. Establishing CERTs is one of many
important ways to promote preparedness in your community.

While there is no single or correct way to educate your community, there are
probably some mistakes to avert. For instance, assuming someone else will share
vital information about terrorism with the public is incorrect and problematic. Also,

failing to coordinate your public education campaign with others may result in the sharing of conflicting and contradictory messages. Finally, educating the public on a one-time basis will ensure that citizens forget what is relayed, which will limit ongoing preparedness activity. Avoiding these mistakes will go a long way to ensuring your community is ready to deal with terrorist attacks.

In The Real World

KnoWhat2Do

In 2007, the North Central Texas Council of Governments produced a public education campaign to reach out to millions of people in the 16-county region surrounding Dallas–Fort Worth. Funded by a grant from the DHS, the KnoWhat2Do initiative included the distribution of calendars, playing cards, DVDs, and brochures to help citizens understand what to do in case of a terrorist attack or other types of disaster. A website, accessed at **www.KnoWhat2Do.com**, has also been created to help people anticipate possible hazards/threats and take measures to protect themselves from harm. KnoWhat2Do is a creative example of reaching out to the community for the purpose of preparedness.

Self-Check

1. A training program does not need to be repeated to ensure success. True or false?
2. The first step in developing an exercise program is to identify weaknesses in the plan or emergency management system. True or false?
3. Which of the following is the BEST example of a first responder?
   a. Mayor
   b. City manager
   c. Police officer
   d. Governor
4. How do Community Emergency Response Teams help increase preparedness beyond normal levels?

Summary

Preparing for terrorism is one of your central responsibilities in homeland security. To help your community get ready for possible terrorist attacks, you will need to comprehend the executive orders and legislation issued by the president and Congress. You should also set the foundation for preparedness by creating an advisory council, passing ordinances, acquiring monetary resources, and establishing an EOC. Writing plans and promoting training, exercises, and public education are other ways to improve the degree of preparedness in your jurisdiction. Setting up a CERT is also helpful to increase post-attack capabilities. These types of preparedness activities are vital if you are to effectively respond to and recover from the effects of terrorism.

Assess Your Understanding

Understand: What Have You Learned?

Go to **www.wiley.com/go/mcentire/homelandsecurity3e** to assess your knowledge of preparedness.

Summary Questions

1. National preparedness is solely the responsibility of the DHS and the FEMA. True or false?

2. A tribal government operates exactly like a state government. True or false?

3. The National Emergency Management Association is mainly composed of emergency management offices from city governments. True or false?

4. Grants always cover the full financial need of a proposed program. True or false?

5. An important consideration for preparedness councils is the interdisciplinary nature of homeland security. True or false?

6. Assistance with writing an emergency operation plan can be sought from homeland security and emergency management personnel in neighboring jurisdictions or from the state. True or false?

7. The paper plan syndrome occurs when a plan provides a false assurance of effective preparation. True or false?

8. Training is meant only for first responders. True or false?

9. It is important to educate the public only once to ensure there will be no confusion about the message. True or false?

10. The process by which government functions are maintained after terrorism is best known as:
    a. Preparedness
    b. Continuity of operation
    c. Prevention
    d. Mitigation

11. Which law serves as the cornerstone of emergency management in the United States?
    a. The Robert T. Stafford Disaster Relief and Emergency Assistance Act
    b. The National Defense Authorization Law
    c. The Comprehensive Homeland Security Act of 2003
    d. The Antiterrorism and Effective Death Penalty Act of 1996

12. A collaborative agreement between jurisdictions that ensures external help is coordinated is best known as:
    a. An ordinance
    b. Mutual aid
    c. Annexes
    d. An emergency operations plan

13. Which of the following is NOT one of the three sections of an emergency operations plan?
    a. Annexes
    b. Basic plan
    c. Appendices
    d. The National Response Plan

14. Training and education about terrorism and how to deal with it should be provided to:
    a. City managers and mayors
    b. Emergency managers
    c. First responders
    d. Everyone

15. Functional exercises are BEST described as:
    a. Scenarios that explore one or a few annexes in a plan
    b. Informal discussions about hypothetical situations
    c. Major scenarios that test the entire response system practice
    d. Scenarios that are studied in an office setting

Applying This Chapter

1. While serving as an emergency manager, it is important that you organize a preparedness council. Who could you get to participate in the council?
2. As a financial officer of your emergency management program, you need to submit a proposed budget to the mayor and city council. What can you do to get as much money as possible for your preparedness program and, over time, increase your budget?
3. Due to your role in homeland security, you have been asked to talk with local governments about the importance of community education. What reasons would you give in support of a strong community education program?
4. You have just completed a full-scale exercise as the new fire chief in your jurisdiction. What steps could you take to properly follow up on the exercise?

Be a Homeland Security Professional

Presentation to the City Council

You are the emergency manager for a local government that is not well prepared for a possible terrorist attack. To accomplish your office's mission of increasing the city's preparedness, you need more people and monetary support. The city council is considering an increase in your budget, but first you must persuade them that such a step is an important and effective use of the taxpayers' money. When you are asked by the council about why preparedness is important, how could you respond?

Assignment: Federal and State Governments

Write a one-page paper comparing and contrasting the roles of federal and state governments in preparedness. Be as thorough as possible.

Creating Ordinances

You were just hired as the emergency manager for a local government. You are aware that ordinances are a necessary component in justifying and executing measures aimed at increasing preparedness. Such regulations indicate what should and should not be done in preparing for terrorism. What methods could you utilize to facilitate the development of ordinances that take into account these issues?

Key Terms

Annexes A portion of the emergency operations plan that discusses specific hazards or functions that will need to be addressed if an event takes place

Appendices Additional information at the end of the emergency operations plan, which includes resource and contact lists, maps, standard operating procedures, and checklists

Basic plan An overview of the entire emergency operations plan

Code/Ordinance An authoritative order or law issued by a county or municipal government

Community Emergency Response Team (CERT) A group of citizens who receive basic training response operations

Continuity of operation The maintenance of government functionality after terrorist attacks through the identification of leader succession, alternate work sites, and resumption of operational practices

Emergency Management Accreditation Program A standard-based assessment and certification initiative for local and state emergency management agencies

Emergency Management Assistance Compact (EMAC) An agreement among states to render assistance to one another in times of disaster

Emergency manager A local government official in charge of disaster mitigation, preparedness, response, and recovery

Emergency operations center (EOC) A location from which disaster response and recovery activities can be overseen and managed

Emergency operations plan (EOP) A document that describes what may be anticipated in terms of homeland security and emergency management and how best to react

Exercises Drills and mock events that test the knowledge and skills of those in charge of or react to attacks

First responders The first official government responders in the field including police, firefighters, and emergency medical technicians

Full-scale exercises Major scenarios that test many functions or the entire response system

Functional exercises Practice scenarios that explore one or a few of the annexes in the plan

Grants Funds given to local governments to support or enhance homeland security and emergency management programs

Homeland Security Exercise and Evaluation Program (HSEEP) A federal program that provides guiding principles for exercises

Local Emergency Planning Committees (LEPCs) Preparedness councils promoted in the 1980s to help communities prepare for hazardous materials releases

Mutual aid A collaborative agreement between jurisdictions when external help is warranted

National Emergency Management Association A professional association of state emergency management agencies

National Incident Management System (NIMS) A comprehensive national approach for incident management in the United States

National Response Framework (NRF) The successor to the National Response Plan, a document that describes the principles, roles, and structures of response and recovery operations

National Response Plan (NRP) A document that describes the procedures for responding to all types of hazards with a multidisciplinary perspective

Paper plan syndrome An attitude that assumes that having a plan ensures you are prepared to deal with terrorism and other types of disasters

Post-Katrina Emergency Management Reform Act A law that specifies ways to avert the slow and disjointed federal response that was witnessed after the catastrophe in New Orleans, Louisiana

Preparedness Concerted efforts to improve response and recovery capabilities

Preparedness council A group of individuals who come together in an advisory role to share recommendations for local government policy and assist emergency managers with program administration

SWAT Special Weapons and Tactics relating to well-trained police forces

Tabletop exercises Informal discussions about hypothetical scenarios that occur in an office setting

Threat and Hazard Identification and Risk Assessment (THIRA) A comprehensive study that outlines what could happen along with possible consequences

Training Information sharing in classroom or field settings to help familiarize people with plans and protocols

Tribal government One of the 574 federally recognized tribes that have relative autonomy in the United States

References

Alexander, D. (2017). *How to Write an Emergency Plan*. Edinburgh, Scotland: Dunedin Academic Press.

Auf der Heide, E. (1989). *Disaster Response: Principles for Preparation and Coordination*. St. Louis, MO: C.V. Mosby.

Buck, D.A., Trainor, J.E., and Aguirre, B.E. (2006). A critical evaluation of the incident command system and NIMS. *Journal of Homeland Security and Emergency Management* 3 (1): 1–27.

Clarke, L.B. (1999). *Mission Improbable: Using Fantasy Documents to Tame Disaster*. Chicago, IL: University of Chicago Press.

Daines, G.E. (1991). Planning, training, and exercising. In: *Emergency Management: Principles and Practices for Local Government* (ed. T.E. Drabek and G.G. Hoetmer), 161–200. Washington, DC: International City/County Management Association.

Department of Homeland Security. (2004). National Incident Management System. www.dhs.gov. (Accessed November 10, 2006).

Department of Homeland Security. (2010). Developing and Maintaining Emergency Operations Plans: Comprehensive Preparedness Guide (CPG) 101. https://www.fema.gov/media-library-data/20130726-1828-25045-0014/cpg_101_comprehensive_preparedness_guide_developing_and_maintaining_emergency_operations_plans_2010.pdf. (Accessed August 2017).

Donahue, T.A. (2014). *National Security and Preparedness: Issues, Development and Analyses.* New York: Nova Science Publishers, Inc.

Dynes, R. (1994). Community emergency planning: false assumptions and inappropriate analogies. *International Journal of Mass Emergencies and Disasters* 12 (2): 141–158.

Frankie, M.E. and D.M. Simpson. (2004). Community response to Hurricane Isabel: an examination of community emergency response team (CERT) organization in Virginia. *Quick Response Report 170*. Boulder, CO: Natural Hazards Center, University of Colorado at Boulder.

Gillsepie, D.F. and C.L. Streeter. (1987). Conceptualizing and measuring disaster preparedness. *International Journal of Mass Emergencies and Disasters* 5 (2): 155–176.

Godschalk, D.R. (1991). Disaster mitigation and hazard management. In: *Emergency Management: Principles and Practices for Local Government* (ed. T.E. Drabek and G.G. Hoetmer), 131–160. Washington, DC: International City/County Management Association.

Goralnick, E., Van Trimpont, F., and Cali, P. (2017). Preparing for the next terrorism attack: lessons from Paris, Brussels, and Boston. *JAMA Surgery* 152 (5): 419–420.

Jensen, J. (2008). NIMS in action: a case study of the system's use and utility. Quick Response Report #203. Boulder, CO: Natural Hazards Center, University of Colorado at Boulder. http://www.colorado.edu//hazards/research/qr/qr203.pdf. (Accessed January 31, 2014).

Jensen, J. (2009). NIMS in rural America. *International Journal of Mass Emergencies and Disasters.* 27 (3): 218–249.

Jensen, J. (2011). The current NIMS implementation behavior of United States counties. *Journal of Homeland Security and Emergency Management* 8 (1): Article 20.

Kirschenbaum, A. (2002). Disaster preparedness: a conceptual and empirical reevaluation. *International Journal of Mass Emergencies and Disasters* 20 (1): 5–28.

Lindell, M.K. (1994). Are local emergency planning committees effective in developing community disaster preparedness? *International Journal of Mass Emergencies and Disasters* 5 (2): 137–153.

McEntire, D.A. (2009). Emergency management grant administration for local government. *IQ Service Report* 41 (2). Washington, D.C.: ICMA.

McEntire, D.A. (2022). *Disaster Response and Recovery: Strategies and Tactics for Resilience.* Hoboken, NJ: Wiley.

McEntire, D.A. and A. Myers. (2004). Preparing communities for disasters: issues and processes for government readiness. *Disaster Prevention and Management* 13 (2): 140–152.

Perry, R.W. (1995). The structure and function of emergency operations centres. *Disaster Prevention and Management* 4 (5): 37–41.

Perry, R.W. (2004). Disaster exercise outcomes for professional emergency personnel and citizen volunteers. *Journal of Contingencies and Crisis Management* 7 (2): 186–200.

Peterson, D.M., and R.W. Perry. (1999). The impacts of disaster exercises on participants. *Disaster Prevention and Management* 8 (4): 241–255.

Quarantelli, E.L. (1984). *Organizational Behavior in Disasters and Implications for Disaster Planning.* Washington, DC: Federal Emergency Management Agency.

Richardson, L. (2007). *What Terrorists Want: Understanding the Enemy, Containing the Threat.* New York: Random House.

Roud, E., Gausdal, A.H., Asgary, A., and Carlstrom, E. (2020). Outcome of collaborative emergency exercises: differences between full-scale and tabletop exercises. *Journal of Crisis Management* 29 (2): 170–184.

Scanlon, J. (1994). The role of EOCs in emergency management: a comparison of American and Canadian experience. *International Journal of Mass Emergencies and Disasters* 12 (1): 51–75.

Tierney, K.J. (2006). Recent developments in U.S. homeland security policies and their implications for the management of extreme events. In: *Handbook of Disaster Research* (ed. H. Rodrigues, E.L. Quarantelli, and R.R. Dynes), 405–412. New York: Springer.

Wickham, E.D., Bathke, D., Abdel-Monem, T., Bernadt, T., Bulling, D., Pylik-Zilling, I., Stiles, C., and Wall, N. (2019). Conducting a drought-specific THIRA (threat and hazard identification and risk assessment): a powerful tool for integrating all-hazard mitigation and drought planning efforts to increase drought mitigation quality. *International Journal of Disaster Risk Reduction* 39: Article 101227.

CHAPTER 12

Responding to Attacks

Important Functions and Coordination Mechanisms

DO YOU ALREADY KNOW?

- How to define convergence and emergence
- Recommendations to perform a variety of response functions
- Why coordination is important for post-disaster operations

For additional questions to assess your current knowledge of response operations, go to **www.wiley.com/go/mcentire/homelandsecurity3e**

WHAT YOU WILL LEARN

12.1 Investigation priorities and how human behavior relates to response operations

12.2 The importance of warnings, evacuations, and sheltering

12.3 Ways to coordinate responses to terrorist attacks

WHAT YOU WILL BE ABLE TO DO

- Predict necessary response activities and evaluate their effectiveness
- Perform necessary functions to care for the victims of terrorist attacks
- Set up an incident command system and manage an EOC

Introduction

When terrorists threaten or attacks occur, it is vital that you anticipate needs and work to address the problems that your jurisdiction will encounter. This is no easy task. In fact, there are countless actions that must be implemented right before and immediately after a terrorist attack. For instance, and if needed, you will want to work with intelligence agencies and law enforcement personnel to thwart terrorists through ongoing investigations and apprehension. If attacks cannot be prevented, you should be aware of the likely convergence of resources at the scene along with the altruistic roles of everyday people. You will also need to fulfill many priorities including securing the scene of the attack and conducting various operations ranging from search and rescue, emergency medical care, and decontamination to crime-scene investigation. Where possible or when required, you may also need to warn citizens of impending attacks or evacuate them away from harm. Because so many organizations are involved in these functions, you will need to harness various approaches to coordination. The incident command system and emergency operations centers (EOCs) are two tools that can help you manage the many responsibilities you will have in the aftermath of a terrorist attack.

12.1 Investigation and Apprehension

One of the most important activities of homeland security is to thwart the attack before it begins or while it is unfolding. This proactive measure requires lots of information from intelligence sources, as noted in Chapter 8. However, while preventing an attack internationally requires the Central Intelligence Agency (CIA) and military involvement, it is the Federal Bureau of Investigation (FBI) and local law enforcement officials who play a crucial role in domestic investigation and apprehension (see Figure 12-1). If there is a tip or any evidence that indicates an attack is imminent, federal investigative agencies and the local police will analyze evidence and complete further investigations in an attempt to apprehend potential terrorists or neutralize active threats. This is clearly seen in the response to the Boston Marathon bombing.

After the Boston Marathon bombing, the FBI took the lead law enforcement role but received the support of the CIA and the National Counterterrorism Center, among others. The dissemination of information online by law enforcement personnel kept the public informed about the bombing and aided in eventually solving the case itself. The Boston Police Department and FBI collected video surveillance and photographs of the suspects from cameras at nearby businesses. The FBI also called for and accepted digital uploads of pictures from the scene from the public via social media sites like Facebook and Twitter. After sorting through an exorbitant amount of information, FBI personnel were able to identify the Tsarnaev brothers as the responsible parties on April 18. The FBI released these details to the media and through multiple online sources. This initiated a massive manhunt for the perpetrators that would be accompanied by an execution of the police officer, hostage taking, a shootout with law enforcement personnel, the death of Tamerlan, and the capture of Dzhokhar.

First, the brothers fatally shot a police officer who was sitting in his vehicle at the Massachusetts Institute of Technology. The brothers then carjacked a civilian, taking the driver as a hostage. While detaining the hostage, Tamerlan confessed that

FIGURE 12-1 The FBI is one of many agencies that attempt to find terrorists, prevent attacks, or bring perpetrators to justice. *Source: US Department of Homeland Security.*

the brothers were responsible for the marathon bombing and the death of the police officer. The hostage was forced to withdraw money from an ATM, (presumably for gas money) but luckily the hostage escaped at a convenience store and called 911.

Law enforcement personnel were able to locate the brothers (who were driving separate vehicles) shortly thereafter in Watertown, and a shootout with the police ensued after midnight. In the firefight, two police officers were wounded, and one ended up passing away from his injuries almost a year later. During the exchange of gunfire, Tamerlan was shot multiple times and was subsequently run over and killed by his brother who was driving the other car. Dzhokhar soon abandoned the vehicle and escaped on foot. While law enforcement pursued Dzhokhar, additional information was released to the public through the media and online. Watertown residents were told to shelter in place and avoid leaving their homes. Everyone in the search area was put on lockdown as the manhunt continued for several hours.

Dzhokhar was discovered on the evening of April 19 when a Watertown citizen saw a man covered in blood hiding inside his boat that was parked in his backyard. The citizen notified the police, and Dzhokhar was finally taken into custody. The apprehension was fortunate in that interrogation of Dzhokhar later revealed that the brothers were headed to New York City after the attack in Boston to bomb Times Square. The trial of Dzhokhar revealed that the brothers were self-radicalized but influenced by propaganda from an Al-Qaeda affiliate. Their motive for this attack was to avenge Muslims who were affected by the U.S. wars in Afghanistan and Iraq. Dzhokhar was eventually sentenced to death by lethal injection, but his case continues to be reviewed in court proceedings at the time of this writing.

CAREER OPPORTUNITY | Police Officer

There are countless individuals and organizations that respond to terrorist attacks, disasters, and other criminal activities. Police officers, detectives, sheriffs, and those working with the highway patrol are essential because they help with a plethora of actions ranging from warning, evacuation, and search and rescue to investigation, apprehension, and incarceration.

Those serving in law enforcement have dangerous, stressful, and physically demanding work responsibilities. Therefore, these personnel are always required to attend police training academies and to understand their duties and responsibilities according to the jurisdictions they serve. This training includes courses about the law, driving under emergency conditions, de-escalation techniques, hand-to-hand combat maneuvers, and firearm proficiency.

Most police departments and law enforcement agencies hire individuals who are at least 21 years old and are able to pass rigorous physical tests and a thorough background check. A high school diploma is always required, but a college degree may be optional depending on the hiring agency.

The employment outlook for those working in law enforcement is somewhat complex. On the one hand, there has been a movement among some politicians and various public groups on the political left to "defund the police" in the aftermath of the death of George Floyd. Funding for some agencies has been cut back, and certain local, state, and federal politicians have been very critical of law enforcement personnel due to a history of actual or perceived racism. As a result, the negative sentiment and the accompanying policy decisions have caused some officers to quit. Meanwhile, the verbal attacks on law enforcement and the lack of police officers have emboldened criminal behavior. Consequently, there are now many openings for police officers and the need for their service has never been greater.

For example, the job outlook is as fast or faster than normal employment growth. There were more than 800,000 jobs open last year, and over the next ten years these positions will increase by about 3%. Median pay is about $70,000, and this will likely increase because of the difficulty of hiring and retaining employees due to the somewhat hostile political environment that exists right now. Nevertheless, police officers perform a vital role in homeland security and they will be essential if terrorism and crime are to be reduced.

For more information, see Adapted from U.S. Bureau of Labor Statistics **https://www.bls.gov/ooh/protective-service/police-and-detectives.htm**.

12.1.1 Human Behavior After Terrorist Attacks

Police officers and law enforcement agencies will investigate the incident (see Figure 12-2). However, they are not the only individuals and organizations to respond to a terrorist attack. Additional first responders and citizens will assist with post-incident operations. In the vast majority of cases, official governmental first responders or volunteer departments will arrive to suppress fires or treat the wounded. This occurs after they are notified by citizens who call emergency numbers (e.g., 911). In this sense, the title of first responder is somewhat of a misnomer. It is actually ordinary people who will typically be present on the scene before police, fire, and emergency medical personnel arrive. Citizens are almost always first to react because they are located everywhere—at home, in their cars, at work, in the

FIGURE 12-2 Numerous law enforcement officials will confront terrorists and initiate an investigation into the event. *Source: Federal Bureau of Investigation (FBI).*

shopping mall, at the movie theater, in sports stadiums, running errands in government offices, and so on. Every day people will generally see or hear of the event first. They will notify official responders through dispatch centers and do what they can to help victims until official personnel arrive. The behavior of responders and ordinary people has been described by sociologists as "convergence" and "emergence."

Convergence is the flow of people and resources to the scene of an emergency or disaster (Kendra and Wachtendorf 2003). When victims have been impacted by a major incident, bystanders will stop what they are doing and may go to the location of the attack. In some cases, they will bring needed supplies with them (e.g., a fire extinguisher, a first-aid kit, or any other resource that might be useful). Their goal is to proceed to the focal point of the incident to provide assistance to those in need. In turn, first responders and government officials will be notified and also arrive on the scene. Until this happens, citizens will engage in new types of behaviors, which is called emergence.

Emergence is the appearance of altruistic behavior that is unfamiliar to the participants (Drabek and McEntire 2002). In emergency or disaster situations, people will take on new roles and interact cooperatively with strangers (e.g., providing basic first aid). They will also develop new relationships that often end when emergency needs have been addressed (e.g., groups that form to address victim donation needs).

The terrorist attacks on 9/11 provide vivid examples of both convergence and emergence. For instance, people from around the United States and the world sent supplies for first responders and monetary support for the victims' families. A whole host of organizations brought equipment, supplies, and personnel to Ground Zero. Also, occupants from different floors in the World Trade Center worked together to evacuate the disabled or injured. Carrying someone down the stairs or treating victims is something people don't normally do on a daily basis. But, in a disaster like 9/11, citizens will take on new responsibilities to care for others.

To be sure, the activities of everyday people and official responders will vary significantly, depending on the type of attack that occurs. For instance, people will run, hide, or even fight with a mass shooter. A mass shooting will also necessitate heavy police involvement and tactical EMS. **Tactical emergency medical services** is the name given to a team of paramedics who are armed and trained in weapons use. Tactical emergency services teams have been given greater attention since the mass shootings at Columbine High School, Virginia Tech, the Pulse nightclub, and the community center in San Bernardino. In other cases, an intentionally set fire may require firefighters from multiple stations and arson investigators. A bombing, in contrast, will cause the Bureau of Alcohol, Tabaco and Firearms to become involved in determining what type of explosive was used. A biological attack will necessitate the involvement of public health officials. Thus, since the type of terrorist incident can vary dramatically, it is also imperative that you understand how to respond to any contingency effectively.

12.1.2   Safety and Security

As can be seen in the cases earlier, terrorist attacks will produce property destruction, injury, and death. For these reasons, there will be a strong inclination for people to rush into the area to help victims. It is important, however, that this temptation is carefully considered and executed—at least initially. The first priority, if possible, is to assess the situation from afar to determine what has occurred and how to respond safely. The process of evaluating the nature of the attack site is known as a **size-up**. This quick assessment is imperative because the location of terrorist attacks is inherently dangerous. Citizens and first responders can be injured by glass, twisted metal, falling debris, unstable buildings, broken gas lines that catch on fire, and many other hazards. Knowing what conditions may be encountered is central to the safety of official and nonofficial responders (Levy 2014).

There are other potential dangers that you must be concerned about as well. It is possible that terrorists may detonate additional bombs (i.e., **secondary devices**) to add to the disruption and fear (see Figure 12-3). By setting off numerous explosions, terrorists attempt to kill those who are trying to aid initial victims. This only adds to the casualty count and creates compounded problems for those trying to react to the mayhem. As an example, Eric Rudolph, a terrorist who opposed abortion, used secondary devices when he attacked an abortion clinic in 1998. **Dirty bombs**—explosives combined with hazardous materials (e.g., chlorine or nuclear material)—have been used in Iraq and could also be utilized in the United States in the future. While the blast area is generally limited to a certain geographic area, chemical fumes or radioactive material can be transported by wind far from the scene of an attack. Someone should therefore be given the task of monitoring safety concerns and other issues such as air quality. Known as **situational awareness**, this keen perception of dangerous working conditions is an important skill for those responding to terrorist attacks. You should expect that terrorists will do all they can to hinder response operations.

FIGURE 12-3 The scene of a terrorist attack can be extremely dangerous. *Source: US Department of Homeland Security.*

There are at least three principles to remember to keep emergency responders out of harm's way at the scene of a terrorist attack (FEMA 1999). First, police, fire, and EMS personnel should consider if they should keep away from the scene of an attack, if possible, to minimize exposure to bombs or dangerous chemicals. Second, responding personnel should not stay at the location of a terrorist attack for an extended length of time. Besides facing dangerous materials, fatigue resulting from

In The Real World

Responding to 9/11

The attacks on the World Trade Center in New York illustrate the variety of functions that have to be performed when terrorism occurs. Ground Zero and the area surrounding it was a dangerous area due to fires, unstable debris piles, broken glass, and twisted metal. The area was cordoned off and blocked by fences. The nature of the disaster required that additional equipment and instructions be given to first responders. Because a secondary attack by terrorists could not be ruled out, anyone wishing to enter the area had to be carefully screened. Only those with legitimate reasons for entering Ground Zero were permitted to do so once they were approved and given proper identification documents. USAR teams arrived from around the nation to find the injured and deceased. 9/11 also produced the largest crime scene in America. Evidence from more than a 16-block area had to be collected by numerous agencies. One of the major concerns was the presence of hazardous materials (e.g., dust, soot, dangerous chemicals). First responders were adversely affected by breathing the contaminated air around Ground Zero. Many required long-term medical care to treat their symptoms and several individuals died as a result. The major lesson from 9/11 is that those working in homeland security and emergency management will be preoccupied with many responsibilities after terrorist attacks occur.

extensive work periods at the location can lead to many accidents and injuries. While this is not always feasible for those involved in response operations, the goal should be promoted where possible. Finally, if responders must enter a dangerous area, they should have the proper personal protective equipment. Fire gear and hazardous materials suits may be needed to ensure survivability in hostile conditions and protect long-term health. Therefore, time, distance, and shielding are central ways to keep responders safe when reacting to terrorist attacks.

In addition to these recommendations, there is a need for site security. As soon as is feasible, law enforcement and/or other personnel should gain control over the scene. This may include the use of squad cars to block off roads as well as barricades, fences, and police or National Guard units on foot patrol. It would be wise for trained individuals to sweep the area for other secondary devices. The main priority is to prevent further attacks so that additional lives can be spared, and a comprehensive response can take place without further constraints. While site security is essential, the scene must not be impermeable. Other responders, public works employees, and contractors may need to enter the area to accomplish vital post-disaster missions. For this reason, a check-in system can be established by locating tables, chairs, and personnel at the site entrance. The intentions, legitimacy, and qualifications of those wishing to enter the scene can then be determined by checking ID cards, reviewing licenses, or making phone calls to sponsoring organizations. Any donations coming into the area should also be carefully checked to ensure they do not include bombs or weapons of mass destruction. This can limit the probability of secondary attacks and protect the safety of emergency responders.

12.1.3 Search and Rescue

While safety and site security issues are major priorities, they are only means to an end. Safety and security will enhance your ability to care for the victims, which ranks among your ultimate goals after a terrorist attack. One of the first steps you will need to take is to participate in search-and-rescue operations. **Search and rescue (SAR)** is defined as "response activities undertaken to find disaster victims and remove them from danger or confinement" (McEntire 2007, p. 142). There are many different types of SAR ranging from swift water to wildland contexts. In relation to terrorism, SAR includes finding the victims under rubble and extracting them to a safe location.

SAR will be undertaken initially by emergent groups as well as some police officers and emergency medical personnel. Firefighters, who have specialized knowledge for this important function, are most likely to be involved in this activity. Firefighters have personal protective equipment and tools such as helmets, goggles, dust masks, and other gear (e.g., saws and jacks) which allow them to search for and retrieve victims. Local fire departments may be insufficient, however. In major events, national Urban Search and Rescue (USAR) teams may be required. FEMA has nearly 30 such teams made up of firefighters, engineers, doctors, and paramedics. These units can be activated by within hours and are transported by C-130 aircraft and/or bus to the scene of an attack. USAR teams are especially valuable when destroyed areas are extensive or when SAR operations will be required over long periods. Such teams can be rotated periodically to keep rescuers fresh, alert, safe, and productive. USAR teams were used extensively after the Oklahoma City bombing and after the 9/11 attacks on the World Trade Center. Their value is especially notable in the first hours and days after terrorist attacks, and then wanes over time (Wenger 1989).

12.1.4 Medical Care and Triage

As victims are extracted from the scene of a terrorist attack or around damaged buildings, it will be necessary to provide them with necessary medical care (see Figure 12-4). Much of the medical attention will be provided by citizens (similar to those engaging in SAR operations). When firefighters and paramedics arrive, these professional responders will take over. These individuals have expertise and supplies that citizens will not possess. Treatment by these official responders will vary depending on the nature and extent of the injuries.

For instance, some individuals may suffer from minor cuts and bruises or smoke inhalation. Others will have broken bones or life-threatening injuries (e.g., damaged internal organs). If the number and extent of injuries are limited, first responders will be able to handle the load easily. They will stop the bleeding, provide oxygen, or immobilize fractures. If there are many casualties and serious injuries, the emergency medical system can be severely taxed. In this case, triage may need to be implemented.

Triage is the assessment, sorting, treatment, and transportation of the injured in such a way as to maximize limited medical resources (Mayer 1997). At the scene of a terrorist attack, there may be large numbers of victims. Because of the quantity of casualties in comparison to available emergency medical technicians or paramedics, choices will have to be made about who will receive care first. For instance, those with minor injuries and those with fatal injuries will be treated last or not at all. Attention will be given to those who require immediate help and have a strong chance of survival. This sorting will help determine who should be sent to hospitals. The practice of triage sounds cruel or inhumane, but the reality of terrorist attacks often dictates the extent to which medical personnel can assist everyone. Difficult choices have to be made to do the most good for the most number of people.

Another major advantage of triage is that it may limit the number of people who go to the hospital for further treatment. In most disasters (terrorist attacks included) people will often be taken to the hospital by friends, coworkers, and neighbors. That

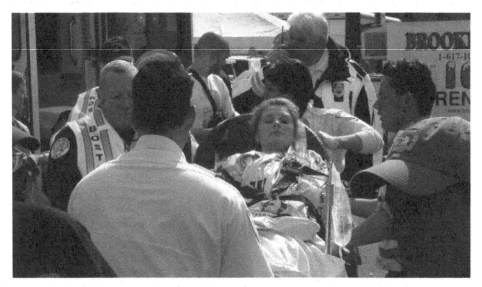

FIGURE 12-4 An important priority after a terrorist attack is emergency medical care.
Source: US Department of Homeland Security.

is to say, they are **self-referred**, walking wounded or ambulatory, meaning they arrive at the hospital whether they require immediate care or not. This often clogs the hospital with patients who may not always have critical injuries or medical needs. Therefore, hospitals also have to practice triage, thereby limiting the chance that nurses and doctors will be overwhelmed with patients who do not require immediate or lifesaving help. Interestingly, this is a continuation of a practice that occurs every day in the hospital. Nurses and doctors in the ER always determine who should be admitted immediately or who can wait for treatment.

12.1.5 Decontamination

Regardless of how patients arrive at the hospital, it may be necessary to clean them before they receive additional treatment. Patients may be covered in hazardous materials that could result in harm to themselves, firefighters, paramedics, and medical staff in hospitals. If a hospital receives contaminated patients or even contaminated gurneys, its operations could be jeopardized. **Decontamination**, or the removal of hazardous materials from victims through clothing removal and the washing of bodies, must therefore be performed at the scene, at a field hospital, or before entering a permanent medical facility (see Figure 12-5). If victims are not decontaminated, the chemicals on them may adversely impact physicians and nurses as was the case after the Sarin gas attacks in Tokyo in 1995.

The process of decontamination is technical and must be followed meticulously. When responding to a terrorist incident involving hazardous materials, three zones should be identified near the attack site. A **hot zone** is the area that has been contaminated by a terrorist attack. A **warm zone** is the location where victims are washed.

FIGURE 12-5 Decontamination is vital to prevent the transmission of dangerous chemicals or hazardous materials. *Source: US Department of Homeland Security.*

A **cold zone** is the uncontaminated area. It is where responders and victims may enter and leave the area. When setting up and cordoning off these zones, it is important that wind direction be taken into consideration. The zones should be set up in such a way that the wind blows toward the hot zone. This will ensure that hazardous materials are not sent back in the direction of responders and decontaminated individuals.

Decontamination in the warm zone is perhaps the most important, and this must take into account several important principles. Those involved in decontamination must have the proper protective gear and training. Inadequate hazardous materials suits, incorrect breathing apparatus, and mistakes in operations can lead to incapacitation and the loss of more lives. Another important priority is to protect the privacy of those being decontaminated. Decontamination tents or colored plastic sheeting can be hung to shield undressing, washing, and dressing areas. To properly clean victims, clothing should be removed. Known as "dry decon," this process may remove up to 85% or 95% of contaminants. If needed, water, mild soap, brushes, and/or sponges can also be used to remove hazardous materials. As this occurs, contaminated runoff can be captured in children's plastic swim pools or other devices made especially for this purpose. Affected clothing or gurneys with hazardous residues should also be properly washed or disposed of. The collection of contaminated liquids can then be treated in an environmentally sensitive way. When the job has been completed, responders should also be decontaminated before they leave the scene. This will ensure their safety as well. When finished with the decontamination process, new clothing or medical gowns can be given to victims or responders. At this point, victims can be transported, receive further medical treatment, or be admitted into hospitals.

Once again, hospital personnel should also ensure decontamination has occurred before a patient is admitted in these medical facilities. In fact, the Joint Commission requires hospitals to have a process to decontaminate victims. This includes a location to wash off patients as well as the availability of personal protective equipment and training of staff. Whether at the site of the attack or at the hospital, remediation teams and the Environmental Protection Agency may need to be contacted and involved to determine the impact on our natural resources and what can be done to clean up hazardous chemicals at the building facilities.

12.1.6 Closing the Investigation

Another major priority after a terrorist attack is to finalize the investigation (McEntire, Robinson, and Weber 2001). After life safety issues have been addressed, additional attention can be shifted toward subsequent law enforcement activities (i.e., final investigation, reporting, and prosecution). To apprehend or prosecute terrorists, evidence must be collected. It is vital that you remember that anything at the scene of a terrorist attack—debris, a body, or other materials and items—may provide evidence. For this reason, responders should be aware that the scene of an attack should be protected to the fullest extent possible. Unauthorized people should not be allowed to enter the area, or they should be asked to leave or operate with minimal disruption in mind. Evidence should be meticulously recorded and stored for future court proceedings. Photos can also be taken of the scene, and maps can be drawn to assist with the investigation. Other measures, including information gathering from witnesses, or the use of laser maps can help you piece together critical facts and data that will be needed in legal trials.

In The Real World

San Bernardino Shooting

On December 2, 2015, Syed Farook and his wife Tashfeen Malik entered the Inland Regional Center and shot several people with AR-15 rifles during a holiday party. Farook was present in the meeting but left at about 10:36 a.m. A short time later, Farook and his wife entered the facility wearing dark clothing. They discharged between 100 and 225 rounds into the room. The shooting led to the death of 14 people and injured another 22. 911 was contacted immediately, and the police responded within four minutes after being notified. The first officer on scene had no tactical gear nor did the first team, which was assembled a short time later. When numbers were sufficient, law enforcement officials encountered a chaotic scene with a blaring fire alarm, heavy gunpowder odor, malfunctioning lights, and water coming out of the fire sprinklers. Police also found numerous victims and others hiding to save their lives. Police officers began to stabilize the scene and address the needs of victims and survivors.

Paramedics stationed themselves at a golf course across the street and began providing emergency medical care to those injured. Four hospitals received victims as they were transported from the scene of that attack. The San Bernardino SWAT team arrived within 11 minutes of the call, as they were training only a few miles away at the time of the incident. Roads were shut down to clear the area of commuters, and buses were used to pick up witnesses and take them to the Rock Church for interviews to begin the investigation.

During conversations with witnesses, one of the employees noted that he saw Farook leave the party and believed it was he who returned. This tip led police to the couple's home. Other departments rushed to assist in the apprehension, including the FBI and the Department of Homeland Security, which provided aircraft surveillance in the area in search of the rented black SUV the suspects used to flee the scene. When police arrived at the couple's home in Redlands, the SUV was seen leaving an alleyway. The police followed the vehicle, and Malik shot at the officers through the back window. Syed exited the vehicle and began shooting at the police as well. Armored police vehicles surrounded the perpetrator's SUV, and the shootout continued. The exchange of over 500 rounds lasted 6½ minutes and involved 23 officers. Both Farook and Malik were killed by police, about four hours after the terrorist attack began.

Back at the scene and at Farook's house, several bombs were found. Robots were used to dispose of the three explosive devices that were discovered. Farook's friend and former neighbor, Enrique Marquez, Jr., also called 911 to implicate Farook in the incident. On December 17, 2015, Marquez was arrested in connection with terrorism. He was charged for his role in purchasing the two rifles that Farook and Malik used in the shooting, as well as the acquisition of materials used in the pipe bombs. Marquez claimed he did not know their plans to attack the building in San Bernardino. However, it was later revealed that Marquez conspired with Farook to execute other terrorist attacks that were never carried out.

Subsequent investigation revealed that Malik posted allegiance to al-Baghdadi, a terrorist leader, on social media. Although the shooting was a terrible tragedy, this incident is a prime example of how a quick response by law enforcement personnel limited further impact. Lessons learned revealed that there is a need for cross-training among police and fire departments. The importance of tactical EMS, the need for automatic weapons, and the benefit of body armor with plate carrier systems were also noted in after-action discussions.

All of these measures could help you identify who committed the attacks, prevent further terrorist acts, and facilitate successful prosecution. Cases have been solved because of seemingly insignificant clues. For instance, the serial numbers of blown-up vehicles have helped law enforcement officials track down the terrorists who rented or owned them. As an example, a 2007 terrorist plot in London was thwarted because a terrorist parked a car illegally. When police went to move the vehicle, they detected explosives and were able to track down the terrorists involved in the diabolical plan. Alternatively, legal battles may be lost on a technicality because evidence has not been carefully collected and recorded with a clear chain of custody. Investigation conducted by law enforcement personnel is one of the things that separates terrorist responses from other types of disaster operations.

Self-Check

1. Usually, the very first people to help out with a post-disaster operation are not first responders but ordinary citizens. True or false?
2. Emergence is the flow of people and resources to the scene of an emergency or disaster. True or false?
3. The process of assessing and evaluating the situation at an attack site is best known as:
    a. Triage
    b. Search and rescue
    c. Situational awareness
    d. A size-up
4. Why should responders be aware that the attack scene should be protected to the fullest extent possible?

12.2 Other Crucial Functions

There are many other important functions that may need to be addressed before, during, or after terrorist attacks occur. If possible, warnings should be issued, and information must be shared with the public. People must be evacuated out of dangerous areas and sheltered if required. These measures can keep citizens safe and minimize the impact of terrorism.

12.2.1 Warning and Public Information

Warnings are notifications sent out to the public so they can take protective measures. Warnings have been issued several times since 9/11. For instance, warnings were given due to possible attacks against the financial district in New York as well as when terrorists attempted to smuggle explosives on airliners crossing the Atlantic Ocean from England to the United States. Unfortunately, most terrorist attacks will not allow advanced notification. Terrorists often make threats against their enemies, but they rarely explain exact details about what they intend to do. The element of surprise is one of their greatest strengths. It is true that intelligence officers and law

FIGURE 12-6 Sharing information with the public via the media can help your community respond successfully to a terrorist attack. *Source: US Department of Homeland Security.*

enforcement agencies may at times intercept communications and obtain terrorist plans. However, this information may be sketchy or incomplete and require corroboration. In addition, there is always the difficulty of knowing what to believe and how much to share with the public.

On the one hand, the intelligence community may wish to avoid sharing sensitive information in a warning because it may compromise the safety of their agents or make terrorists aware of their tracking methods. This, in turn, will harm future intelligence efforts. On the other hand, sharing information with the public could help avert attacks or lead to the capture of terrorists. For instance, if terrorists threaten to attack airports, relaying this information to the public could help them be more aware of the activities taking place around them while traveling. Since this dilemma may never be completely resolved, authorities should use their best judgment about what to do. The particulars of the situation will probably dictate the best course of action to follow.

Regardless of whether or not a warning is possible or desirable, it is imperative that officials in emergency management and homeland security communicate often to the public after a terrorist attack (see Figure 12-6). Citizens will want to know:

- What happened?
- Who was responsible for the attack?
- What are the impacts?
- Are they still in danger?
- What steps should they take to protect themselves?
- What is the government doing?
- What should they do if they need assistance?

To notify people of impending attacks or to answer their questions, emergency management and homeland security officials should rely heavily on existing warning systems (McEntire 2007). **Weather radios** are electronic devices that receive information from the National Weather Service to warn people of approaching severe weather. If needed, these warning systems could be used to relay information about terrorist attacks. Alternatively, the **emergency alert system** can be activated. This warning system interrupts TV and radio programs and provides an announcement about what is taking place and what people should do for protection. An **Integrated Public Alert & Warning System (IPAWS)** can also be used to shared information with the public through cell phones. This system requires periodic testing according to legislation passed in 2015. The most recent test at the time of this writing was on October 4, 2023, The drawback of these systems is that not everyone has a weather radio, is watching TV, listening to the radio, or is in possession of a cell phone.

In addition to the warning systems described earlier, some cities have **reverse 911 systems,** which are essentially computerized phone messages sent rapidly to large numbers of people in a designated area. Reverse 911, such as the one operated by Everbridge, can relay detailed information to thousands of citizens and businesses within a specific area code or zip code. It allows you to target a specific group of people with a detailed message. The drawback of reverse 911 systems is that they are expensive and require constant updating and data entry. The good news is that 911 systems are increasingly able to connect with cell phones.

Another common method of communicating with the public is through the media and/or social media. Before and after terrorist attacks, emergency management and homeland security officials should interact frequently with reporters representing the television, radio, the Internet, and print media. Press conferences can be held periodically to make sure information is getting to the public. Because the media wants lots of information, it is vital that you meet their requests to the best of your ability. Failing to do so will likely result in them seeking knowledge elsewhere or reporting inaccurate information. In addition, anytime you hear any report that is incorrect, it should be brought to the attention of the media and corrected or clarified. The rule is to provide clear, consistent, and repeated information by a credible authority (Quarantelli 1990). Whenever these rules are not followed, public information will be ineffective and even counterproductive. Relying on a **public information officer**, a city employee who specializes in working with the media, can help you share information successfully with the media and the citizens you are trying to inform.

12.2.2 Evacuation and Sheltering

Depending on the consequences of the attack, you may need to evacuate your citizens and shelter them in safer locations. Terrorist attacks will result in buildings that have been gutted by fire or become unstable structures. Along these lines, a bomb may release toxic chemicals into the atmosphere on a temporary basis. A dirty bomb—an explosive device laden with radiation—could contaminate the environment for an extended period of time. Worse yet, the detonation of the nuclear device would level a large city and leave it uninhabitable for decades. Those victims that survive would need to leave or risk radiation contamination.

While not the same as terrorist attacks, events like Chernobyl illustrate this point clearly. The leak of radiation from the reactor made the area uninhabitable since the release occurred. This is a good reminder that you may need to evacuate the citizens in your community after a terrorist attack.

Evacuation is the movement of people away from hazardous areas or situations. When attacks threaten the lives or well-being of individuals, an evacuation request should be made. This decision should not be taken lightly since "unnecessary evacuations are expensive, disruptive, and unpopular" (Baker 1990, p. 3). However, evacuation for short or long periods may be required before or after terrorist attacks. It is therefore important that you notify people of the need to evacuate and provide clear instructions about when and how they will leave. In addition, after a major attack, you may need to make transportation arrangements for those without vehicles. Buses, trains, and planes can help you evacuate large numbers of people. As neighborhoods or communities are evacuated, you will want to have sufficient law enforcement personnel to help monitor and direct traffic. Contraflow plans—reversing the flow of transportation arteries as is done before hurricane landfall—can help to speed up the evacuation process. This will enable a quicker and safer exit from the location of the terrorist attack. However, it is important to recognize that some people may not evacuate even when warned or advised to do so. This reality has been witnessed in many disasters, including Hurricane Katrina. In some cases, people simply ignore requests to leave. In other cases, individuals and families may not have the means to travel to a safer area.

If people are leaving their homes and neighborhoods, they will logically require a place to go. Some individuals and families will stay in hotels, while others will visit friends and families in other areas. Different people will not have financial resources or supportive networks that can help them find a place to stay. In other cases, the number of evacuees is so large that this puts an extreme burden on receiving communities. In these cases, sheltering will be required elsewhere.

Sheltering is the location of relocation of individuals in places of safety and refuge (Mileti, Sorensen, and O'Brien 1992). It includes not only a roof overhead but also other life-sustaining activities. There are a number of factors that need to be considered when opening shelters:

- The number of evacuees versus the number of shelters needed to house them
- Occupancy rates in relation to the size of the buildings or rooms
- Electrical supply
- Sleeping arrangements (e.g., beds, cots, sleeping bags, pillows, blankets)
- Food and water (i.e., mass care arrangements)
- Bathroom and shower facilities
- Medical care
- Law enforcement presence
- Records of who is staying in the shelter

Churches and organizations like the American Red Cross and Salvation Army are frequently involved in sheltering operations. Faith-based agencies and volunteer groups can help you understand the variety of factors that must be taken into account when establishing and running shelters.

When possible, you will want to encourage those occupying shelters to find temporary or permanent housing (e.g., apartments or homes). The time frame of this may depend largely on whether or not evacuees can return home or start a new life elsewhere. Because some individuals and families will lack resources, the transition from an evacuee status to a returned or permanent resident can be long and challenging. You may need to provide government assistance or financial support from nonprofit organizations to make this transition more likely.

In The Real World

Sheltering First Responders and Other Emergency Management Personnel

When sheltering is discussed by emergency managers, politicians, or the media, the impression is given that this service is only directed toward the victims of terrorist attacks and disasters. While this is often true, it ignores the fact that those responding to the event may also need to place to stay. After the terrorist attacks on 9/11, hundreds of emergency workers had to be sheltered in the Jacob Javits Convention Center in New York City. Cots and bedding were acquired for this purpose. Shower facilities and food were also required to care for those responding to the terrorist attack. The sheltering operations at the Javits Center went on for an extended period of time and were vital to the success of the recovery activities at Ground Zero. Sometimes the need to shelter first responders and other personnel can limit the ability to house the victims of disasters. After the Maui fire occurred in Lahaina, Maui, there was some concern that FEMA personnel would occupy hotel rooms that could be used to care for the many individuals and families that lost their homes in the devastating blaze. Those involved in homeland security and emergency management must consider the sheltering needs of victims and responders alike.

Self-Check

1. Most terrorist attacks will not allow advanced warning. True or false?
2. The most common method of communicating with the public is through the emergency alert system. True or false?
3. The movement of people away from hazardous areas or situations is best known as:
   a. A warning
   b. Sheltering
   c. An evacuation
   d. The emergency alert system
4. Why is there a dilemma in the intelligence community about whether or not to share sensitive information?

12.3 Coordination Mechanisms

As can be seen, there are numerous activities that must be performed after terrorist attacks. They range from site security and SAR to evacuation and sheltering. Because of this wide array of responsibilities, there are countless organizations involved in incident response. First responders will take care of life safety issues and investigate. State and federal government officials will arrive to provide security and investigate. Public works will be involved in damage assessment and debris removal. Public information officers will work closely with the media to provide information to citizens. Public health may assist with medical needs along with hospitals. Businesses can provide resources and nonprofit organizations will assist with sheltering. As can be seen, there are innumerable individuals and organizations involved in response operations after terrorist attacks.

Due to this disparate set of activities and actors, coordination becomes imperative. **Coordination** is defined as cooperative efforts to pursue common goals in the wake of terrorist attacks. Such goals may include investigating leads, protecting life, assisting victims, and minimizing social disruption. Coordination helps to identify who will be in charge of vital functions, what collective problems exist, and how they will be overcome. These joint endeavors help to limit gaps in service and promote efficiency and effectiveness in response operations. Conversely, the lack of coordination may result in "an inability to determine priorities, misunderstanding among organizations, failure to fully utilize equipment and personnel, overly taxed organizations, delays in service, omission of essential tasks, duplication of effort, safety problems, and counterproductive activity among other things" (McEntire 2007, p. 293).

12.3.1 The Incident Command System

One of the best ways to promote coordination among field-level personnel is to employ the **incident command system (ICS)**. The ICS is "a set of personnel, policies, [and] procedures ... integrated into a common organizational structure designed to improve emergency response operations of all types and complexities" (Irwin 1989, p. 134). It was developed in California after responses to forest fires witnessed several problems including poor communications, lack of joint planning, and inadequate resource management. ICS helps to overcome these challenges and manage organizations involved in response operations.

ICS is typically a field organization mechanism and is based on an incident commander and various supporting officers (see Figure 12-7). It also includes four organizational sections and a number of widely accepted principles. While ICS can help promote coordination, first responders must be aware of its potential weaknesses if response operations are to be successful.

Under the organizational structure of ICS, the position of incident command will be established. **Incident command** is the on-scene leader or leaders for field operations (often fire fighters, police officers, or their respective chiefs or captains). When a terrorist attack occurs, incident command will be established by the first person on the scene and later taken over by those with more expertise or higher authority. Incident command may also include more than one commander. In other words, a variety of individuals may meet to make joint decisions about response priorities and methods.

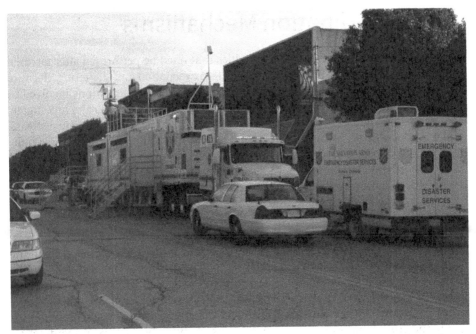

FIGURE 12-7 Incident Command Vehicles like this one may facilitate coordination at the scene. *Source: US Department of Homeland Security.*

In The Real World

Gedaper

When you or the incident commander(s) arrives on the scene, you should take several steps to accurately assess the situation. The National Fire Academy recommends the acronym GEDAPER (FEMA 1999, p. 43). GEDAPER includes:

- Gathering information about the event
- Estimating the potential impact of the attack
- Determining response goals
- Assessing tactical options and resources at hand
- Planning and implementing response actions
- Evaluating progress
- Reviewing results

By following these recommendations, you will be more likely to protect yourself and successfully respond.

The incident commander/commanders work(s) closely with three officers. These officers are attached laterally to incident command and include the public information officer, the safety officer, and the liaison officer. The **public information officer** gathers information for the incident commander/commanders and shares information with the media. The **safety officer** evaluates the dangers at the scene and makes sure everyone is operating according to safety policies. The **liaison officer** serves as the link between the incident commander/commanders and other organizations.

Together, the incident commander/commanders and officers oversee the entire field response operation.

The incident commander/commanders and officers do not have to do everything, however. They have four organizational sections below them to assist them in responding to terrorist attacks. These sections include planning, operations, logistics, and finance/administration. Each section may include one person or scores of people to fulfill important preplanning and post-event functions.

- **Planning** is the section in charge of collecting information about the terrorist attack, including operational priorities given to them by the incident commander. It determines what has happened and what may occur and identifies a strategy to accomplish response goals.
- **Operations** is the name given to the section that is in charge of implementing the strategy and tactics to satisfy the objectives of the incident commander and the planning section. It includes a number of activities ranging from fire suppression and triage to SAR and decontamination.
- **Logistics** is the section that supports operations. It acquires people, equipment, and other resources needed by those responding to the attack.
- **Finance/administration** is the final section under ICS. This section tracks the expenses associated with response operations and logistics.

As can be seen, ICS helps to organize leadership, personnel, and key functions in response operations. It also promotes coordination and effectiveness because it is based on commonly accepted principles. For instance, ICS encourages people to use common terminology when communicating with others; jargon and "ten" codes (e.g., 10-4) are avoided. ICS likewise allows for expansion of the organizational structure based on the nature and scope of the incident; it may be very simple or include additional layers (e.g., divisions, branches, and strike teams). ICS principles are geared toward the integration of communications; frequencies are assigned and clearly identified. Unity of command is promoted under ICS; this implies that each person reports to one supervisor only. Unified command is also an important priority in ICS; all major organizations may be involved in joint decision making. Another principle is consolidated incident action plans. These written documents help guide operations over a 12-hour period. Three other principles are a manageable span of control, designated incident facilities, and comprehensive resource management. The span of control implies that each supervisor should have between three and seven people to manage and oversee. Designated incident facilities refer to the desire to make everyone aware of the incident command post, staging areas, camps, helibases, and so on. Comprehensive resource management suggests that all resources, whether human or material, must be checked in and carefully tracked during response operations. Such principles are believed to help promote coordination among field personnel.

12.3.2 Strengths and Weakness of ICS

ICS has both advantages and disadvantages (McEntire 2007, p. 329). On the one hand, ICS may help promote collaboration among key leaders, increase safety among responders, and enhance communication among many organizations. Realistic management processes and improved use of resources are other benefits of ICS. On the other hand, ICS may not resolve all of the challenges inherent in response operations, and it can even exacerbate some (Dynes 1994; Neal and Phillips 1995). Critics argue that incident command may be too rigid for the dynamic nature of post-disaster operations and fail to appreciate the need for collaboration instead of control. Some feel that first-response organizations do not work well with other organizations and that ICS becomes less important in larger terrorist attacks. Regardless of this controversy, ICS remains the principal system for field response operations. Those working under ICS will be more successful if they are aware of the potential pitfalls (e.g., the dangers of focusing too much on technological solutions instead of addressing organizational communication problems or the problems of stressing who should be in charge vs. how organizations can work together harmoniously).

12.3.3 Utilization of Emergency Operations Centers

Because terrorist incidents will most likely include more than a field response, other organizational layers will be required. This is especially true when terrorist attacks are large and have significant consequences. In other words, ICS is helpful for dealing with field operations at a specific location or multiple locations (e.g., SAR, medical care, and decontamination). Incident command activities will also be needed to oversee the larger picture of terrorist attacks. This may result in the establishment of **area command** (which supervises several incident command posts) and **multiagency coordination centers (MACCs)** (which supervises incident command across several jurisdictions). If national assets are required, a **joint field office (JFO)** will be established. The JFO is an incident command organization with federal personnel (and state and local officials on certain occasions). Their purpose is to provide resource support to incident command teams at lower levels. Area command, MACCs, and JFO are known in emergency management as EOCs.

However, in contrast to incident command posts, EOCs are more concerned with the broader issues pertaining to incident response and disaster recovery. As noted in Chapter 11, EOCs are the centralized locations where information is gathered, processed, and acted upon by key decision makers. However, multiple EOCs will be activated in the jurisdiction, in neighboring communities, in regional organizations, at the state level, and even among numerous federal government agencies. Several EOCs can therefore be running and even interacting at the same time.

When working in an EOC, you must ensure that numerous organizations are represented. This may include police, fire, and emergency medical departments. However, EOCs will also incorporate all major city departments and even major businesses and nonprofit organizations. These participants will assist all community-wide response operations such as public information, evacuation, and sheltering. In terrorist attacks, public health and intelligence officials should logically be given a presence in EOCs. This will aid with bioterrorism attacks and investigation processes.

EOCs can and have been organized under the principles of incident command. However, in most cases, EOCs are laid out based on organizational units or according to functions (e.g., Emergency Support Functions such as warning, evacuation, public information, debris management, and resource management). Furthermore, there have been EOCs organized by the categories of life safety, social services, and infrastructure. Thus, the main point to remember is that EOCs can be organized in dramatically different ways and based on the needs and interests of those involved. Nevertheless, all EOCs should endeavor to facilitate response coordination.

In The Real World

Responsibilities of Local Departments and Organizations

When responding to a terrorist attack, it is imperative that you are aware of the other important participants. In cities and counties, this may include several organizations as noted in the following chart (see Perry 2003):

Department/organization	Roles and responsibilities
Fire	• Isolate impact area and set up perimeter • Position equipment and responders upwind, uphill, and upstream from the incident site • Assess downwind hazards and implement evacuation or shelter-in-place decisions • Identify agenda and adjust scene layout if required • Respond to victim needs with appropriate PPE • Decontaminate all victims, responders, and equipment as needed
EMS	• Implement mass casualty triage procedures • Provide medical treatment as dictated by the incident • Transport victims to definitive care facilities • Determine mental health impact and treat accordingly
Police	• Share preliminary intelligence data with incident command and the EOC • Notify and interact with the FBI • Deploy law enforcement personnel, including bomb squads and tactical operations teams • Assure incident security for first responders • Collect and control evidence • Apprehend and assume custody of suspects at the scene
Hospitals	• Implement lockdown of the facility to ensure security • Decontaminate and triage all arriving patients • Track patients, including their symptoms, and communicate with public health officials • Decide where to treat patients (internally or externally) • Treat as dictated by the nature of injuries
Public health	• Conduct surveillance for evidence of epidemics • Identify and control agent • Determine and implement protective measures for the population, including immunizations or prophylactic medicines • Work with police to implement quarantines if needed

(continued)

In The Real World *(Continued)*

Department/organization	Roles and responsibilities
Coroners	• Receive human remains • Safeguard personal property • Identify the deceased and notify the next of kin • Prepare and complete the file for each decedent • Photograph, fingerprint, and collect DNA specimens as appropriate • Provide death certificates • Coordinate and release remains for final disposition

12.3.4 EOC Management

In addition to facilitating coordination, one of your major priorities in the EOC is to acquire and manage resources. For instance, if the incident commander and logistics section requests additional body bags, the EOC may be tasked with the responsibility of obtaining and transporting them to the right location. E.L. Quarantelli (1979) notes that EOCs also help determine response policies, host visitors, and keep records.

Because EOCs are in charge of so many functions, they can become very noisy and stressful. The nature of response operations is also dynamic because the impacts of terrorism can unfold and change quickly. For this reason, it may be necessary to have everyone stop what they are doing every few hours and report on the key issues they are dealing with. This is somewhat similar to the planning function of ICS. In both cases, briefings will help ensure that everyone is up to speed on what is taking place and identify what yet needs to be done.

Finally, those working in EOCs may be required to work long hours under emotionally draining circumstances. Breaks should be taken periodically, and healthy food should be supplied to keep energy levels up. Shifts should be designated so that employee burnout does not occur. When shift transitions take place, transfer of command briefings should be given to the fresh crew. While these principles are geared toward EOCs, these same guidelines can also be applied to those working in field response.

Self-Check

1. Under the incident command system (ICS), incident command is always given to a single on-scene leader. True or false?

2. There is only one EOC operating during a terrorist attack. True or false?

3. Under the strategy of ICS, who serves as a link between incident command and other organizations?

   a. The information officer

   b. The safety officer

   c. The liaison officer

   d. The emergency manager

4. Why is coordination an important aspect of responding to attacks?

Summary

This chapter illustrated that the first priority is to investigate and apprehend terrorists before they launch attacks. However, if terrorism cannot be prevented, it is important to remember that successful response operations do not occur by chance. They require you to be familiar with convergent and emergent behavior. In addition, effective reactions to terrorist attacks include giving priority to the safety of first responders as well as the decontamination of victims. At times, you may need to warn citizens before an attack or evacuate them to safer areas and shelter them after terrorism occurs. To help you with your responsibilities, ICS and EOCs can be utilized. By anticipating your responsibility after a terrorist attack, you will be better able to apprehend and prosecute terrorists, protect life, and collaborate effectively with others.

Assess Your Understanding

Understand: What Have You Learned?

Go to **www.wiley.com/go/mcentire/homelandsecurity3e** to assess your knowledge of response operations.

Summary Questions

1. In emergency or disaster situations, people will take on new roles and interact cooperatively with strangers. True or false?
2. When providing medical care at an attack site, those with minor or fatal injuries are treated last or not at all. True or false?
3. Triage is the process by which harmful chemicals are removed from victims' clothes and bodies. True or false?
4. Sheltering is the location of people to places of safety and refuge. True or false?
5. Weather radios are never used to relay information on terrorist attacks. True or false?
6. IPAWS is a warning system that distributes information through the television only. True or false?
7. ICS is the principal system to coordinate field response operations. True or false?
8. EOCs have a narrower scope in disaster response than ICS. True or false?
9. Logistics is the ICS section that supports operations by acquiring and distributing people, equipment, and other needed resources. True or false?
10. Initial search and rescues are often undertaken by:
    a. Firefighters
    b. Urban Search and Rescue teams
    c. Emergent groups
    d. Police officers

11. An important way to prevent overloading of hospitals after a terrorist attack is BEST known as:

    a. Triage

    b. Situational awareness

    c. A size-up

    d. Decontamination

12. All of the following are drawbacks of reverse 911 systems, EXCEPT that such systems:

    a. Are able to reach large numbers of people in an expedited manner

    b. Call only homes and not businesses

    c. Require constant updating of data

    d. Are expensive

13. To better share information with the media after an attack, it could be most helpful to obtain communication assistance from:

    a. An emergency manager

    b. A public information officer

    c. A homeland security official

    d. A preparedness committee

14. Under the strategy of ICS, who evaluates the dangers at an attack scene?

    a. The information officer

    b. The liaison officer

    c. The safety officer

    d. The emergency manager

15. The cooperative effort to identify common goals in the wake of a terrorist attack is BEST known as:

    a. ICS

    b. Logistics

    c. Planning

    d. Coordination

Applying This Chapter

1. As a homeland security expert, you have been asked to speak at an emergency management seminar on the importance of safety in responding at a terrorist attack site. What could you say about the importance of a size-up?

2. You are in charge of medical operations and have been asked to speak at a press conference about the response to the attack. One reporter asks you about the process and importance of decontamination. How could you respond?

3. As an emergency manager, you are concerned about how best to communicate with the public. How could you effectively utilize the media for this goal?

4. Your duty with the Red Cross is to deal with sheltering in the case of an evacuation. What are some factors that you would need to be considered if sheltering was required?

Be a Homeland Security Professional

Advising a Local Government

You are a consultant on emergency management. A local government has asked you for suggestions on how to keep their first responders safe in the event of a terrorist attack. What recommendations could you provide?

Assignment: Convergence and Emergence

Write a one-page paper describing the concepts of convergence and emergence and how these processes factor into the response to terrorist attacks. Be as thorough as possible.

Managing an EOC

You are an emergency manager. Your office has recently created an EOC. When an EOC opens up to respond to a disaster, the work environment is often noisy, fast-paced, and exhausting. What management strategies could you use to maintain high levels of communication and staff energy in the event your EOC responds to an attack?

Key Terms

Area command   An ICS organization that supervises several incident command posts

Cold zone   The uncontaminated area where responders and victims may enter and leave

Convergence   The flow of people and resources to the scene of an emergency or disaster

Coordination   Cooperative efforts to pursue common goals in the wake of terrorist attacks

Decontamination   The removal of hazardous materials from victims through clothing removal and the washing of bodies

Dirty bombs   Explosive devices laden with dangerous chemicals or radioactive material

Emergence   The appearance of altruistic behavior that is unfamiliar to the participants

Emergency alert system   An announcement that interrupts TV and radio programs and relays information about what is taking place and what people should do for protection

Evacuation   The movement of people away from hazardous areas or situations

Finance/administration   The final section under ICS. This section tracks the expenses associated with response operations and logistics

Hot zone   The area contaminated by the terrorist attack

Incident command   The on-scene leader or leaders in the incident command post

Incident command system (ICS)   A set of personnel and procedures that helps facilitate coordination among first responders

Integrated Public Alert & Warning System (IPAWS)   A warning system that shares information with the public through cell phones

Joint field office (JFO)   An incident command organization with federal personnel (and state and local officials on certain occasions)

Liaison officer   The person who serves as the link between the incident commander(s) and other organizations

Logistics   A section that supports operations. It acquires people, equipment, and other resources needed by those responding to the attack

Multiagency coordination centers (MACCs)   An ICS organization level that supervises incident command across several jurisdictions

Operations   The name given to the section under ICS that is in charge of implementing the strategy created by those in planning

Planning   The section under ICS in charge of collecting information about the terrorist attack, including operational priorities

Public information officer   The person who gathers information for the incident commander(s) and shares information with the media, or a city employee who specializes in working with the media

Reverse 911 systems   Computerized messages sent over phone lines rapidly to anyone in a designated area

Safety officer   The person who evaluates the dangers at the scene and makes sure everyone is operating according to safety policies

Search and rescue (SAR)   Response activities undertaken to find disaster victims and remove them from danger or confinement

Secondary devices   The detonation of other bombs to add to the disruption and fear of the initial attack

Self-referred   Patients who arrive at the hospital whether they require immediate care or not

Sheltering   The location of individuals in places of safety and refuge

Situational awareness   Continual monitoring of safety concerns at the scene of a terrorist attack

Size-up   The process of evaluating the nature of the attack site

Tactical emergency medical services   The name given to a team of paramedics that are armed and trained in weapons use

Triage   The assessment, sorting, and treatment of the injured in such a way as to maximize limited resources

Warm zone   The location where victims are washed. It is located between the hot and cold zones

Warnings   Notifications sent out to the public so they can take protective measures

Weather radios   Electronic devices that receive information from the National Weather Service to warn people of approaching severe weather

References

Baker, E.J. (1990). Evacuation decision making and public response in Hurricane Hugo in South Carolina. *Quick Response Research Report No. 39.* Boulder, CO: Natural Hazards Research and Applications Information Center, University of Colorado.

Drabek, T.E. and D.A. McEntire. (2002). Emergent phenomena and the sociology of disaster: lessons, trends and opportunities from the research literature. *Disaster Prevention and Management* 12 (2): 97–112.

Dynes, R.R. (1994). Community emergency planning: false assumptions and inappropriate analogies. *International Journal of Mass Emergencies and Disasters* 12 (2): 141–158.

FEMA (1999). *Emergency Response to Terrorism.* Independent Study Course. Washington, DC: FEMA.

Irwin, R.L. (1989). The incident command system. In: *Disaster Response: Principles of Preparedness and Coordination* (ed. E.Auf der Heide), 133–161. St. Louis, MO: C.V. Mosby Company.

Kendra, J.M. and T. Wachtendorf. (2003). Reconsidering convergence and converger legitimacy in response to the world trade center disaster. In: *Terrorism and Disaster: New Threats, New Ideas,* Research in Social Problems, vol. 11 (ed. L.Clarke), 97–122. New York: Elsevier.

Levy, J. (2014). *The First Responder's Guide to Hazmat & Terrorism Emergency Response.* Firebelle Productions. Campbell, CA.

Mayer, T.A. (1997). Triage: history and horizons. *Topics in Emergency Medicine* 19 (2): 1–11.

McEntire, D.A. (2007). *Disaster Response and Recovery: Strategies and Tactics for Resilience.* New York: Wiley.

McEntire, D.A., Robinson, R.J., and Weber, R.T. (2001). Managing the threat of terrorism. *IQ Rep* 33 (12). Washington, DC: International City/County Management Association.

Mileti, D.S., Sorensen, J.H., and O'Brien, P.W. (1992). Toward an explanation of mass care shelter use in evacuations. *International Journal of Mass Emergencies and Disasters* 10 (1): 25–42.

Neal, D.M. and B.D. Phillips. (1995). Effective emergency management: reconsidering the bureaucratic approach. *Disasters* 19 (4): 327–337.

Perry, R.W. (2003). Municipal terrorism management in the United States. *Disaster Prevention and Management* 12 (3): 190–202.

Quarantelli, E.L. (1979). *Studies in Disaster Response and Planning.* Newark, DE: Disaster Research Center, University of Delaware.

Quarantelli, E.L. (1990). The Warning Process and Evacuation Behavior: The Research Evidence. *Preliminary Paper No. 148.* Newark, DE: Disaster Research Center, University of Delaware.

Wenger, D. (1989). The study of volunteer and emergency organizational response in search rescue: approaches and issues for future research. *HRRC Publication 5P.* College Station, TX: Hazard Reduction and Recovery Center. Texas A&M University.

CHAPTER 13

Recovering From Impacts

Short- and Long-Term Measures

DO YOU ALREADY KNOW?

- How to assess impacts and damages resulting from terrorism

- Ways to deal with debris and mass fatalities

- What makes a relief operation effective

For additional questions to assess your current knowledge of recovery activities, go to **www.wiley.com/go/mcentire/homelandsecurity3e**

WHAT YOU WILL LEARN

13.1 **The importance of disaster declarations**

13.2 **The key functions of disaster recovery**

13.3 **Types of assistance available after terrorist attacks**

WHAT YOU WILL BE ABLE TO DO

- Assess the consequences of terrorist attacks

- Justify the need for outside assistance and support the emotionally traumatized

- Assemble an effective relief operation

Introduction

Reacting to terrorist attacks includes much more than initial lifesaving measures, as important as those are. If you are employed in homeland security or emergency management, you should also be aware of the steps that must be taken to facilitate and promote recovery. For instance, you should understand the urgency of assessing damages and the steps required to declare a disaster. You must be able to deal with mass fatalities, debris, and psychological issues resulting from terrorist attacks. You should likewise be aware of the different types of assistance that can be provided to both state and local governments as well as citizens. Applying the novel approaches to recovery as mentioned in this chapter can also enable your community to rebound quickly from the consequences of terrorist attacks.

13.1 Initial Recovery Steps

If you are to promote recovery after a terrorist attack, you will need to understand the impact of the event. It may also be in your best interest to seek help from the state or federal government by acknowledging the limited capabilities you may have to react to the attack. However, assessing damages and declaring a disaster are the initial steps that must be taken to start your recovery after terrorism occurs.

13.1.1 Impact and Damage Assessments

One of the first things you will need to do to facilitate recovery is to assess the effects of an attack. As the emergency period begins to wane and after you have taken care of the most pressing lifesaving operations, you must quickly begin to think about short- and long-term recovery issues. To comprehend what needs to be done, you must consider the broad impacts of the attack and carry out a specific evaluation of damages. An **impact assessment** includes a very broad evaluation of the number of deaths and degree of social disruption caused by terrorists. Homeland security personnel and emergency managers may gather data about general impacts from first responders, hospitals, media reports, and other organizations. For instance, a bombing could illustrate housing shortages and the need for crisis counseling, and so on. One of the very specific types of impact evaluations is a damage assessment. **Damage assessment** is a survey of physical destruction and related economic losses. In most cases, damage assessment should identify major rebuilding and financial needs. Therefore, the assessment of damages is really your first priority to foster recovery (Oaks 1990, p. 6). Without an assessment, the community or nation will not be able to quickly and successfully overcome the negative consequences of terrorist attacks.

The types and methods of damage assessment can be described in simple terms, but this function is actually a very complicated process. There are typically three types of damage assessments as well as different ways of completing them (McEntire 2002). A **rapid assessment** is a quick survey of impacts. It is designed to gain fast comprehension of the scope of the attack so immediate needs can be met and additional help can be summoned. This type of assessment is undertaken by numerous local government officials. **A preliminary damage assessment (PDA)** is a more detailed evaluation of the destruction and financial losses which typically take place within days or weeks after the attack. The goal of this assessment is to determine if and to what degree state and federal help is warranted and in what ways. A PDA is performed by the affected community as well as state and federal officials. This type of assistance is labeled as "preliminary" because it occurs before the federal government verifies needs and provides relief and financial assistance.

A **technical assessment** is an evaluation by engineers, and it points out expected methods for rebuilding along with anticipated costs. It identifies what materials and labor will be required for demolition and/or reconstruction. This assessment is completed by engineers, insurance agents, contractors, and Federal Emergency Management Agency (FEMA) employees. It is a more thorough type of assessment than the PDA.

The three types of damage assessments are not accomplished in the same fashion. The rapid assessment may be completed by having first responders or public works personnel drive near the attack site to view damages or by having public officials fly overhead to gain an aerial perspective. In other cases, a rapid assessment may demand that the damaged area be toured on foot. Such walkthroughs (often by engineers) are most likely to be used for preliminary and technical damage assessments. At times, PDAs and technical assessments may require those involved to talk to victims, impacted businesses, and community leaders to tally deaths, estimate economic losses, and determine societal disturbance. The important point to remember is that the goals of the assessment will determine how it is conducted. Safety should also be a top priority.

13.1.2 Damage Assessment Concerns and Procedures

When your jurisdiction is completing damage assessments, it will be imperative that you recognize the extreme danger of the attack site. Sharp glass and twisted metal may be present from bombings. Fires and unstable structures are also associated with explosions. Hazardous materials could be located in and around the area of the terrorist attack. There are many other dangers that could cause injury or death to those who are involved in damage assessment. For these reasons, access to the site of the attack should be carefully controlled. You should also do all you can to verify that the damage assessment is accurate and complete. Recovery will be slowed down if your evaluation is incorrect or performed in a superficial manner.

To ensure that your damage assessments are successful, it is a good idea to hold a meeting to discuss who will evaluate the site of the attack and how it will be accomplished (McEntire 2022). Participants can be trained in safety precautions and given assignments based on geographic location or the type of damage to be tallied. For instance, those doing the assessment can be told to search a particular floor in a building or visit a block in a particular portion of the downtown area. Assessment teams can also be advised if the building lacks structural integrity or if there is the possibility of gas leaks. Communication devices and protective equipment should also be distributed at this meeting. This will facilitate the sharing of information in case people have questions or reduce the probability that someone will get hurt. Cameras and forms should likewise be given to those assessing damages along with guidelines on when reports are to be turned in.

At the scene of the attack, you will need to determine if buildings are safe, sanitary, and secure. In other words, you will need to discover if the building poses a risk, if it can be inhabited, and if its windows and doors can be locked. Depending

In The Real World

The Complexity of Assessments and Use of Technology

Impact and damage assessments involve countless organizations. First responders may evaluate the situation at the scene of an attack. EMS and hospital personnel will provide the number of people who have been killed or injured in the incident. Utility companies will identify if power and water systems are damaged. Public works may explore how a terrorist event affected roads or bridges. Insurance agencies will examine damages at businesses. Citizens will report if the attack had an impact on their homes and property. Engineers will share specific details relating to recommended construction methods and materials. Nonprofit organizations will share details about the special needs of disaster victims. County, state, and federal departments will collect, verify, and distribute information about impacts and damages with each other and with politicians. Some of these individuals and organizations act immediately after an attack and others engage in activities during the initial hours and days, within weeks, or even over the following months.

The type of technology used in impact and damage assessments is also diverse. This includes flood gauges, satellite images, drones, laser mapping, digital cameras, Survey 123/ARC GIS, Crisis Track software, WebEOC, city websites, social media posts, computer spreadsheets, and many other modern tools and equipment. Those serving in homeland security and emergency management should be aware of the participants involved in impact and damage assessments. They should also understand how technology can benefit the collection of data after a terrorist attack.

---

on the degree of damage, you may need to categorize structures in one of three ways. Buildings designated as "green" are habitable and occupants are allowed to return. "Yellow" structures have known or anticipated safety concerns and should only be entered by well-trained and properly equipped individuals. Buildings labeled as "red" are unsafe and should be condemned and destroyed. Once the assessments are completed, the information should be collected and reviewed for accuracy, compiled with other data, and then given to state and federal authorities. This information will influence what additional measures must be taken to promote recovery.

13.1.3 Declaring a Disaster and Seeking Help

All terrorist attacks will include the involvement of many responding organizations, especially law enforcement agencies and the FBI. These departments will arrive on the scene and support local and state efforts to determine who was responsible and to investigate perpetrators for apprehension and prosecution. In addition, the president may declare a national emergency and other leaders will be present to advocate for victims (see Figure 13-1). This has an overtone of wartime situations and brings attention to concerns about ongoing security and desperate immediate needs. A national emergency makes the public aware of what has happened and mobilizes other federal agencies to support the affected jurisdiction. However, depending on the findings of your impact and damage assessments, you might need to issue a statement acknowledging the extent of the issues you are facing. For this reason, you should know the process of declaring a disaster at the local level.

FIGURE 13-1 Political leaders are often present when a disaster is declared or to remember the victims of attacks. *Source: © National Guard.*

A **disaster declaration** is an acknowledgment of the severity of the event and it reveals that outside response and recovery assistance is required. It is one of the many expectations that must be fulfilled to obtain state or federal funds for rebuilding. Without this acknowledgment and justification, you may not get help from other levels of government.

After an attack occurs, it will be important for you to initiate your response and recovery operations. As mentioned, you will determine the negative consequences of the event through your damage assessment process along with your ability to handle such challenges. If the municipal and county governments determine that the event will overstretch their abilities, a disaster is declared. This usually includes a formal statement recognizing the death and damages, societal impact, economic losses, and the need for outside assistance. Disaster declarations may also discuss if the damage is covered by insurance if the attack traumatized the community, and if the consequences result in unemployment or evacuation and sheltering. At this point, state emergency management personnel will contact you to discuss the situation and may send representatives to verify the impact (or do so virtually with digital photos and videos). If the state feels it may also be unable to deal with the consequences effectively, it will also declare a disaster and relay this information to the FEMA regional office responsible for their area. If the officials in one of the ten regional offices around the nation concur with the state's assessment, the declaration will be forwarded to the FEMA administrator and to the president. A large or significant event will result in a presidential disaster declaration. This action will free up funds and mobilize the federal government to the aid of the affected community. However, if the event is less serious, the process of declaration could theoretically be stopped at any point. Because of the political nature of terrorism and homeland security, it is likely that the federal government will be involved in some fashion in any and all attacks against our nation. This should not discount the fact that there will be many functions that will require the attention of local governments.

In The Real World

Declaring a Disaster

When Timothy McVeigh blew up the Murrah Federal Building in Oklahoma City, first responders initially thought the explosion was a result of a broken gas line. A short time later, it became apparent that the devastation was the outcome of an intentional attack. After consulting with local officials, the governor of Oklahoma called the Regional Director of FEMA Region VI in Denton, Texas. The Regional Director then notified James Lee Witt, the Director of FEMA in Washington, D.C. By that afternoon, President Clinton was made aware of the terrorist attack in Oklahoma. He addressed the nation publicly and declared a state of emergency. This freed up funds, personnel, and federal resources so the government could support the response operations at and around the Murrah Federal Building. Declaring an emergency or disaster is often the first step toward getting the help you need.

Self-Check

1. A survey of physical destruction and economic losses is known as a damage assessment. True or false?

2. A rapid assessment is generally undertaken by federal officials. True or false?

3. In assessing damage, the color "red" indicates:
   a. The building is safe.
   b. The building is sanitary and secure.
   c. The building is completely destroyed, unsafe, or beyond repair.
   d. The building should be entered with caution.

4. An acknowledgment of the severity of an event that requires outside response and recovery assistance is best known as:
   a. A rapid assessment
   b. A preliminary damage assessment
   c. A disaster declaration
   d. A windshield assessment

13.2 Key Recovery Functions

Besides assessing damages and declaring a disaster, there are many other activities that must be undertaken if you are to promote recovery for your community. You may have to deal with mass fatalities, clean up debris, and address emotional issues. These are some of the major priorities to be addressed during recovery operations.

13.2.1 Mass Fatality Management

Terrorism may produce a large number of deaths. Even relatively small attacks around the world may result in numerous fatalities. For instance, Israel has experienced continual threats of violence since it was founded and many people have been killed each year. Larger attacks, such as Al-Qaeda's bombings in Africa on August 7, 1998, claimed even more lives. In Nairobi (Kenya) and Dares Salaam (Tanzania), 212 people died when U.S. embassies were blown up with explosives in cars parked near the buildings. In addition, on October 12, 2002, over 200 people died when Jemaah Islamiyah initiated three bombings in the tourist area of Bali, Indonesia. Iraq and Syria have been plagued with numerous ongoing attacks with multiple fatalities over the past decade.

In the United States, 168 people died on April 19, 1995, when Timothy McVeigh parked a Ryder truck laden with explosives near the Murrah Federal Building. And, on September 11, 2001, an appalling 2,974 people died when 19 terrorists hijacked planes and flew them into the World Trade Center (WTC), the Pentagon, and a field in Pennsylvania. These numbers are significant, but experts anticipate greater losses in the future if terrorists use nuclear or biological weapons. It is no exaggeration to suggest that terrorist attacks could result in thousands, hundreds of thousands, and even millions of fatalities.

Because of the significant numbers of victims and the possibility of even more consequential attacks down the road, it will be imperative for you to understand the challenges of and recommendations for dealing with a mass fatality incident. A **mass fatality incident** is an attack that creates so many deaths that the processing of remains stretches government agencies or is beyond the ability of local government. In other words, the large number of deceased is greater in comparison to the personnel that are available to collect, identify, and bury them. Besides the large quantity of dead, there are other challenges associated with mass fatality incidents. While dead bodies do not normally pose a threat to public health, a terrorist attack involving weapons of mass destruction can complicate mass fatality management. Some of the bodies may be contaminated and that can adversely affect those trying to process remains and bury deceased victims. In addition, well-intentioned citizens may move bodies in an attempt to help public officials (Scanlon 1998). This may complicate investigation and record keeping (because bodies are no longer located at the scene of the attack). A third problem deals with the heavy emotions of survivors. Family members will be understandably distraught over their losses, and those working in mass fatality management should be sensitive to the situation. In fact, the loss of life could be so disturbing that it may also impact responders emotionally. Finally, it is also possible that the remains may not be identified. Some bodies can be obliterated in attacks and beyond recognition. After 9/11, searchers found 19,893 separate body parts. However, the remains of 1,268 individuals were not found (Hampson and Moore 2003). DNA is typically useful in determining who people are, but it is not 100% successful as we discovered after the WTC attacks.

To process large numbers of bodies for burial or cremation, several steps must be taken (Hooft, Noji, and Van De Voorde 1989):

1. If possible, the location of bodies should be recorded. This will help with the investigation. ID tags on the deceased and maps of their location can assist with this objective.

2. The remains should be stored for processing. This may occur at hospitals or county morgues. Innovative ideas can help you when bodies outstrip storage facilities. Bodies have been stored in unmarked refrigerated trucks or even at ice skating rinks.

3. Clothing, jewelry, and other items (e.g., hats or wallets) should be removed and saved for family members. The height, weight, gender, and other identifying features (e.g., tattoos, mustaches, dentures, and cavities) must be recorded. Fingerprints and pictures of the body should also be taken to facilitate identification.

4. After bodies have been identified, they can be returned to families for burial. If remains are not claimed for whatever reason, they can be buried in recorded graves.

5. Respect for the wishes of the surviving family members (e.g., cultural burial practices) should be ensured.

Should mass fatality incidents warrant substantial outside involvement, a **Disaster Mortuary Operations Response Team (DMORT)** can be requested. A DMORT is a group of private citizens from around the nation who may be activated by the federal government to assist with mass fatality incidents. They include funeral directors, medical examiners, coroners, pathologists, and medical record technicians. Their purpose is to recover bodies and issue death certificates. These teams can be sent to any location and can assist communities with their portable morgue units. DMORTs can augment the capabilities of any community impacted by a terrorist attack. They will be a vital asset if fatalities outstretch local resources.

13.2.2 Debris Management

Because of the nature of the most common types of terrorist attacks (i.e., bombings), there could be a great deal of rubble that will need to be cleaned up (see Figure 13-2). This will include concrete, glass, metal, wood, wiring, and other construction materials. The quantity of debris can be overwhelming. For example, when the Murrah

FIGURE 13-2 Debris from 9/11: Debris is an important, but often neglected, function that must be addressed after a terrorist attack. *Source: US Department of Homeland Security.*

Federal Building was demolished after the Oklahoma City bombing, an average of 800 tons of debris was removed on a daily basis. The amount of debris at the WTC is even more noteworthy. As many as 10 major buildings were destroyed, which left behind 1.2 million tons of debris (McEntire, Robinson, and Weber 2003, p. 451). It took several months working around the clock to remove this large quantity of debris. This brings up the important concept of debris management. **Debris management** is the removal, storage, disposal, or recycling of rubble produced from terrorist attacks. It is an extremely important function to facilitate recovery.

Debris management after a terrorist attack sounds like a straightforward activity in theory, but it is actually a very complicated issue. There are a variety of factors that must be taken into account when dealing with rubble. First, trained personnel with adequate gear and heavy equipment will be needed to remove debris. People may not be able to help if they do not have gloves, hard hats, steel-toed boots, shovels, backhoes, cranes, or dump trucks. Debris can also pose a danger to those trying to remove it. After the Oklahoma City bombing, a large piece of concrete broke loose from an upper floor and fell on a nurse who was helping with search-and-rescue operations. She was killed instantly. There are numerous cases where debris has created injuries among those trying to clean it up. Cuts, bruises, and crushing from debris are all possible at the site of a terrorist attack. Safety should be a top priority when removing debris.

Debris management operations are also problematic in that debris may contain evidence as well as human remains. Sorting of debris must be painstakingly careful. Law enforcement detectives and coroners/medical examiners may need to be involved in the debris removal process to find evidence, full corpses, or body parts. Another challenge relates to where to take the debris. Because of the large quantity, debris may need to be stored somewhere temporarily until the waste can be recycled or buried. This will require a large holding area. Eventually, debris will need to be sent to a designated landfill, burned, or disposed of in other ways. The location and disposal of debris bring up another difficulty. The expense associated with moving and burying debris can be enormous. Local governments may need to seek federal assistance for debris management. Community officials will also need to monitor contractors to make sure they are being honest. Some companies

In The Real World

Debris from the WTC

The terrorist attacks on 9/11 resulted in the collapse of the WTC towers. When these buildings came down, at least eight other buildings were completely destroyed or substantially damaged. The pile of rubble that was produced was equivalent to 1.2 million tons of debris. To deal with this large amount of debris, public officials divided the 16-acre WTC site into quadrants. They then assigned a number of contractors to remove, ship, and dispose of debris. Cranes, backhoes, trucks, and even barges were used in the process. A unique challenge was the sorting of debris to search for and collect evidence, bodies, body parts, and personal belongings. In spite of the monumental undertaking, debris was removed much quicker than initially anticipated. In May 2002, Ground Zero was emptied of steel beams, broken concrete, and other demolished building materials.

involved in debris management have defrauded the government for their services. Continual oversight will be needed. As you deal with debris after a terrorist attack, organizations like the FEMA can assist you with helpful advice. The Environmental Protection Agency can also help you dispose of debris in an environmentally friendly manner.

13.2.3 Emotional Issues

Besides the physical impacts that can be readily seen, terrorism may create notable distress in emotions and mental well-being. The victims of terrorism and even witnesses recognize the toll of attacks in terms of injuries, fatalities, and economic disruption. The death of a loved one, resulting disabilities, and monetary losses could be unbearable to certain individuals in some cases. But many people are most troubled by the fact that the incident is an intentional act undertaken to instill fear. The violence is directed against the ordinary citizens. A common response is to ask "Why was this done and how could someone do this?" These issues and questions bring up a major function that you should address during recovery.

Crisis counseling is the treatment of psychological problems that may arise from the stress produced by terrorism. It includes active listening, sympathetic understanding, and emotional support (both direct and indirect) for those affected by terrorism. These victims may include first responders who experience critical incident stress. **Critical incident stress (CIS)** is defined as the inability of emergency service personnel to cope with the trauma that is experienced while on the job. For instance, a firefighter may do all he/she can to save a child after a terrorist attack, but ultimately be unsuccessful. This experience, along with the sights, smells, and sounds at the attack site, may be emotionally burdensome over time.

Alternatively, citizens may suffer from symptoms of **post-traumatic stress disorder (PTSD)**. PTSD is the clinical diagnosis for individuals who become depressed due to a traumatic event that they personally witness or experience indirectly in their lives. As seen after 9/11, terrorism certainly constitutes such an experience. Many people were psychologically disturbed by the terrorist attacks on this day.

Victims experiencing CIS or PTSD may exhibit a number of symptoms. Besides depression, they may gain or lose weight, become angry, abuse alcohol and drugs, have headaches, and experience mood swings. Facial twitches, social withdrawal, sleeplessness, and flashbacks are also common signs indicating that someone has been affected psychologically. If a person has difficulty coping with everyday stress and does not have a strong support network, they are considered to be vulnerable to CIS or PTSD.

After a terrorist attack occurs, homeland security and emergency management personnel may set up clinical treatments to help those who suffer psychologically. There are two strategies for dealing with the emotional problems created by terrorism. A **defusing** is a short, unstructured meeting to allow a person to discuss an experience as soon as it takes place. It is commonly provided to first responders as they wrap up at the scene so they can vent and unload frustration among colleagues and peers. Defusing is one way to relieve pent-up stress and reduce its compounding effects over time. A **debriefing** is a recurring and more in-depth discussion designed to redirect harmful thinking and develop improved coping mechanisms. Therapists reiterate the fact that humans have normal reactions to disappointing events and offer practical suggestions for stress management. The goal is to allow people to talk

---

CAREER OPPORTUNITY | Psychological Counselor

Terrorist attacks, disasters, and other disturbing events may result in CIS and post-traumatic stress. These situations may impact various employees in homeland security and emergency management, as well as the public at large.

Psychologists and mental health counselors are needed to help people work through emotional problems due to dangerous work conditions, the loss of a loved one, the destruction of property, or being a victim of or even a witness to a violent crime.

Those serving as psychological counselors need to have a master's or PhD in psychology. They work for first-response organizations, social service agencies, the American Red Cross, or as independent counselors.

The median annual income for psychologists was over $85,000 in 2022. Because of a variety of social, cultural, political, and economic challenges that exist and will likely worsen in the future, the demand for psychologists will increase by 6% over the next decade.

For more information, see **https://www.bls.gov/ooh/life-physical-and-social-science/psychologists.htm**.

---

after the attack, provide victims support, and help them find ways to deal with or overcome negative psychological impacts.

The success rate of defusings and debriefings is debated among scholars (Barnett-Queen and Bergmann 1989; Mitchell 1988). Some research seems to indicate that crisis counseling helps people recover more quickly, but other evidence suggests that repeated exposure to a traumatic event through frequent discussions could be counterproductive. While this issue remains to be resolved, you can seek the most appropriate help for your first responders and citizens. Police departments, fire departments, and the American Red Cross are organizations that have experience in dealing with CIS and PTSD. Specialists in these organizations can provide advice to help your community recover emotionally from terrorist attacks.

In The Real World

Memorials

After terrorist attacks, communities frequently honor the victims of the incident. For instance, a memorial was created in Oklahoma City after the bombing of Murrah Federal Building. It includes a reflection pool and an empty chair for each of the victims. The Flight 93 memorial in Pennsylvania incorporates a visitor's center and plaques to remember the victims of this hijacking. The largest and most visible memorial is in New York City. The 9/11 memorial museum is building on the footings of the twin towers at the WTC. It includes artifacts from the site of the attack, video and audio recordings of the response, pictures of the victims, and other notable information about the event. The goal of these solemn monuments is to remember those who perished and educate people about the nefarious impacts of terrorism. Such steps also help the community to recover from the impacts of the attacks.

Self-Check

1. Experts anticipate that there will be less loss of life from terrorist attacks in the future. True or false?

2. Defusing is a recurring and in-depth discussion to improve coping mechanisms for those suffering from emotional problems after an attack. True or false?

3. The removal, storage, and disposal of rubble produced by a terrorist attack is BEST known as:

   a. Defusing

   b. Debriefing

   c. Rapid assessment

   d. Debris management

4. What challenges are associated with mass fatality incidents?

13.3 The Importance of Disaster Assistance

As individual victims and the entire community seek to begin the process of recovery after a terrorist attack, aid will arrive from unofficial and official sources. Unofficial sources include volunteers and donations from concerned citizens. Official sources of assistance will come from organizations like FEMA. In either case, it will be your responsibility to harness these resources and implement innovative strategies to promote recovery.

13.3.1 Volunteer and Donation Management

After a terrorist attack, citizens will experience a deep and sincere interest in helping victims and the impacted jurisdiction. A study of volunteer behavior after 9/11 reveals that people sympathize with those affected and desire to find ways to assist them in coping with the aftermath. For instance, a woman who wanted to donate blood after the WTC towers came down stated, "There needs to be some positive that comes out of it, and the only way it's positive is if I and other people make it positive" (Lowe and Fothergill 2003, p. 299). This strong feeling will motivate people to volunteer their time, talent, and energy for altruistic causes.

On 9/11, spontaneous volunteers arrived at the WTC to participate in ad-hoc search-and-rescue operations, provide basic first aid, cheer on first responders, serve food to those removing debris, and give massages to tired emergency workers. Others offered translation services or were willing to perform any other duty that might be required after such a devastating event (e.g., clerical duties, crisis counseling, and transportation). In events such as 9/11, it is not uncommon for there to be hundreds or even thousands of volunteers. While not everyone was able to assist because of the dangerous conditions, the arrival of so many volunteers can be truly impressive—and perhaps overwhelming.

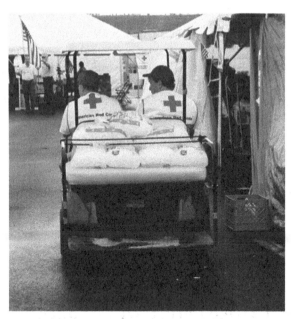

FIGURE 13-3 People will donate goods and supplies to the Red Cross and other organizations for distribution to victims. *Source: US Department of Homeland Security.*

The influx of significant quantities of volunteers is matched and even outpaced by the number of donated items that pour into a community after a terrorist attack (see Figure 13-3). The recovery efforts after the Oklahoma City bombing provide a good case in point (Oklahoma Department of Civil Emergency Management 1996). Southwestern Bell donated cell phones and a cell tower to help those responding to the event. Fast-food companies brought in meals to feed emergency workers. Clothing, pharmaceuticals, water, and other supplies and equipment arrived from around the state, region, nation, and world. Besides these **in-kind donations** (i.e., physical donations including food, water, clothing, supplies, and equipment), money was also sent to Oklahoma City. Hundreds of thousands of dollars were sent to care for the victims and their families. Donations of money and supplies will occur after almost every significant terrorist attack. Experience suggests that large quantities of donations will poor into affected communities.

This mass assault of volunteers and donations, as it has been described by Thompson and Hawkes (1962), can create several problems for those working in emergency management and homeland security. For example, there can be too many volunteers in comparison with the number of individuals needed to manage them effectively. Some volunteers may be untrained or lack the skills you require after a terrorist attack. Another difficulty is keeping volunteers safe in spite of the dangerous conditions of the attack site. Some volunteers may also become frustrated if you do not put them to work immediately or if they are given tasks that they feel are not "making a difference."

To overcome the challenges associated with those who want to help, you should implement a coherent strategy for volunteer and donation management. **Volunteer management** is the harnessing of volunteers to take advantage of their potential contributions while averting potential negative consequences. Successful volunteer management often requires an online registration process or a physical registration center. A **volunteer registration center** is a location where citizens fill out forms noting their skills and share other information that can help them when making assignments. While giving responsibilities to volunteers, it is wise to review some basic safety rules and briefly train them on what they will be doing. You will then need to evaluate their progress and care for them as needed (e.g., provide food, water, or necessary equipment). Without ensuring the safety and well-being of volunteers, injury and death can result. In addition, liability will be increased if volunteers are not carefully supervised. No community wants to be sued if insufficient protections are not in place to protect volunteers. Although too many uncoordinated volunteers can be a headache, their strength in numbers and desire to serve can have a dramatic and positive impact on recovery operations.

The problems resulting from excessive and unrequested donations may also be addressed through effective donations management (Neal 1994). **Donation management** is the collection, sorting, and distribution of goods and money for the benefit of victims of terrorist attacks. It includes recognizing your needs after an attack and relaying your specific requests to the public through the media. Once received at designated areas (e.g., a staging area or a warehouse), you will need to have a team in place to help sort and disperse needed items. Faith-based groups and nonprofit organization programs can be very helpful in this respect. Tracking donations with the use of computers can also help you determine what you are lacking, what is coming in, and where it should be sent.

An even better way to avoid the hassle of donations is to request monetary contributions. Cash funds prevent unwanted items from arriving, eliminate some of the labor needed to deal with them, speed up assistance to victims, and help the local economy (assuming goods and services are purchased in the affected or nearby areas). If in-kind and monetary donations are not getting to the appropriate individuals, an unmet needs committee may be established. An **unmet needs committee** is a group of concerned citizens and community leaders who work together to collect donations and address the long-term needs of victims (Wedel and Baker 1998). They play a positive role in helping victims recover and rebound after terrorist attacks.

13.3.2 Individual and Public Assistance

When a terrorist attack results in a presidential emergency and/or disaster declaration, citizens and communities alike may receive federal disaster assistance. There are two types of disaster assistance. **Individual assistance** provides relief to citizens and businesses impacted by terrorist attacks. **Public assistance** makes aid available to government entities that have been affected by terrorism. Both types of assistance have special programs and requirements.

Individual assistance includes loans at low interest rates as well as grants that do not have to be repaid. Some of the loans are directed toward homeowners and renters who have lost personal property. In this case, up to $200,000 can be provided for primary residences and $40,000 to replace personal property. Other loans are for businesses. Loans up to $1,500,000 can be obtained to repair or replace structures and machinery. If the business has suffered economic hardship because of the attack, another $1,500,000 can be acquired. Small grants up to $27,000 can also be obtained by individuals and families who do not qualify for loans. This money can be used to provide temporary housing in mobile homes, help with hotel/motel stays, or repair or replace destroyed property.

Individuals and families may also obtain needed services after a terrorist attack. Federal and state assistance can help victims understand the types of programs that are available, file and settle insurance claims, protect against price gouging and fraud, and obtain counseling from attorneys regarding legal issues and contracts. Food stamps, unemployment assistance, and social security and veterans' benefits can all be given to victims of terrorist attacks.

To obtain individual assistance, victims of terrorism should call the National Processing Service Center at 1 (800) 621-FEMA. The **National Processing Service Center** is a FEMA office set up to help victims apply for federal assistance programs. Call takers will input information about the victim's needs into a computer. Alternatively, the victim may go to a Disaster Recovery Center for help. A **Disaster Recovery Center (DRC)** is a temporary facility near the attack location where victims can seek information about federal assistance programs (see Figure 13-4). It includes phone and Internet access to federal disaster programs. The selection of the location of the DRC is a local and state responsibility. However, the DRC is usually set up in a building that can accommodate tables, chairs, phones, copiers, and parking. Federal officials will help to publicize and run the DRC. Alternatively, victims can submit information about disaster losses online through FEMA's website. Regardless of how citizens apply for assistance, the information they provide will be reviewed and verified by FEMA employees and contractors. Eventually, decisions will be made about loans, grants, services, and benefits. When warranted, funds will be distributed to individuals, families, and businesses.

If a local or state government, native American tribe, or private nonprofit organization is adversely impacted by a terrorist attack, it may also qualify for public assistance. Public assistance is divided into two types: emergency assistance and permanent assistance. **Emergency assistance** is financial help given to local governments to take care of immediate needs. It may include monetary and technical support for debris removal and the implementation of various safety precautions. **Permanent assistance** is financial aid for the repairing of publicly owned critical infrastructure and key assets. It may include the help to reconstruct roads, bridges, and lights; dikes, levees, and dams; public buildings and equipment; water, gas, and sewage systems; or parks, airports, and recreational facilities.

FIGURE 13-4 Victims of large terrorist attacks may be able to seek assistance at a FEMA Disaster Recovery Center. *Source: US Department of Homeland Security.*

If individual or public assistance is required, community leaders will need to declare a disaster and share damage assessment findings with federal officials. If assistance is deemed justified, a **kickoff meeting** will be held. A kickoff meeting is a gathering of local, state, and federal officials for the purpose of explaining public assistance programs in detail (e.g., application materials and deadlines). This meeting will help answer questions about public assistance and begin the process of obtaining outside aid to promote recovery.

The oversight of federal public assistance programs will take place in a **Joint Field Office (JFO)**. The JFO is a temporary office that is created with local, state, tribal, and federal representatives who will manage the paperwork regarding public assistance. Those working in the JFO will work to ensure that recovery projects are being completed as promised, check for fraudulent activities, and consider the special circumstances surrounding historic buildings, environmental concerns, and rebuilding with mitigation in mind. However, this does not mean that the process will be smooth or easy. Some jurisdictions and states could be highly critical of FEMA and the federal government after disasters and terrorist attacks. They assert that the process is bureaucratic, inflexible, and even wasteful. Politicians have therefore called for the dissolution of FEMA at times or advocated for the distribution of recovery funds directly to local and state governments (thereby bypassing FEMA). Recovery can therefore be a contentious process as people seek resources or attempt to control the outcome of disaster assistance and rebuilding efforts.

13.3.3 Novel Approaches

As can be seen there are many different aspects and elements associated with recovery operations. However, performing the functions listed above will not necessarily ensure that long-term issues will be successfully addressed. Innovative ideas are needed to help the community rebound quickly from terrorist attacks. One exemplary case is from Manchester, England (Bathos, Williams, and Russel 1999).

On June 15, 1996, the Irish Republican Army called officials in this city and advised them of a bomb that would be detonated shortly. A warning was issued, and efforts were made to evacuate the area of approximately 80,000 people. About 40 minutes later, a 3,000-lb. fertilizer bomb in an illegally parked Ford van exploded. Fortunately, no one was killed in the bombing. Nevertheless, over 200 people were injured. At least 12 buildings were damaged, and many of these had such severe structural problems that they had to be demolished. As a result, 672 businesses were displaced, and nearly 50,000 m² of office space was lost. Residents in the area were forced to leave their homes due to fears about the effects of infrastructure damage. The event provided both a test for responders and an opportunity for those involved in recovery.

When the emergency period involving fire suppression and lifesaving activities ended, attention soon focused on long-term concerns. In accordance with the community disaster plan, the area was cordoned off. This "protected people from physical danger, helped preserve criminal evidence at the disaster scene, and [was central to] the city-council's re-occupancy strategy" (Bathos, Williams, and Russel 1999, p. 222). On the next day, a task force was created and made up of local leaders, police officials, and representatives from the private sector. This group toured the area and devised a methodical strategy to help the community recover.

The city's architectural department was put in charge of assessing damages and determining what could be done to make the area safe enough for further recovery activities (e.g., demolition of buildings). As structures were deemed or made safe, business owners and citizens were allowed back into the structures to retrieve belongings and resume normal activities. Over the next several days, police presence diminished, and companies hired their own security staff to monitor the destroyed buildings. The cordoned-off area was reduced to five buildings.

During the first week, a major challenge for city officials was the large number of people (between 5,000 and 10,000) who arrived at city hall wanting information or assistance after the bombing. At first, there was no system in place to address citizen concerns. In time, however, various meetings were held for owners and occupants in different geographic areas. This helped to provide details about getting back into damaged buildings, making them safe, and the process of recovery.

The city council also helped large businesses relocate to available retail property so the private sector could resume normal operations. A massive media campaign was initiated to help citizens understand what was taking place and how they could assist in the recovery efforts. By the end of the first month, the city created a committee to oversee the rehabilitation of the area. Three priorities became evident.

First, the mayor created a fund and asked people to donate to it to help those experiencing hardship from the attack. Over £2.5 million was raised for victims over an 18-month period. Second, the deputy prime minister announced an international urban planning competition in July to generate a vision and plan for recovery. The decision on the winning design was made public 16 weeks later. Finally, an organization called the Manchester Millennium Task Force was created to oversee the rebuilding of the area. It was put in charge of recovery activities until they were completed in spring 2000.

A review of the reaction to this bombing reveals several positive features. Prior planning and training helped to ensure a quick response to the event. The oversight of post-attack functions was also successful due to the use of a designated emergency

For Example

Unmet Needs After the Oklahoma City Bombing

Because Timothy McVeigh's terrorist attack on the Murrah Federal Building produced so many deaths and injuries, there was an almost overwhelming sense of responsibility for the victims and their families. Within a short time, the community shifted from short-term emergency needs to long-term concerns. A committee was formed with the participation of city, business, and nonprofit organizations (Wedel and Baker 1998). Its goal was to collect monetary and in-kind donations and ensure that the donations were directed to those in need. Cars, tuition scholarships, and many other resources were given to those in need. The unmet needs committee also tracked relief assistance to make sure that it was not duplicated by other organizations or abused by disaster victims. The superb handling of donations and the treatment of victims were labeled the "Oklahoma Standard." However, in time, the distribution of resources has taken a and a minority of victims are claiming the resources are not being used as intended.

operations center. Police were effective in evacuating most people out of the area and maintaining a security presence after the attack occurred. Although determining who should have access to the area was problematic, the City Architect Department was soon made responsible for issuing passes to the area. The leadership and innovation of city officials helped to channel resources and ideas to facilitate recovery. While the rehabilitation of the area was not problem-free, the aftermath of the Manchester bombing illustrates the benefit of organizations working together to overcome post-attack problems. Partnerships and cooperation among various parties were cited as reasons for a speedy recovery.

Self-Check

1. A gathering of local, state, and federal officials for the purpose of explaining public assistance programs is best known as a kickoff meeting. True or false?
2. Contributions of money to help out victims of a terrorist attack are also known as in-kind donations. True or false?
3. Financial aid for repairing publicly owned critical infrastructure and key assets is BEST known as:
   a. Permanent assistance
   b. Emergency assistance
   c. Individual assistance
   d. In-kind donations
4. What problems can volunteers create in the recovery process?

Summary

If a terrorist attack occurs in your jurisdiction, you will need to perform a variety of recovery measures. After assessing damages, you will need to ensure that a state of emergency or disaster is declared. It will also be imperative that other functions are addressed, including mass fatality management, the disposal of debris, and the provision of emotional support for those who have been emotionally impacted by the event. As you begin to address long-term issues, you will want to ensure that your jurisdiction applies for public and individual assistance. The impacts of terrorist attacks can be alleviated through donations and volunteer management. Innovative approaches can speed up the time it takes for recovery and help your community rebound from disturbing events.

Assess Your Understanding

Understand: What Have You Learned

Go to **www.wiley.com/go/mcentire/homelandsecurity3e** to assess your knowledge of recovery.

Summary Questions

1. An impact assessment is no different than a damage assessment. True or false?
2. A preliminary damage assessment is a quick survey of attack-related impacts undertaken by local government officials. True or false?
3. A technical assessment identifies what materials and labor will be required for demolition and/or construction. True or false?
4. If a local government declares a disaster, the federal government must provide aid. True or false?
5. If an attack causes so many deaths that a local government is unable to collect, identify, and bury all of the deceased, it is called a mass fatality incident. True or false?
6. DNA is generally, but not always, 100% successful in helping to identify remains after an attack. True or false?
7. The Environmental Protection Agency is one of the organizations that can assist local governments in debris management. True or false?
8. After a terrorist attack, people are afraid and will not want to volunteer or help out. True or false?
9. Only local or state governments can qualify for public assistance after a terrorist attack. True or false?
10. If a mass fatality incident warrants substantial outside help in dealing with remains, then a request can be made for:
    a. A Disaster Mortuary Operations Response Team
    b. Tactical emergency medical services
    c. Crisis counseling
    d. Defusing
11. Critical incident stress is BEST described as:
    a. The process of determining which buildings are safe after an attack
    b. A clinical diagnosis for individuals suffering from depression or trauma
    c. The inability of first responders to cope with trauma from work
    d. A treatment for psychological problems associated with attacks
12. A short, unstructured meeting to allow a first responder to discuss job-related experiences and reduce emotional stress is BEST known as:
    a. A damage assessment
    b. Defusing
    c. Debriefing
    d. Decontamination
13. The oversight of federal public assistance programs after a terrorist attack takes place in a:
    a. National Processing Service Center
    b. Joint Field Office
    c. Disaster Recovery Center
    d. Volunteer Registration Center

14. Financial help given to local governments for immediate needs, such as debris removal, is BEST known as:

    a. Permanent assistance

    b. In-kind donations

    c. Individual assistance

    d. Emergency assistance

15. Relief provided to citizens and businesses affected by terrorist attacks is BEST known as:

    a. Permanent assistance

    b. Emergency assistance

    c. Individual assistance

    d. Public assistance

Applying This Chapter

1. As a homeland security expert, you are asked by a local government official about the process by which a municipality can obtain federal assistance after a terrorist attack. How could you respond?

2. When having lunch with friends, you mention that you are studying emotional issues related to terrorist attacks in your homeland security class. Your friends are curious about how authorities deal with the stress faced by first responders. How could you answer their question?

3. As a homeland security expert, a local government has asked you about the potential resources that may become available for their community in the event of a terrorist attack. They are curious about the different types of disaster assistance. How could you respond?

4. As an emergency manager, you have been asked by your organization to come up with some strategies for volunteer management in case of a terrorist attack. What are some recommendations that you could provide?

Be a Homeland Security Professional

Planning for Debris Management

You are an emergency manager with a local government. In the event of a terrorist attack, there could be a large quantity of debris that needs to be removed, stored, disposed, or recycled. You have been asked to prepare a report on this aspect of recovery in case of an attack. What concerns would you address in your report on debris management?

Assignment: Damage Assessments

Write a two-page paper discussing the concerns you might want to address if you were to complete an impact or damage assessment. Be as thorough as possible.

Addressing Donations Management

You are an employee in an emergency management office. To be prepared for recovery in case of a terrorist attack, your personnel should be ready to handle the large influx of donations that usually accompanies such an event. Without proper management, donations can become excessive or irrelevant. What recommendations could you provide on how to effectively manage donations?

Key Terms

Crisis counseling   The treatment of psychological problems that may arise from the stress produced by terrorism

Critical incident stress (CIS)   The inability of emergency service personnel to cope with the trauma that is experienced while on the job

Damage assessment   A detailed survey of physical destruction, economic losses, deaths, social disruption, and recovery needs resulting from a terrorist attack

Debriefing   A recurring and more in-depth discussion designed to redirect harmful thinking and develop improved coping mechanisms

Debris management   The removal, storage, disposal, or recycling of rubble produced from terrorist attacks

Defusing   A short, unstructured meeting to allow a person to discuss an experience as soon as it takes place

Disaster declaration   An acknowledgment of the severity of the event and that outside response and recovery assistance is required

Disaster Mortuary Operations Response (DMORT)   A group of private citizens from around that nation who may be activated by the federal government to assist with mass fatality incidents

Disaster Recovery Center (DRC)   A temporary facility near the attack location where victims can seek information about federal assistance programs

Donations management   The collection, sorting, and distribution of goods and money for the benefit of victims of terrorist attacks

Emergency assistance   Financial help given to local governments to take care of immediate needs such as debris removal or safety precautions

Impact assessment   A broad evaluation of the number of deaths and degree of social disruption caused by terrorists

Individual assistance   Relief programs for citizens and businesses impacted by terrorist attacks

In-kind donations   Physical donations including food, water, clothing, supplies, and equipment

Joint Field Office (JFO)   A temporary office that is created with local, state, and federal representatives who will manage the paperwork regarding public assistance

Kickoff meeting   A gathering of local, state, and federal officials for the purpose of explaining public assistance programs in detail

Mass fatality incident   An attack that creates so many deaths that the processing of remains is beyond the ability of the local government

National Processing Service Center   A FEMA office set up to help victims apply for federal assistance programs

Permanent assistance   Financial aid for the repairing of publicly owned critical infrastructure and key assets

Post-traumatic stress disorder (PTSD)   The clinical diagnosis for individuals who become depressed due to a traumatic event in their lives

Preliminary damage assessment (PDA)   A more detailed assessment of impacts that typically takes place within days or weeks of the event; it determines the possibility and extent of outside assistance

Public assistance   Relief programs that make aid available to government entities that have been affected by terrorism

Rapid assessment   A quick survey of impacts designed to gain an appreciation of the scope of the attack

Technical assessment   A survey of damages that points out methods and costs for rebuilding

Unmet needs committee   A group of concerned citizens and community leaders who work together to collect donations and address the long-term needs of victims

Volunteer management   The harnessing of volunteers to take advantage of their potential contributions while averting potential negative consequences

Volunteer registration center   The location where citizens fill out forms noting their skills and other information that can help you when making assignments

References

Barnett-Queen, T. and L.H. Bergmann (1989). Counseling and critical incident stress. *The Voice* (August/September), pp. 15–18.

Bathos, S., Williams, G., and Russel, L. (1999). Crisis management to controlled recovery: the emergency planning response to the bombing of the Manchester city centre. *Disasters* 23 (3): 217–233.

Hampson, R. and M.T. Moore. (2003). Two years after Sept. 11, NYC couple to Bury Son. *USA Today* (Thursday, September 4), pp. 1A–2A.

Hooft, P.J., Noji, E.K., and Van De Voorde, H.P. (1989). Fatality management in mass casualty incidents. *Forensic Science International* 40: 3–14.

Lowe, S. and A. Fothergill (2003). A need to help: emergent volunteer behavior after September 11th. In: *Beyond September 11th: An Account of Post-Disaster Research* (ed. L. Jacquelyn), 293–314. Boulder, CO: Natural Hazards Research and Applications Information Center, University of Colorado.

McEntire, D.A. (2002). Understanding and improving damage assessment. *IAEM Bulletin* (May), pp. 9, 12.

McEntire, D.A., Robinson, R.J., and Weber, R.T. (2003). Business responses to the world trade center disaster: a study of corporate roles, functions and interaction with the public sector. In: *Beyond September 11th: An Account of Post-Disaster Research* (ed. L. Jacquelyn), 431–457. Boulder, CO: Natural Hazards Research and Applications Information Center, University of Colorado.

Mitchell, J.K. (1988). Stress: the history, status and future of critical incident stress debriefings. *JEMS* 13 (11): 47–52.

Neal, D.M. (1994). The consequences of unrequested donations: the case of hurricane Andrew. *Disaster Management* 6 (1): 23–28.

Oaks, S.D. (1990). The damage assessment process: an overview. In: *The Loma Prieta Earthquake: Studies of Short-Term Impacts. Program on Environment and Behavior Monograph #50* (ed. R. Bolin), 6–16. Boulder, CO: Institute of Behavioral Science, University of Colorado.

Oklahoma Department of Civil Emergency Management (1996). Donations Management Case Study of the Alfred P. Murrah Federal Building Bombing, 19 April 1995 in Oklahoma City, OK: Summary and Lessons Learned. Oklahoma City, OK.

Scanlon, J. (1998). Dealing with mass death after a community catastrophe: handling bodies after the 1917 Halifax explosion. *Disaster Prevention and Management* 7 (4): 288–304.

Thompson, J. and R. Hawkes. (1962). Disaster community organization and administrative process. In: *Man and Society in Disaster* (ed. G. Baker and D. Chapman), 268–300. New York: Basic Books.

Wedel, K.R. and D.R. Baker. (1998). After the Oklahoma City bombing: a case study of the resource coordination committee. *International Journal of Mass Emergencies and Disasters* 16 (3): 333–362.

CHAPTER 14

Assessing Significant Threats

WMD and Cyberterrorism

DO YOU ALREADY KNOW?

- Future threats that will confront homeland security
- The dangers of radiological weapons
- How to minimize the spread of nuclear weapons
- Ways to respond to biological weapons
- Steps to prevent the use of chemical weapons
- How to define cyberterrorism

For additional questions to assess your current knowledge of the significant threats, go to **www.wiley.com/go/mcentire/homelandsecurity3e**

WHAT YOU WILL LEARN	WHAT YOU WILL BE ABLE TO DO
14.1 Serious risks facing our nation	• Predict new threats confronting the United States
14.2 The threat of radiological weapons	• Estimate the outcome of dirty bombs
14.3 The potential impact of nuclear weapons	• Evaluate the consequences of nuclear terrorism
14.4 What to do about biological weapons	• Select ways to respond to biological attacks
14.5 The types of chemical weapons	• Plan how to respond to chemical weapons
14.6 The vulnerabilities associated with cyberterrorism	• Implement measures to prevent and minimize cyberattacks

Introduction

If you are to prevent terrorist attacks and minimize their consequences, you must anticipate future threats. In the following chapter, you will learn about the probability of terrorist attacks involving radiological, nuclear, biological, and chemical weapons. You will gain knowledge about the threat these weapons pose, the ability of terrorists to acquire such capabilities, and the effects of their use. You will also learn what steps can be taken to minimize these types of attacks or respond effectively should they occur. In addition, the nature of cyberterrorism will be explored along with a number of measures to prevent these types of attacks against the nation. Knowing the information presented in this chapter will be imperative as terrorist attacks will likely be more problematic in the future.

14.1 The Future of Terrorism and WMD

To succeed in homeland security, it is vital that you are aware of the innumerable concerns about future threats. There are new terrorist groups appearing each day, and their motivations for attacks are becoming more intense and entrenched. For instance, left-wing groups became more vocal a few years ago, and even people like school teachers and Madonna have asserted that they would like to shoot President Trump and blow up the White House. Meanwhile, right-wing domestic terrorist groups have also become a very concerning threat, more so than they have been in the past. This is to say nothing about the many other international and domestic terrorists that want to do us harm.

In addition, terrorism may manifest itself in recurring or new ways in the future. For instance, the possibility of suicide bombings in crowded public areas remains an ongoing concern. Terrorists employ this tactic around the world, and it is likely this type of terrorism will likely be witnessed in the United States. Also, the disturbing impacts of mass shootings in Florida, California, and Nevada will likely lead to more these types of slayings. Terrorists can take out large numbers of people with automatic weapons. All of these threats have major implications for homeland security policy and programs.

Of course, a prevalent focus over the past few decades has been on radical Islamic terrorists. They have been constantly fighting American troops in Afghanistan, Iraq, and Syria, among other locations. Attacks from groups like ISIS and HAMAS have become more brutal, and the impacts may have broader and more negative consequences. These individuals and groups have also vowed to bring the fight to the United States. Radical Islamic terrorists desire to attack the West directly by launching attacks on our own soil. The porous border in recent years may permit these types of attacks in the United States according to the FBI.

Countries, including Iran and Syria, have increased their support of terrorism (financially and in other ways). In addition, we have learned again that Iran trains terrorists within its borders, funds operations in the Gaza Strip and Lebanon, and has sent its proxy soldiers and weapons into Iraq to be employed against American soldiers or in Israel to exterminate the Jewish people (see Figure 14-1). An additional concern is that Iran is developing nuclear technology. While the leaders of this country maintain they are advancing their nuclear capabilities to meet future energy

FIGURE 14-1 The Ayatollah Khamenei is an example of an Iranian leader who has espoused revolution and violence. *Source: Unknown author/Wikimedia Commons.*

needs, it is also possible that highly enriched uranium could be used for belligerent purposes. If this occurs, Israel or the United States could be targeted or blackmailed by Iran. Furthermore, it is possible that Iran could share nuclear materials or weapons with terrorist organizations. Such an action would theoretically make it more difficult to prevent attacks or place blame. In any case, Iran is regarded by many to be a major threat in the Middle East and to the United States. Its leaders have clearly stated their desire to attack Israel, and this country is no fan of the United States. In response to this situation, President Obama focused on diplomatic solutions. But

In The Real World

Al-Qaeda, ISIS, and WMD

Research illustrates that terrorist groups like Al-Qaeda and ISIS have taken and may continue to pursue numerous measures to acquire and use weapons of mass destruction (WMDs). Dunn's study (2008) revealed that Al-Qaeda was working diligently to obtain WMDs. For instance, it is believed the Al-Qaeda operatives attempted to purchase radiological material in Russia and elsewhere. Concrete evidence suggests that Al-Qaeda contacted scientists in Pakistan to learn more about nuclear weapons. Training manuals seized in Afghanistan reveal that labs were beginning to develop ricin and botulinum toxin. Al-Qaeda ran experiments to test the impact of cyanide on dogs in 2001.

ISIS is another group that covets WMD, and it appears that they have used them against U.S. troops in Syria. ISIS also established a unit to seek and develop chemical weapons. The *Washington Times* reported that ISIS fired a shell with a dangerous chemical in Iraq where U.S. troops were operating. The shell landed within the security perimeter, and tests confirmed mustard gas. It is also believed that ISIS may have acquired iridium-192 from a storage facility in southern Iraq. This radioactive isotope is highly dangerous and could injure or kill those who come in contact with it. Belgian police also disrupted an ISIS plot to obtain radioactive materials in that country. It appears that terrorists are intent on obtaining and using WMDs.

many feel this approach is paving the way for Iran to obtain nuclear weapons. It appears that diplomacy has not really worked as intended. In light of this situation, the Trump administration therefore implemented new sanctions against Iran. Later on President Biden reversed Trump's policies reinstituted some of the political preferences of President Obama. It will be interesting to see how American policy unfolds in the future as threats evolve. Nevertheless, continued conflict, terrorism, and even all-out war are not out of the question. If things turn more violent, the use of nuclear weapons cannot be ruled out.

The possibility of Iran developing nuclear weapons brings up the most pressing concern for future homeland security: weapons of mass destruction. As noted in earlier chapters, **weapons of mass destruction (WMDs)** are weaponry that will create major carnage, destruction, and disruption when utilized. WMD's are not only possible but also probable in the future. Many scholars and policy experts agree that forthcoming attacks will rely on WMDs going forward (Gurr and Cole 2000; Kelley 2014; Mahan and Griset 2013).

There are numerous arguments why WMDs could be the future of terrorism. *America's Achilles' Heel*, a study by three terrorism experts (Falkenrath, Newman, and Thayer 1998, pp. 5–6), reveals that terrorists may seek and use WMDs for the following reasons:

- **Massive casualties**. WMDs may result in an overwhelming number of injuries and casualties. Hundreds, thousands, and even millions of people could be adversely impacted by WMDs. While the number of fatalities was limited, over 5,000 people sought medical care after the 1995 Sarin gas attack on a Tokyo subway in Japan. The potential for mass death is certainly a concern going forward.

- **Degraded response capabilities**. Terrorism involving WMDs will have at least three consequences for first responders. First, police, fire, and emergency medical service personnel will be overwhelmed by the demands placed upon them by the initial wave of victims. Second, countless first responders could be numbered among the victims of these types of attacks because of secondary devices or unanticipated exposure and contamination. Third, responding to WMD attacks will require a great deal of specialized knowledge and skill. Unfortunately, this expertise is not as widespread as it needs to be due to the technical nature of WMD and insufficient training programs or funding sources.

- **Contamination**. The use of WMD has a significant negative consequence on the environment. Because of the dangers associated with WMDs, the impacted area may require complex and lengthy remediation efforts. In certain cases, the location of the attack may be uninhabitable for days, weeks, months, and even years. For instance, the accidental release of radiation from a nuclear power plant in Chernobyl in 1986 illustrates this potential in a vivid manner. Decades after the incident, the area is still considered hazardous and is therefore officially declared uninhabitable. Some studies suggest this area in the Ukraine won't be completely safe for hundreds or thousands of years.

- **Economic damage**. An attack involving WMDs will have extensive financial consequences. Direct financial expenses will include the loss of buildings, property, and infrastructure as well as the costs associated with response and recovery. Indirect economic losses are inevitable and could result from disruptions in business transactions, astronomical insurance payouts or unsettled claims, and resulting unemployment. The terrorist attacks on 9/11 cost at least $40 billion; future costs could be unimaginable.

In The Real World

WMD Identifiers

When responding to terrorist attacks, first responders and others should pay special attention to WMD identifiers. FEMA has identified several of them (1999, pp. 25–26):

Biological

- Unusual numbers of sick or dying people or animals
- Dissemination of unscheduled and unusual sprays, especially outdoors and/or at night
- Abandoned spray devices with no distinct odors

Nuclear

- Presence of Department of Transportation placards and labels
- Monitoring devices

Incendiary

- Multiple fires
- Remains of incendiary devices
- Odors of accelerants
- Heavy burning
- Fire volume

Chemical

- Massive onset of similar symptoms in a large group of people
- Mass fatalities
- Hazardous materials or lab equipment that are not relevant to the location
- Exposed individuals reporting unusual odors and tastes
- Explosions dispersing liquids, mists, or gases
- Detonations that destroy a package or the bomb device alone
- Unscheduled dissemination of an unusual spray
- Abandoned spray devices
- Numerous dead animals, fish, and birds
- Absence of insect life in a warm climate
- Mass casualties without obvious trauma
- Distinct pattern of casualties and common symptoms
- Civilian panic in potential target areas, for example, government buildings, public assemblies, and subway systems

Explosives

- Large-scale building damage
- Blown-out windows
- Scattered debris
- Victims with shrapnel-induced trauma
- Appearance of shock-like symptoms
- Damage to eardrums

- **Psychological impact**. Because the outcome of any WMD attack is likely to be significant in so many ways, victims and others witnessing this type of terrorism may feel intense fear. Anxiety, sleeplessness, stress, and other emotional tolls are probable. While most people are resilient after major events, terrorism adds psychological distress because it is intentionally caused.

- **Political change**. The use of WMD could result in a significant transformation of governmental systems. The freedoms we enjoy could be seriously curtailed in an attempt to respond effectively after an attack occurs or prevent future attacks. Isolationism or vengeance in foreign policy is another possibility if terrorists use WMD. The organizational and policy changes after 9/11 are indicative of how societies can sometimes change after major terrorist attacks. New laws were passed, and the creation of the Department of Homeland Security resulted in the most sweeping transformation of government in half a century.

Of course, not all types of WMD will have each of these impacts or to the same degree. This is because WMDs may range from crude bombs made of common household cleaning products and more elaborate and large-scale explosives composed of fuel and fertilizers acquired from commercial outlets. However, the most feared WMDs include radiological, nuclear, biological, and chemical weapons. For this reason, each of these will be discussed in turn.

Self-Check

1. Scholars and practitioners are increasingly fearful of the use of WMD in the future. True or false?
2. WMD produces injuries and deaths, but not social disruption. True or false?
3. Responding to WMD attacks is difficult because:
   a. Responders are likely to be victims.
   b. There will be a large number of victims.
   c. Responses to WMD require technical expertise.
   d. All of the above.
4. Discuss three reasons why terrorists may use WMD in future attacks.

14.2 Radiological Weapons

Radiological weapons spread dangerous radiological material but do not result in a nuclear explosion. Instead, **radiological dispersion devices (RDDs)** as they are known are made out of a combination of conventional explosives and nuclear materials. RDDs are commonly known as **dirty bombs**. Terrorism involving radiological material can also occur by simply exposing it to people and the environment. In other words, terrorists could place a container with radiological material in a crowded area and then remove the lid. This would affect all those in the immediate vicinity. As will be seen, the alpha, beta, and gamma radiation in these bombs generate additional complications beyond more routine terrorist activities.

Radioactive substances are not readily available to the average person. However, they can be obtained from hospitals, industrial facilities, and elsewhere (e.g., through theft, smuggling, or black market purchases). Philip Purpura, a well-known security expert, asserts that radiological materials are used for various commercial purposes, including food sterilization and the treatment of cancer (2007, p. 74). Unfortunately, the sources for dirty bombs or RDDs are not always secure. "Since 1999 ... federal investigators have documented 1,300 cases of lost, stolen, or abandoned radiological material" in the United States (Purpura 2007, p. 74). There are also over 100 countries that have not adequately controlled radiological substances. The ease of access (in comparison with other types of WMD) as well as their potential negative impacts makes radiological weapons inviting to terrorists.

It is true that radiological weapons are not commonplace—at least when compared with conventional bombings. Nevertheless, there have been confirmed attempts or actual uses of radiological weapons in the United States. José Padilla, a U.S. citizen trained by Al-Qaeda in Afghanistan, was arrested in 2002 when he plotted to use an RDD in Chicago. Although the attack was thwarted, it did generate a significant degree of media interest. Since instilling fear in others is a major motivator for terrorists, the recognition gained by RDDs makes them attractive weapons. For these reasons, it is highly likely that radiological weapons will be used in the future.

In The Real World

Radiological Terrorism

Terrorism involving radiological material is not a theoretical proposition, but an empirical reality. Gavin Cameron, an expert at the Centre for Military and Strategic Studies, observes that radiological terrorism is more common than one might think. In 1974, a man called police and noted that he had placed a nonlethal amount of iodine-131 on a train bound to Rome. The material was stolen as it was being shipped to a hospital. The man took this measure due to his grievance about the treatment of mentally ill patients in Austria. In 1985, a man sent a letter to the Mayor of New York demanding the release of a prisoner. If the city failed to respond adequately, plutonium tri-chloride would be placed in reservoirs serving New York. Although there was no proof that the threat was acted upon, the U.S. Department of Energy did find elevated levels of plutonium in the water. In 1993, a Moscow businessman was killed when the Russian mafia placed gamma-ray-emitting pellets in his office. In 1996, three men attempted to kill officials of the Republican Party by placing radium in the victims' cars, food, and toothpaste. One culprit was found unfit to stand trial (on grounds of insanity), and the other two were sentenced for their participation in the attack. In 1995, the Chechen guerrilla leader, Shamil Basayev, placed radioactive material in Izmailovo Park. Russian authorities were able to find the cesium-137 wrapped in a yellow plastic bag within a case. In 2001, Ivan Ivanov, a Bulgarian businessman, told British officials that he was asked by bin Laden's associates to obtain radiological material. He was also offered $200,000 to acquire fuel rods from the Kozlodui nuclear power plant in Bulgaria. In 2006, Alexander Litvinenko, a former spy, was killed by radiation poisoning in London, England. After the 2015 terrorist attacks in Paris, authorities uncovered video surveillance of a scientist employed at the Belgian Nuclear Research Center. It was feared he was working on radiological weapons. Radiological weapons have been used in the past, and the efforts of terrorists indicate that they are attempting to employ them in the future.

If deployed, radiological weapons could pose serious health hazards to humans. The length of exposure will determine the extent of injuries and deaths. Those who are exposed for a brief time may not suffer any notable consequences. However, a strong dose or prolonged exposure could lead to immediate fatalities or cancer that develops over time. In addition, radiological weapons could contaminate geographic locations for an extended period more of time. Even if the amount of radiological material dispersed is insignificant, the presence of such substances could create social disruption and psychological concern. People will not want to live or work in an area that has been exposed. This could lead to major evacuations, housing shortages, and economic decline around the nation.

The theft of radioactive material from a medical clinic in Brazil illustrates the potential negative outcome of dirty bombs. In 1987, scavengers entered an abandoned medical facility and came across a container with radiological material. They broke the source open, thereby releasing the radioactive material into the environment. As a result, "four people died, more than 100,000 others had to be monitored for contamination, and cleanup costs amounted to tens of millions of dollars" (Ferguson and Lubenau 2008, p. 139).

To prevent the use of radiological weapons, it will be imperative to secure the sources of this material. For instance, each year, 50 radioactive gauges are stolen from the oil industry in Nigeria (Ferguson and Lubenau 2008, p. 141). If such thefts are to be prevented, the material will have to be held under lock and key, monitored by armed guards, or protected by other means. In addition, the issuing of licenses (allowing the legal use of radiological materials) should also be carefully controlled. As an example, Stuart Lee Adelman posed as a professor in 1996 and illegally obtained radiological material. He pled guilty to this fraud and was sentenced to five years in prison (Ferguson and Lubenau 2008, p. 143). It is believed he may have sought the materials to obtain money from terrorists. Regardless of Adelman's motive, terrorists themselves may also seek radiological materials through similar deception. Records should therefore be carefully kept, and shipments should also be meticulously monitored. Homeland security personnel should ensure that steps are taken to prevent terrorists from acquiring radiological material.

If radiological weapons are used by terrorists, it will be imperative to rely on the expertise of highly trained individuals. Normally, this will require military personnel or other specialists who have knowledge and understanding of how to isolate and clean up radiological material. The Department of Defense has 58 **Civil Support Teams** around the nation, with more on the way (see Figure 14-2). These teams can be quickly activated and mobilized to respond to this type of event anywhere in the country. They assist local and state governments in identifying hazardous agents (including RDDs), determining consequences and appropriate response techniques, and requesting additional support if required. Other military personnel, from U.S. Northern Command (NORTHCOM), may also be activated to assist when circumstances warrant their assistance. If an attack occurs, medical personnel and coroners will also be needed to treat the injured and dead. A major priority is to wash those affected by the radiation and "purge inhaled or ingested materials" (Purpura 2007, p. 305). The disposal of the deceased must be completed in the proper manner (assuring that others are not contaminated in the process).

To be prepared for terrorism involving radiological materials, monitoring devices should be placed strategically around cities to detect the presence of the weapon. This is especially important in that RDDs may or may not have an obvious and associated explosion. The lack of forewarning or the desire to avert drawing

FIGURE 14-2 Civil Support Teams can help local jurisdictions deal with attacks involving weapons of mass destruction. *Source: © U.S. Department of Homeland Security.*

In The Real World

Treating Radiation Poisoning

Three medical experts in the Dallas–Fort Worth area (Elvin Adams, Ira Nemeth, and John White) assert that radiological poisoning can be minimized in some cases. If a person has been exposed to radioactive U-235, an antidote of baking soda and water can be administered to eliminate the radioactive isotope from the body. If cesium has been used in an attack, the pigment Prussian blue can bind the cesium and carry it out in the stool. If iodine-131 is present in the environment, the administration of potassium iodide will help. Large doses are required to saturate the thyroid to block exposure to radioactive I-131. If given to the victim quickly, some forms of cancer can be prevented. The most critical treatment to prevent dangerous internal exposure is to prevent ingestion of a radiological agent. The use of an N-95 mask will potentially minimize the inhaling of the agent. When drinking is required, using a straw to minimize ingestion of materials on the lips will reduce significantly the number of radioisotopes taken into the body.

attention to this type of attack means that these weapons can be employed covertly. In addition, radiological materials cannot be detected by human senses because they are colorless and odorless. The only way to know that they have been used is through the symptoms of victims or constant monitoring of the environment. Having adequate equipment will also be necessary for those responding to RDDs. This may include Geiger counters, personal protective clothing covering the entire body, and a breathing apparatus. Keeping a safe distance from the radiological materials or limiting time of exposure is another recommendation when responding to this type of terrorist attack.

Self-Check

1. Examples of radiological weapons include dirty bombs and RDDs. True or false?
2. There have been no cases of radiological weapons used in the past. True or false?
3. Which country illustrates the potential deaths and economic impact of radiological terrorism?
   a. Nigeria
   b. Brazil
   c. Colombia
   d. Iran
4. What experts will be needed to react effectively to radiological terrorism?

14.3   Nuclear Weapons

Nuclear weapons generate radiological material as dirty bombs or RDDs do. However, **nuclear weapons** produce massive explosions due to the release of vast amounts of energy through the process of nuclear fission or fusion. Most nuclear weapons are placed in the tip of rockets and cruise missiles. However, portable nuclear weapons are also possible. In fact, many experts in homeland security fear the use of a **suitcase bomb**. A suitcase bomb is a portable weapon that can be carried or rolled to the target location.

Nuclear weapons are possessed by major powers including China, France, Great Britain, Russia, and the United States. India, Israel, North Korea, and Pakistan also have nuclear weapons, and it is believed that Iran is actively seeking to develop them. While it would be difficult for a terrorist organization to acquire highly enriched material for nuclear bombs, it is not impossible to rule out their success in obtaining them. Theft, blackmail, and bribery are all potential ways of acquiring plutonium or other nuclear material. And, existing nuclear powers could share nuclear materials with terrorist organizations. Once attained, terrorists with sufficient knowledge and technological sophistication sometimes may theoretically build nuclear weapons. Information on their construction is accessible in scientific journals, in academic books, or even on the Internet. It is also true that states possessing nuclear weapons may also simply give them to known terrorists. This is why the United States and others are currently concerned about Iran and its nuclear energy program.

FIGURE 14-3 The atomic bomb in Japan was devastating. Today's nuclear weapons are far superior and will inflict even more death and destruction if utilized. *Source: Unknown author/ Wikimedia Commons / Public domain.*

Should nuclear weapons be acquired and used in a terrorist attack, the impacts are almost unimaginable (see Figure 14-3). The devastation produced by the nuclear weapons at the end of World War II was imposing. America's use of the A-bomb killed approximately 130,000 people in Hiroshima and 65,000 more in Nagasaki. But today's technology is far superior. The destruction and death of a modern nuclear weapon would certainly be more notable.

The explosion of a nuclear weapon produces a mushroom cloud that can be seen from miles and miles away. In the blink of an eye, people and property in the immediate area (perhaps a 15-mile radius or greater) would simply be vaporized. Those on the outskirts would be killed by the intense heat or injured by flying debris and blinded by the extreme light produced by the explosion. Sickness and death from radiological poisoning would also occur, requiring a major medical and mass fatality response. The infrastructure would be severely damaged, and the environment would be contaminated for decades. Living near the blast or fallout zone would be impossible due to the health consequences of radiation. Because of the loss of business and impact on the stock market, serious economic repercussions would be inevitable. Government agencies would cease to exist or function in the immediate area, and nearby jurisdictions would be severely hampered or overburdened. The results truly bring up images of Armageddon.

To prevent the use of nuclear weapons by terrorists, those involved in homeland security must work closely with intelligence officials, foreign governments, military agencies, and multilateral institutions to avert proliferation. **Proliferation** is the acquisition, sharing, and spread of nuclear weapons and materials to those who do not currently possess them. The **Non-Proliferation Treaty (NPT)** is an international regime designed to prevent nuclear states from giving nuclear weapons or materials to those who do not possess them. It was initiated in 1968 due to the Cold War and has been signed by over 180 governments around the world. It is administered by the **International Atomic Energy Agency (IAEA)**.

Unfortunately, not everyone has signed the NPT, and others may not adhere to IAEA requirements. As an example, Iran has disregarded the wishes of the NPT repeatedly and hindered the activities of the IAEA and others. Furthermore, terrorist organizations are not signatories of the treaty because they are not nation-states. They may seek to develop nuclear weapons outside of international law. Therefore, the loopholes in the NPT and the desire of others to acquire nuclear materials and technology limit 100% compliance. In addition, one radical Saudi cleric states that it would be morally permissible to use a nuclear bomb against the United States to stop American actions against Muslims (Bunn and Wier 2008, p. 125).

Bunn and Wier's chapter in *Weapons of Mass Destruction and Terrorism* dispels seven myths pertaining to nuclear terrorism (2008):

- Some terrorists do want to use nuclear weapons against the United States and others. Osama bin Laden declared this to be a major priority before he was killed.

- Acquiring nuclear material is not impossible. Bribery and theft are only two of the many ways terrorists may acquire highly enriched uranium or plutonium.

- Building a nuclear weapon is not improbable. The U.S. Office of Technology Assessment asserts that a small group of individuals could build such a device if they were sufficiently committed.

- Setting off a nuclear weapon is not as unlikely as is commonly believed. Some of the older weapons in the possession of Russia do not have permissive action links (electronic locks requiring coded access).

- State sponsorship in obtaining nuclear weapons is not a requirement. As long as nuclear materials can be acquired, individuals and groups will be able to manufacture nuclear weapons alone.

- Smuggling a nuclear device into the United States is likely to be achieved. It would be very difficult for border officials to guarantee that a nuclear weapon or its component parts could be shipped and assembled because of the vast quantity of goods that enter our nation each day through legal and illegal means.

- A strong military offensive cannot always stop terrorists from using nuclear weapons. Counterterrorism operations have not prevented terrorist attacks in the past, and there is no reason to believe they will be 100% successful in the future.

For these and other reasons, it is wise to prepare for a terrorist attack involving a nuclear weapon. Getting ready for this scenario is likely to be undertaken especially by the military and medical communities. Personnel in the armed forces have special training on nuclear weapons in addition to personnel protective gear and radiological monitoring devices. Doctors, nurses, and paramedics will also be needed to care for the injured. Those exposed to radiation will lose hair and experience the destruction of the cells that are vital for the brain, heart, and other internal organs. There is no way to reverse the effects of this type of radiation exposure, although symptoms can be relieved through various medical treatments.

Because of the dangers of radiation, government leaders will also need to plan ways to evacuate hundreds of thousands of people from the impacted area. Such an evacuation could take place if a credible threat is issued or after a nuclear detonation has occurred. Doing so would require detailed and comprehensive evacuation plans that take into account all exit routes and all methods of transportation. Contraflow strategies with the use of countless law enforcement personnel would be needed along with others to monitor vehicle breakdowns and fuel shortages.

In The Real World

Future Possibilities of Nuclear Terrorism

Research reveals differing opinions about the future possibility of nuclear terrorism (Maerli, Schaper, and Barnaby 1998). Some assume that terrorists will attack a U.S. city as soon as they are able to acquire nuclear weapons. Others assert that other types of WMD are more likely to be employed due to the technical requirements associated with developing nuclear weapons. While the risks of nuclear terrorism may be low, the consequences should warn us about complacency or a false sense of security. If terrorists cannot obtain nuclear weapons from supporting governments, they may be able to make crude weapons such as those that destroyed Hiroshima and Nagasaki during World War II. In addition, one nuclear physicist warns if terrorists are able to obtain nuclear material, "even a high school kid could make a bomb in short order" (Alvarez in Maerli, Schaper, and Barnaby 1998, p. 115). While this may or may not be true, one additional concern about nuclear weapons is an electromagnetic pulse. An electromagnetic pulse is a brief burst of energy that may have devastating consequences. Detonating a nuclear weapon high above the United States in the atmosphere may damage all types of electronic equipment. Power-generating stations, water systems, industrial plants, vehicles, and computers may cease to function. The resulting disruption could result in hundreds of thousands or millions of deaths and would take years and decades to recover.

However, many vehicles will be rendered useless due to the electromagnetic pulse associated with nuclear weapons.

Any mass evacuation will also necessitate large quantities of sheltering and housing. Hurricane Katrina resulted in one of largest evacuations in U.S. history. The destruction of New Orleans and surrounding cities caused over one million people to migrate to other locations. Every state received evacuees. This led to the need for short-term shelters operated by organizations like the American Red Cross. The integration of evacuees into long-term residential housing will also be required after terrorist attacks, which will lead to severe shortages in the housing market and place significant burdens on the construction industry. This is to say nothing about the need to help evacuees find schools, obtain jobs, and gain a sense of belonging in their new community. Evacuation is a short-term action, but it has long-term consequences for those involved.

Self-Check

1. The only threat of nuclear terrorism comes from missiles and rockets. True or false?
2. Pakistan currently possesses a nuclear weapon. True or false?
3. Which of the following is not associated with nuclear weapons?
    a. Radiation sickness
    b. Vaporization of people in the immediate detonation area
    c. Immunity through vaccination programs
    d. Injuries due to fires and flying debris
4. What is the Non-Proliferation Treaty?

14.4 Biological Weapons

Another major concern among those in homeland security is biological weapons (Guillemin 2005). **Biological weapons** are living organisms—or agents produced by living organisms—that may be used in terrorist attacks. There are three types of biological agents that can be converted into WMDs:

- **Bacteria** are single-cell organisms that cause disease in plants, animals, and humans. Anthrax is an example of such pathogens that can be reproduced on their own.

- A **virus** is a microscopic genetic particle that infects the cells of living organisms. It lives inside a host cell and cannot multiply outside of this location. Ebola, smallpox, and AIDS are well-known viruses.

- A **toxin** is a poison that is produced by plants or animals. They vary in strength but can destroy blood cells, tissues, and the central nervous system. Ricin is frequently mentioned as a toxin.

Developing and using biological weapons in attacks is ironically straightforward and difficult at the same time (Falkenrath, Newman, and Thayer 1998). On the one hand, converting biological agents into weapons is relatively simple. A seed stock of the agent can be acquired and then it must be produced in bulk. Information to accomplish this objective is widely available in the scientific literature and within the medical community. Equipment to reach this goal is also present in many locations because of numerous research or commercial enterprises engaged in bioengineering. On the other hand, it is admittedly more challenging to store, transport, and employ biological weapons. For instance, there are significant technical hurdles inhibiting the development of aerosol dissemination, unless the perpetrator is willing to utilize less effective methods for spreading the biological agent.

In The Real World

The 2001 Anthrax Crisis

Shortly after the 9/11 attacks, Bruce Edwards Ivins sent four envelopes containing powdered anthrax spores from Trenton, New Jersey (Thomas 2003). The first victim was a photo editor who worked with the National Enquirer. Over the next several weeks, 21 others became ill. Four of these individuals succumbed to the effects of inhaling anthrax and died. Because antibiotics can stop infection and treat anthrax disease, the government implemented an aggressive prophylactic treatment. Thirty-two thousand people were given antibiotics for 10 days. Of these, 10,000 were also given a 60-day treatment. Although it was impossible to tell who was infected, public health officials and the Center for Disease Control (CDC) believed that this measure may have saved countless lives. As the emergency period ended, the focus shifted to other issues. The post office and other buildings had to be cleaned of any remaining anthrax spores. These facilities had to be shut down for an extended period of time until they were believed to be free of any remaining biological agent. The Federal Bureau of Investigation (FBI) initiated a massive manhunt to determine who was involved in the attacks. Clues eventually led the FBI to Ivins. The anthrax crisis in 2001 illustrates the deadly and disruptive potential of bioterrorism along with the difficulty of prosecution.

The Impacts of biological weapons will vary dramatically, depending on the type of agent used and the method of distribution (see table below). **Category A agents** pose a serious risk to people because they are easily transmitted to others, and due to the fact that they result in high mortality rates. Smallpox is an example of these types of biological weapons. **Category B agents** have a moderate chance of contagion and generally result in lower morbidity rates than category A agents. Typhus fever is an example of category B agents. **Category C agents** could theoretically be used for mass dissemination and high morbidity if engineered for that purpose. An example of this type of agent is the hantavirus.

	Category A agents	Category B agents	Category C agents
Example	Plague	Viral encephalitis	West Nile virus
Transmissibility	High	Moderate	Low
Mortality rates	High	Moderate	Low

The effects of **bioterrorism**—terrorism that employs biological weapons—will also be dependent on whether it is absorbed through cracks in the skin, ingestion, or inhalation. Depending on the agent used, extent of exposure, and health of the victim, symptoms may appear quickly or over time. Signs of infection may include fever, respiratory distress, vomiting and diarrhea, painful lesions, blackened fingers and toes, shock, paralysis, and death in certain cases. Ricin is one example of a biological agent. It is a toxin that is produced by castor bean plants and can be very deadly. While ricin may kill only those who have been directly exposed to it, other types of biological weapons could kill countless people. Respiration (e.g., coughing) and physical contact (e.g., touching pustules) are factors in contagion. The ease of travel today may also spread diseases around the nation and world within days. In 2007, an infected passenger traveled from Europe to the United States even though he was known to have been diagnosed with tuberculosis. In another case, severe acute respiratory syndrome spread rapidly in Canada and Hong Kong as people traveled within cities or to distant locations.

It is vital to recognize that humans may not be the only target of biological attacks, however. David Franz, the director of the National Agricultural Biosecurity Center, notes that hoof-and-mouth disease could easily be used in a terrorist attack (1998, p. 190). It is readily available in the natural environment, it does not pose a risk to humans and it would easily spread through the livestock population. This brings up the concept of agroterrorism. **Agroterrorism**, or terrorism against farming industries and products, must therefore be taken into consideration by those involved in homeland security. Agroterrorism could limit or severely disrupt the supply of food. Death and disease could result if terrorists employed this type of tactic.

Preventing bioterrorism or agroterrorism is similar to efforts to inhibit terrorism involving radiological or nuclear weapons. That is to say, the proliferation of biological agents among states and nonstate actors should be strongly opposed by the international community. The **Biological Weapons Convention (BWC)** has this purpose, but it also suffers from some of the similar weaknesses of the NPT. As an example, Iraq successfully hid its anthrax program from the United Nations until after the first Gulf War. For this reason, George W. Bush's administration observed that the BWC was "inherently unverifiable" in 2001.

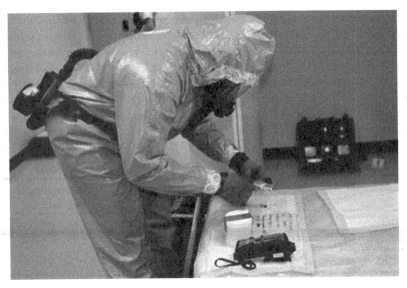

FIGURE 14-4 Deadly diseases can be acquired by terrorists from bio-medical labs.
Source: © National Guard.

Minimizing the threat of bioterrorism will likewise require the close monitoring of biological agents in other ways. Efforts, including the security of facilities and verification of legitimate use, should be taken to limit access to qualified medical personnel and researchers only (see Figure 14-4). Protecting air intakes of buildings and using high-efficiency particulate air filters in large occupancy buildings are other methods to minimize the possible use or impact of bioterrorism. Recurring evaluation of herbs and crops should also be a top priority of those trying to prevent agroterrorism. Inspection personnel in the agricultural and ranching sectors must maintain situational awareness at all times. If cattle and agricultural produce appear to have been tainted in any way, they should be reported to the proper authorities immediately. This may include the U.S. Department of Agriculture, the FBI, or the Department of Homeland Security.

Effective responses to a bioterrorism attack will require a great deal of collaboration among doctors, nurses, public health officials, emergency managers, the National Disaster Medical System (NDMS), and the CDC. It will also necessitate quick distribution of medicines through the strategic national stockpile to state and local governments. The **strategic national stockpile (SNS)** is a cache of medicines in secret locations that can be quickly sent to affected locations around the country. This includes 14 push-pack components such as antibiotics that can easily fill a 747 aircraft. Medicines can also be obtained from the vendor-managed inventory, which includes drugs manufactured and distributed by large pharmaceutical companies. The goal is to break these pallets of supplies down and send them to the **points of distribution (PODs)** as soon as possible. PODs are locations where medicines may be given to victims. PODs may be set up at government buildings, schools, churches, or any location that can handle large numbers of people. Depending on the type of agent, victims may need to be given appropriate shots or pills within as little as 48 hours.

In some cases, vaccines can be administered to prevent the agent from negatively affecting victims. For instance, smallpox vaccinations have been given to soldiers, medical personnel, and first responders around the nation. If a biological attack does take place, the general population may also be given necessary antibiotics, respiratory treatments, or other forms of prophylaxis. This obviously assumes that the biological agent is identified early enough and that medicines are readily available to be dispersed. Unfortunately, it may take days or weeks to ascertain a bioterrorist attack, and some medicines can only be created once the proper strain of the pathogen has been identified.

Planning activities must likewise take into consideration the important process of resource distribution. Preparedness measures must also involve the public. For example, the washing of hands is the single most effective way of reducing the spread of disease. Steps must be taken to isolate infected people and animals from others. Home nursing care, work-at-home programs, and even quarantines may need to be implemented after a biological terrorist attack. Unfortunately, we lack sufficient information on the best methods and effectiveness of quarantines. The last time they were used extensively was during the Spanish flu outbreak in the early 1900s. More studies on how to deal with bioterrorism will be required in the future.

In The Real World

Terrorist Attacks Versus Other Types of Disasters

Responding to a terrorist attack involving WMD is similar, in some ways, to other types of disasters. All events may require emergency medical care, media relations, and donations management. However, there are also some significant differences between WMD events and natural disasters. For instance, there may be no warning for a terrorist attack, whereas some events like hurricanes or tornadoes provide advanced notification. Debris removal should be handled with care after a terrorist attack since destroyed buildings may contain evidence and deadly contaminants. There may be a greater demand for crisis counseling after a terrorist attack because it has been intentionally caused by humans. It is also possible that terrorist attacks involving biological weapons could kill millions of people. People may also overrun the capacity of hospitals because they do not know if they have been affected or not. It is important to be able to compare and contrast the impact of terrorist attacks and other types of disasters so your actions will be appropriate for the situation at hand.

Self-Check

1. Health impacts of biological weapons may include fever, vomiting, paralysis, and death. True or false?

2. A category B agent is more deadly than a category A agent. True or false?

3. Which type of biological weapon is a poison produced by plants or animals?

   a. Bacteria

   b. Toxin

   c. Virus

   d. Category C agents

4. What is the strategic national stockpile?

14.5 Chemical Weapons

Terrorists may also attempt to access and use chemical weapons in the future (see table). Lethal human-made poisons that can be disseminated as gases, liquids, or aerosols are known as **chemical weapons** (Falkenrath, Newman, and Thayer 1998, p. 17). Chemical weapons are divided into five categories:

- **Nerve agents** prevent the transmission of electrical signals in the nervous system. An example of a nerve agent is soman.
- **Vesicants** are blister agents that produce chemical burns on the body. Lewisite is a type of vesicant.
- **Blood agents** prevent the flow of oxygen in the blood. Cyanide is an example of blood agents.
- **Choking agents** inhibit the pulmonary system. Phosgene falls into the category of choking agents.
- **Irritants** are agents that lead to allergic reactions (e.g. tearing, runny nose, or respiratory distress). Mace is a common example of an irritant.

	Example	Impacts
Nerve agents	Sarin	Convulsion of muscles
Vesicants	Mustard gas	Burns/blisters on the skin
Blood agents	Cyanide	Depletion of oxygenated blood
Choking agents	Chlorine	Suffocation
Irritants	Pepper spray	Allergic reactions

Many countries used chemical weapons during World War I, and the consequences were devastating to soldiers on the battlefield. Today, many (but not all) countries have eliminated their stockpiles to avoid aggravating an arms race in this area. However, these same nations could produce chemical weapons at a moment's notice. Others, including Iran and Syria, are believed to be current possessors of chemical weapons. Thus, it is possible that terrorists could obtain chemical weapons from governments around the world. Theft and bribery are, again, possible ways for terrorists to obtain chemical weapons.

Virtually any individual or organization could produce chemical weapons since their precursor materials are available from legitimate commercial suppliers (Falkenrath, Newman, and Thayer 1998, p. 103). It is estimated that the United States produces 300,000 metric tons of cyanide to be used in electroplating, dyeing, printing, and the production of plastics (Tucker 2008, p. 217). Ordinary household chemicals or supplies from mail-order companies could also be used to acquire chemical weapons.

For instance, terrorist organizations like Aum Shinrikyo have developed and utilized chemical weapons in the past (see Figure 14-5). Aum Shinrikyo was a quasi-Buddhist sect composed of disillusioned intellectuals from Japan and Russia during the late 1980s (Tucker 2008, p. 214). Led by Shoko Asahara, this terrorist organization was able to acquire a net worth of $1 billion through legitimate and illegitimate practices. With this money, the terrorist group desired to instigate a major war between the United States and Japan by using chemical weapons. Scientists were subsequently hired to develop this type of weaponry within the organization's

FIGURE 14-5   The Tokyo Subway is one of the world's busiest commuter transport systems, and the Sarin gas attack was the deadliest attack against it to occur in Japan since the end of World War II. *Source: © Flickr/Dick Thomas Johnson.*

"Ministry of Science and Technology." In time, these technicians were successful in producing Sarin, which was later tested on sheep on a remote ranch in Australia. In the mid-1990s, Aum Shinrikyo then used Sarin to kill three judges in Matsumoto. Seven others were later killed with the use of aerosolized Sarin during an attack on a subway in Tokyo. Fortunately, the Sarin gas in this latter attack was diluted and not as potent as it could have been. That attack brought more attention to the use of chemical weapons by terrorist organizations.

Chemical weapons may be composed of conventional explosives that disperse hazardous materials into the atmosphere (e.g., a bomb that disperses chlorine). Alternatively, chemical weapons may be employed by simply releasing chemicals that are normally stored in protective containers (e.g., opening up a valve on a chlorine tanker). Opening a lid and using a spray device to impact others are various means to distribute chemical agents. For instance, chemical weapons could be introduced into a heating, ventilation, and air conditioning system inside a building. Product tampering, injection with a syringe, and the sending of letters or packages with powdery substances are other ways to deliver chemical weapons. Jonathan Tucker, a senior fellow at the Center for Nonproliferation Studies, provides a list of three more types of chemical weaponry:

- Distribution of military-grade chemical warfare agents into the air.
- Use of toxic agents to contaminate water or food supplies.
- Assassination of a specific individual or group with chemical agents (2008, p. 213).

In conjunction with the type of attack, the impact of chemical agents could vary significantly. Some chemicals are persistent (meaning that they evaporate slowly), while others dissipate quickly. Certain chemical agents may lead to minor medical problems, while others lead to immediate death. Depending on the type of chemical agent used and environmental conditions (i.e., temperature, humidity, and wind speed), physical results could include physical annoyance, long-term respiratory or central nervous system damage, and widespread mortality. Environmental problems are also likely consequences of chemical weapons. Water and soil can remain contaminated long after a chemical agent has been used in a particular geographic area.

Much like radiological, nuclear, or biological weapons, the development of chemical weapons can be minimized by carefully controlling the manufacturing, use, transportation, and storage of hazardous materials. Suppliers can also ensure that those requesting chemicals have legitimate and peaceful purposes. Nevertheless, it could be impossible to prevent the acquisition of chemical weapons. Materials and knowledge to make them are omnipresent. Known state possessors could also give them to terrorist organizations. Even industry and transportation methods can be used against us. A bomb could release dangerous chemicals at a manufacturing plant. A tanker truck or railcar could also be penetrated by diverse weapons, thereby releasing toxic materials in the air near a school, business district, or residential area.

The possibility of a chemical weapon attack suggests that much more attention should be given to detection, decontamination, emergency medical response, and environmental reclamation. Planning with the Metropolitan Medical Response System, the Department of Health and Human Services, and the NDMS needs to be improved. More paramedics, doctors, and nurses will be required to treat the victims of chemical weapons. Training of these individuals should focus on ways to identify chemical weapons and respond effectively. For instance, those involved in homeland security should be aware that the FBI has labs to help you identify what agent is being used. In addition, atropine may need to be administered quickly to victims if their lives are to be spared. Along these lines, military personnel and some paramedics may have Mark I autoinjector kits for this purpose. Mark I is a spring-loaded injector that has been developed by the military. It contains atropine and oxime, which can help treat patients affected by nerve agents. Mark I kits and other drugs may only help to save those with moderate exposure, however. Also, such medicines will certainly be lacking in a major attack. For instance, only

For Example

Chemical Terrorism

Between 1960 and 2001, there were 125 incidents of terrorists using chemical weapons (Tucker 2008). Most of these were small in scope and were committed by a wide range of terrorists (e.g., nationalists, religious extremists, antiabortion or environmental groups, left-wing and right-wing organizations). However, it is anticipated that the number of chemical weapons attacks will rise in the future. One case in Russia illustrates how easy it is for scientists to develop chemical weapons. A man named Valery Borzov was fired from the Moscow Scientific Research Institute of Reagents in 1997. To maintain an income, the 40-year-old chemist started to develop blister agents in his own clandestine laboratory. His desire was to sell vials of the poisonous substance for $1,500 each to the Russian mafia or other criminal organizations. When arrested, investigators found recipe books, equipment, and dangerous products in his home. Borzov was found incompetent to stand trial, and he was committed to a mental institution. Police were fortunate to interdict his efforts to make money at the expense of others' lives.

29% of hospitals that responded to a survey had sufficient medicines to care for 50 victims (Tucker 2008, p. 222). Some larger jurisdictions do have CHEMPACK antidotes strategically placed across the city, but it is unclear if funding is adequate to replace them when the drugs reach their expiration date. Organizational resources, including the Disaster Medical Assistance Teams and environmental restoration companies, will need to be deployed quickly after a chemical attack takes place. Chemical weapons, along with other types of WMDs, are all likely to be used by terrorists in the future. It will be crucial that homeland security personnel take measures now to prevent their use or be ready to respond if that cannot be averted.

Self-Check

1. Chemical weapons are disseminated as liquids only. True or false?
2. Responding to chemical weapons may involve decontamination, emergency medical response, and environmental reclamation. True or false?
3. Which type of agent prevents the transfer of electrical signals in the body?
   a. Blood agents
   b. Vesicants
   c. Irritants
   d. Nerve agents
4. Why are attacks involving chemical weapons increasingly likely in the future?

14.6 Cyberterrorism

The WMDs mentioned earlier have received a lot of attention from those working in homeland security over prior decades and in recent years, and for good reason. However, another threat—which has often been underestimated—is cyberterrorism (CEDAT 2008). Cyberterrorism, which is also known as electronic terrorism, is not the same thing as cybercrime or hacking. Cybercrime has the goal of economic gain, while hacking could be equated to electronic vandalism. In contrast to cybercrime or hacking, **cyberterrorism** is defined by the FBI as "premeditated, politically motivated attack against information, computer systems, computer programs, and data which results in violence against non-combatant targets by subnational groups or clandestine agents" (Centre of Excellence Defence Against Terrorism 2008, p. 119). Cyberterrorism is sometimes confused with cyberwarfare in that both nonstate and state actors target cyberspace systems that are composed of computers, networks, servers, routers, fiber optics, and so on. However, cyberterrorists use computer viruses, worms, malware, coded programs, and technology in such a way as to impact people, the infrastructure, and the economy to achieve ideological goals (see Figure 14-6).

Barry Collin, who coined the term cyberterrorism in the 1980s, stated that cyberterrorism occurs at the intersection of the physical and virtual worlds. Along these lines, Pluschinsky suggests that "a cyber-attack involves digitally targeting a computer information system to destroy, damage, or steal data and thereby disrupt or disable telecommunications, health, transportation, finances, utilities, food distribution, and

FIGURE 14-6 A computer may look like an innocent piece of technology, but it can be a very disruptive tool in the hands of a terrorist. *Source: Federal Bureau of Investigation (FBI).*

other critical infrastructure systems. The primary target of a cyber-attack is information, but such attacks can ultimately cause casualties, depending on the system targeted" (Pluschinsky 2006, p. 367).

Today, almost everyone—reporters, researchers, practitioners, and politicians—fears the threat of cyberterrorism. The concern is that terrorists could gain access to computers via networks and disrupt vital services including dams, power generating stations, air traffic control systems, emergency services, and even the self-driving cars of the future. Thomas Friedman painted such a picture. He said, "If someone is able to knock out a handful of key Internet switching and addressing centers in the US, ... here's what happens: many trains will stop running, much air traffic will grind to a halt, power supplies will not be able to shift from one region to another, there will be no e-mail, your doctor's CAT scanner ... won't work" (Nacos 2006, p. 238). Such access and impacts are not farfetched. Students, hackers, and amateur computer programmers have entered secure sites relating to banking, the military, and NASA. Disruption of our critical systems is a very real possibility. Even death could be the consequence of cyberterrorism if critical infrastructure is impacted. Perhaps for this reason, Michael Chertoff noted: "Of the many challenges facing [us in] ... the twenty-first century, one of the most complex and potentially consequential is the threat of large-scale cyberattack against shared information technology and cyber infrastructure, including the Internet" (2009, p. 95).

Examples of cyberterrorism, although not always well known to those outside of the cybersecurity field, are notable. For instance, during the 1990s, Iraq set up over 100 websites around the world. Their goal was to launch denial-of-service attacks against companies from the United States. In Japan, Aum Shinrikyo was able to access the computers of the Metropolitan Police Department, and they developed the capability to track 150 police vehicles. After 9/11, the armed services obtained Al-Qaeda laptops in Afghanistan that revealed significant research on computer systems relating to water systems, nuclear power plants, and sports stadiums. In 2015 and thereafter, France was hit by more than 19,000 cyberattacks after Islamic terrorists attacked the Charlie Hebdo news organization, killing 12 and injuring 11 others. ISIS has similarly attempted to launch cyberattacks on computers that control the U.S. electric power grid.

In The Real World

The Colonial Pipeline attack

The most well-known and largest cyberattack in recent years is the Colonial Pipeline incident. On May 7, 2021, DarkSide hackers—perhaps with Russian knowledge and support—took control of computers that operated the Colonial Pipeline. This pipeline system carries 55% of the gasoline, diesel, and jet fuel used by the northeastern portion of the United States from Texas. The hackers disabled the payment system and demanded 75 bitcoin ($4.4 million) before access to the computer billing programs would be given back to the operators of the Colonial Pipeline. Although the ransom was given to the hackers by the company, the Department of Justice was able to recover much of the payment. However, this did not occur before the disruption to people and the economy could be prevented. Fuel shortages were witnessed at filling stations in Alabama, Florida, Georgia, North Carolina, South Carolina, and Virginia. Some airports and airlines had to temporarily alter or halt their operations. Fuel prices rose and economic losses were felt in this part of the nation and elsewhere. As a result of the incident, President Biden declared a state of emergency as did at least one governor. President Biden also signed Executive Order 14028 on May 12 to increase the security of computers and software for companies that operate this type of fuel transmission system. The Colonial Pipeline attack illustrated the vulnerability of our computer systems and the significant implications of threats to our critical infrastructure.

The appeal of cyberterrorism is owing to several factors. According to Gabriel Weimann (2004), the author of "Cyberterrorism: How Real is the Threat?" cyberterrorism is cheaper than other methods of terrorism because all that is required is a computer and an Internet connection. Cyberterrorism occurs from a distance and also allows anonymity, which could limit apprehension and prosecution. Finally, cyberterrorism expands available targets and has the potential to affect millions of people. For this reason, we must take serious efforts to protect ourselves against cyberterrorism.

CAREER OPPORTUNITY | Cybersecurity Specialists

Our increasing reliance on computer technology, as well as the rise of diverse digital attacks from state and nonstate actors, is leading to an explosion in the demand for cybersecurity specialists.

Cybersecurity specialists—including security analysts, engineers, architects, researchers, policy/compliance experts, and consultants—are experts who identify possible cyber threats, implement measures to prevent them, and respond to those that cannot be stopped. They work at all levels of government and for many different types of large corporations and even smaller companies.

Employees working in this profession typically have cybersecurity certifications such as COMPTIA SEC+ or ISACA. They may also have a bachelor's degree in computer science, information technology, information systems, and software engineering. Annual salaries are often more than $110,000 (median wage).

The job outlook for this profession is likely more robust than any other. In fact, the Bureau of Labor Statistics indicates that this field may grow by more than 32% over the next decade and beyond.

For further information, see **https://www.bls.gov/ooh/computer-and-information-technology/information-security-analysts.htm**

14.6.1 Efforts to Reduce Cyberterrorism

Because of the horrifying potential of cyberterrorism, the U.S. government is taking this threat seriously. New laws have been created as necessary including the Computer Fraud and Abuse Act. In addition, October has been designated each year as Cyber Security Awareness Month. Federal agencies are likewise spending $5 billion annually on cyber-defense. The U.S. government has created a National Cybersecurity and Communications Integration Center. This is a 24 × 7 center composed of officials from the federal government, the intelligence communities, and law enforcement officials to monitor the Internet, respond to threats, and manage incidents.

In addition, the government has proposed a **National Strategy to Secure Cyberspace**. This program is under the Information Analysis and Infrastructure Protection Unit and takes into account home users/small businesses, larger enterprises, critical infrastructure, and national and international Internet. The strategy includes the following priorities:

1. **A National Cyberspace Security Response System.** Efforts need to be taken to assess attacks and work with the private sector to share information. Continuity plans must also be created and exercised.

2. **A National Cyberspace Threat and Vulnerability Reduction Program**. The goal is to enhance law enforcement capabilities to prevent attacks and prosecute terrorists. In addition, there is a need to prioritize research and development to secure emerging systems.

3. **A National Cyberspace Security Awareness and Training Program**. This promotes a national awareness program and fosters training.

4. **Securing Governments' Cyberspace**. The federal government must maintain access for authorized users only and secure federal wireless networks.

5. **National Security and International Cyberspace Security Cooperation**. This objective relates to strengthening cyber-counterterrorism intelligence. It also entails interaction with international organizations to promote global cybersecurity.

As part of the effort to secure cyberspace, President Trump signed into law the Cybersecurity and Infrastructure Security Agency Act on November 16, 2018. This required the federal government to elevate the status of the National Protection and Programs Directorate through the creation of the Cybersecurity & Infrastructure Agency (CISA). The **Cybersecurity & Infrastructure Agency** is a unit within the Department of Homeland Security that is responsible for promoting cyber security in the United States. The mission of this organization is to "lead the national efforts to understand, manage, and reduce risk to cyber and physical infrastructure." CISA personnel consequently work closely with various government departments and industry partners to assess threats, improve training, provide resources, and share tools to enhance cyber security and ensure resilient infrastructure for the future. More information about CISA can be accessed at **https://www.cisa.gov**.

While the National Strategy to Secure Cyberspace and CISA are government-initiated programs and organizations, this should not imply that the average person has no role in cyberterrorism prevention. Robert Jorgensen, a former Director of Cybersecurity at Utah Valley University, asserted that cybersecurity requires a "cyber citizen." In other words, people must recognize that their actions at home or work have a big impact on national vulnerability. Simple efforts to protect passwords and avoid phishing scams are basic steps to limit cyberterrorism in the future.

In addition, the private sector can be a valuable asset in efforts to reduce cyberterrorism. For instance, the National Cyber Security Alliance is a partnership that was created after 9/11. Its goal is to spread information about what people, businesses, and the government can do to limit cyberterrorism. Everyone needs to know how to protect themselves online. Personal responsibility is imperative in the view of this organization. Their motto for the country is "Stop. Think. Connect." There is also a need to engage emergency management to prepare for possible cyberattacks against government organizations and community infrastructure (McEntire 2022).

In The Real World

Examples of Cyberterrorism

If a terrorist can gain access to computers and computer networks, they may be able to wreak havoc on individuals, businesses, and government organizations. Several types of attacks are possible, and a thorough list of these categories can be accessed at **https://www.simplilearn.com/tutorials/cyber-security-tutorial/types-of-cyber-attacks**. Some terrorists may infect computers with malware—whether that be a virus or a worm. A **virus** may overload a single computer through replication or may spread to other computers and cause damage through other means. A **worm** is similar to a virus, but it is capable of reproducing itself over many computer systems. Some attacks are **Trojan horses**, which use what appears to be legitimate software to conceal malware of various types.

Certain cyberterrorism activities may produce what is known as denial of service. A **denial of service** refers to a type of attack where a server is overwhelmed with illegitimate requests to the point where legitimate users cannot access the service.

There are numerous examples of these activities. In 1996, a Swedish hacker tied up phone lines in 11 counties, and this stopped 911 systems for a time. In 1998, two high school students hacked into the Pentagon, NASA, and nuclear sites. These students routed the attack through the United Arab Emirates. If students can create such consequences, it will only be a matter of time before terrorists engage in these types of activities.

Self-Check

1. Cyberterrorism is expected to decline in frequency in the future. True or false?
2. A denial-of-service attack overloads computers with illegitimate requests to the point where legitimate users cannot access the service. True or false?
3. What is another name for cyberterrorism?
   a. Electronic terrorism
   b. Network terrorism
   c. Phishing terrorism
   d. Fiber-optic terrorism
4. Explain the National Strategy to Secure Cyberspace.

Summary

As a participant in homeland security, it is imperative that you prepare for a future that will likely include WMDs. It will be vital for you to comprehend the impacts of dirty bombs and RDDs. You must recognize the possible devastation that may result if nuclear weapons are acquired and utilized by terrorists. Preparing for biological or chemical weapons should be one of your top responsibilities. It will also be necessary that you consider the possibility of cyberterrorism in the future and prepare accordingly. Working with key partners in each of these areas will be essential. Without anticipating future attacks like these and taking preventive and preparatory measures, you will not be successful in your efforts to promote homeland security.

Assess Your Understanding

Understand: What Have You Learned?

Go to **www.wiley.com/go/mcentire/homelandsecurity3e** to assess your knowledge of significant threats.

Summary Questions

1. Terrorists do not intend to use WMD. True or false?
2. Theft of radiological material has been a problem in recent years. True or false?
3. The IAEA is not responsible for the NPT. True or false?
4. The nature and extent of impact from biological weapons are dependent on the type of agent used as well as the health of the victim and the extent of exposure. True or false?
5. Irritants are the most deadly type of chemical agent. True or false?
6. The creation of new response plans should take into account complicated responses to terrorist attacks and other disasters in the nation. True or false?
7. Agroterrorism has no relation to biological agents. True or false?
8. The IAEA is in charge of preventing the spread of chemical weapons. True or false?
9. CISA stands for the Chemical Incident Standards Act. True or false?
10. Radiological agents:
    a. Can be obtained from hospitals, industrial facilities, and the black market
    b. Are not used in dirty bombs
    c. Never pose health threats to humans
    d. Do not necessitate the involvement of the military or expert medical physicians

11. Which of the following causes the least concern among terrorism experts?

    a. A suitcase bomb involving conventional explosives

    b. A declared nuclear possessor giving nuclear weapons or nuclear materials to a terrorist organization

    c. Iran developing nuclear weapons

    d. The United States losing one of its nuclear weapons

12. Which type of biological agent is composed of a single-cell organism that causes diseases in plants, animals, and humans?

    a. Bacteria

    b. Virus

    c. Toxin

    d. Category A agents

13. Which of the following helps to prevent the spread of biological weapons?

    a. IAEA

    b. NPT

    c. BWC

    d. NTW

14. Which type of chemical weapon damages the pulmonary system?

    a. Nerve agents

    b. Vesicants

    c. Blood agents

    d. Choking agents

15. Which of the following is malware that appears to be legitimate software?

    a. Worm

    b. Virus

    c. Trojan horse

    d. Denial of Service

Applying This Chapter

1. While serving as an intelligence analyst for the CIA, you have been asked to testify in a closed session with congressional representatives. A Senator from California wonders why the United States government should be concerned about WMDs today (when the Cold War ended a long time ago). What would you say?

2. During dinner, you tell your family what you learned about radiological weapons from a renowned expert who spoke in your class. They ask you what radiological weapons are. What would you tell them?

3. An official from the State Department expresses concern about the potential for a nuclear attack from a terrorist organization. During the press briefing, a reporter wonders how serious the threat really is. What should the State Department official say?

4. In your role as an analyst for the Department of Defense, your supervisor has requested a report about the types of chemical weapons. What would you include in your report?

5. In your capacity as an employee of the CISA, you are responsible for helping businesses and citizens limit cybersecurity attacks. What are some key messages that you would share with them to protect vital computer systems?

Be a Homeland Security Professional

Nuclear Proliferation

Write a paper on the proliferation of nuclear weapons and how it relates to terrorism. Be sure to explain who is responsible for halting the spread of nuclear weapons around the world.

Bioterrorism

As a seasoned analyst in the Center for Disease Control, you have been asked to identify the risk of bioterrorism. Be sure to explain why bioterrorism is both likely and unlikely.

Cyberterrorism

Imagine that you have been asked to write a report on cyberterrorism. Can you explain what it is? Why is cyberterrorism a concern today? What can be and is being done to prevent this type of attack?

Key Terms

Agroterrorism   Terrorism against farming industries and products
Bacteria   A single-cell organism that causes disease in plants, animals, and humans
Biological weapons   Living organisms or agents produced by living organisms that may be used in terrorist attacks
Biological Weapons Convention (BWC)   An international treaty designed to prevent the proliferation of biological agents around the world
Bioterrorism   Terrorism that employs biological weapons
Blood agents   Chemical weapons that prevent the flow of oxygen in the blood
Category A agents   Biological weapons that pose a serious risk to people because they are easily transmitted to others, and they result in high mortality rates
Category B agents   Biological weapons that have a moderate chance of contagion and generally result in lower morbidity rates than category A agents
Category C agents   Biological weapons that could be used for mass dissemination and high morbidity if engineered for that purpose
Chemical weapons   Lethal human-made poisons that can be disseminated as gases, liquids, or aerosols
Choking agents   Chemical weapons that cause respiratory distress
Civil Support Teams   Specialized military units that assist local and state governments that have been affected by WMDs
Cyberterrorism   Premeditated, politically motivated attack against information, computer systems, computer programs, and data that result in violence against noncombatant targets by subnational groups or clandestine agents
Cybersecurity & Infrastructure Agency   A unit within the Department of Homeland Security that is responsible for promoting cyber security in the United States
Denial of service   A type of attack where a server is overwhelmed with illegitimate requests to the point where legitimate users cannot access the service

International Atomic Energy Agency (IAEA)   The international organization responsible for the enforcement of the Non-Proliferation Treaty

Irritants   Agents that lead to allergic reactions

National Strategy to Secure Cyberspace   A component under the Information Analysis and Infrastructure Protection Unit and takes into account home users/ small business, larger enterprises, critical infrastructure, and national and international Internets

Nerve agents   Chemical weapons that prevent the transmission of electrical signals in the nervous system

Non-Proliferation Treaty (NPT)   An international regime designed to prevent nuclear states from giving nuclear weapons or materials to those who do not possess them

Nuclear weapons   Weapons that produce massive explosions due to the release of vast amounts of energy through fission or fusion

Points of distribution (PODs)   Locations where medicines may be given to victims

Proliferation   The acquisition, sharing, and spread of nuclear weapons and materials to those who do not currently possess them

Radiological dispersion devices (RDDs)   See **radiological weapons**

Radiological weapons   Weapons that spread dangerous radiological material but do not result in a nuclear explosion. Radiological weapons are also known as radiological dispersion devices (RDDs) or dirty bombs

Strategic national stockpile (SNS)   A cache of medicines in secret locations that can be quickly sent to affected locations around the nation

Suitcase bomb   Portable nuclear weapons that can be carried or rolled to the target location

Toxin   A poison that is produced by plants or animals

Trojan horses   The use of what appears to be legitimate software to conceal malware of various types

Vesicants   Blister agents that produce chemical burns

Virus (biological)   A microscopic genetic particle that infects the cells of living organisms but cannot multiply outside a host cell

Virus (computer)   Malware that may overload or a single computer through replication or may spread to other computers and cause damage through other means

Weapons of mass destruction (WMDs)   Weaponry that will create major injuries, carnage, destruction, and disruption when utilized

Worm   Malware similar to a virus, but is capable of reproducing itself over many computer systems

References

Bunn, M. and A. Wier. (2008). The seven myths of nuclear terrorism. In: *Weapons of Mass Destruction and Terrorism* (ed. R.D. Howard and J.J.F. Forest), 125–137. New York: McGraw-Hill.

Center of Excellence Defense Against Terrorism (CEDAT). (2008). *Responses to Cyber Terrorism*. NATO Security Through Science Series. Amsterdam: IOS Press.

Chertoff, M. (2009). *Homeland Security*. Philadelphia, PA: University of Pennsylvania Press.

Dunn, L.A. (2008). Can Al Qaeda be deterred from using nuclear weapons? In: *Weapons of Mass Destruction and Terrorism* (ed. R.D. Howard and J.J.F. Forest), 295–316. New York: McGraw-Hill.

Falkenrath, R.A., Newman, R.D., and Thayer, B.A. (1998). *America's Achilles' Heel: Nuclear, Biological and Chemical Terrorism and Covert Attack*. Cambridge, MA: MIT Press.

FEMA. (1999). *Emergency Response to Terrorism*. Washington, DC: Independent Study Course.

Ferguson, C.D. and J.O. Lubenau. (2008). Securing US radiological sources. In: *Weapons of Mass Destruction and Terrorism* (ed. R.D. Howard and J.J.F. Forest), 139–166. New York: McGraw-Hill.

Franz, D. (1998). International Biological Warfare Threat in CONU. Posture Statement for the Joint Committee on Judiciary and Intelligence. United States Senate, Second Session, 105th Congress, International Biological Warfare Threats in CONUS (March 4). https://fas.org/irp/congress/1998_hr/s980304-franz.htm

Guillemin, J. (2005). *Biological Weapons: From the Invention of State-Sponsored Programs to Contemporary Bioterrorism*. New York: Columbia University Press.

Gurr, N. and Cole, B. (2000). *The New Face of Terrorism: Threats from Weapons of Mass Destruction*. New York: I.B. Tauri.

Kelley, M. (2014). *Terrorism and the Growing Threat of Weapons of Mass Destruction*. Anchor: Hamburg.

Maerli, M.B., Schaper, A., and Barnaby, F. (1998). The characteristics of nuclear terrorist weapons. In: *Weapons of Mass Destruction and Terrorism* (ed. R.D. Howard and J.J.F. Forest), 100–124. New York: McGraw-Hill.

Mahan, S. and P.L. Griset. (2013). *Terrorism in Perspective*. Thousand Oaks, CA: Sage.

McEntire, D.A. (2022). Cyberattacks and their implications for emergency management. *Journal of Emergency Management* 20 (1): 7–8.

Nacos, B.L. (2006). *Terrorism and Counterterrorism: Understanding Threats and Responses in the Post-9/11 World*. New York: Penguin Academics.

Purpura, P.P. (2007). *Terrorism and Homeland Security: An Introduction with Applications*. Burlington, MA: Butterworth-Heineman.

Thomas, P. (2003). *The Anthrax Attacks. The Century Foundation's Homeland Security Project Working Group on the Public's Need to Know*. New York: The Century Foundation.

Tucker, J.B. (2008). Chemical terrorism: assessing threats and responses. In: *Weapons of Mass Destruction and Terrorism* (ed. R.D. Howard and J.J.F. Forest), 213–226. New York: McGraw-Hill.

Weimann, Gabriel. 2004. Cyberterrorism: How Real Is the Threat? Special Report, May 13. United States Institute of Peace.

CHAPTER 15

Evaluating Other Pressing Problems

Criminal Activity, Social Disturbances, Pandemics, and Climate Change

DO YOU ALREADY KNOW?

- The various categories of criminal activity
- How to define a social disturbance
- What an epidemic is versus a pandemic
- The relation of global warming to climate change

For additional questions to assess your current knowledge of how to evaluate other pressing problems, go to **www.wiley.com/go/mcentire/homelandsecurity3e**

WHAT YOU WILL LEARN

15.1 Reasons why rising crime is of great concern to law enforcement agencies

15.2 The difference between a protest and a riot

15.3 How viruses like COVID-19 can impact physical health and the economy

15.4 Trends in global temperatures and the politics of climate change

WHAT YOU WILL BE ABLE TO DO

- Argue ways to prevent mass shootings, the smuggling of drugs, and human trafficking
- Plan and undertake activities to prepare for possible riots
- Implement measures to address public health emergencies
- Develop recommendations to reverse or stop global warming

Introduction

As this book illustrates, homeland security is mainly known for the attention it gives to preventing terrorism, preparing for and responding to disasters, and promoting safety in various sectors of the economy. Although this focus is warranted, those working within the homeland security profession are increasingly concerned about mass shootings that appear to have increased over the past several decades. There is also a great deal of worry about the rise of general crime along with organized criminal behavior including the drug trade and human trafficking. Events over the past few years have brought attention to social disturbances, riots, and public health emergencies like COVID-19. Global warming and climate change have also become of central priority to those working in homeland security. Each of these issues poses a threat to people's immediate safety, their ongoing well-being, and even their long-term survival. With this in mind, the following chapter explores other threats that are not always front and center when people think about the responsibilities of homeland security. The chapter also discusses measures to deal with these growing problems in an efficacious manner. It illustrates that law enforcement agencies, public health organizations, and environmental specialists play important roles in homeland security.

15.1 Criminal Activity

Most people understand that there are warranted concerns about the prevalence of mass shootings in modern society. There is also a widespread belief that criminal activity is increasing in the United States, and this sentiment seems to be justified. Statistics reveal that general property crime and other types of violent attacks have risen in many locations throughout our nation in recent years. Political leaders and their constituents are now focusing a great deal of attention on organized crime. The production and distribution of illicit drugs along with the disturbing trend in human trafficking are also topics that are of relevance to those working in homeland security.

15.1.1 Mass Shootings

An attack that is committed by a perpetrator with a pistol, shotgun, or rifle which injures and kills multiple individuals is known as a **mass shooting**. These types of events are increasingly viewed as a concern by those involved in homeland security since they have been far too common in the United States for the past 40 years. As an example, on July 18, 1984, a man named James Huberty entered a McDonald's restaurant and began a campaign of "hunting humans." He wounded 19 individuals and killed 21. This was followed by other mass shootings that decade in the workplace and at various shopping centers.

In the 1990s, the number of mass shootings expanded dramatically with incidents in Alabama, Arkansas, Kentucky, and Mississippi. These appalling acts of violence often occurred at schools and churches and perhaps set the precedent for many other notable shootings in the 2000s. For instance, in 2002, John Allen Muhammad and Lee Boyd Malvo engaged in the D.C./Beltway Sniper attacks in which they killed

17 people and injured 10 others over a 10-month period. In 2012, two well-publicized mass shootings took place in Aurora, Colorado, and Newtown, Connecticut. In the first case, a young man named James Holmes gained access to the Century 16 movie theater during the premiere of *The Dark Knight Rises*. He killed 12 people and injured another 58. In the second event, Adam Lanza entered Sandy Hook Elementary School and murdered 20 children and 6 adults.

In The Real World

Columbine and Uvalde School Shootings

After a year of extensive planning and preparations, two teenagers—Eric Harris and Dylan Klebold—entered Columbine High School in Jefferson County, Colorado, to seek revenge on their peers for alleged social mistreatment. These disturbed students placed and detonated several bombs in and around this school and then fired their weapons as they entered the school and walked to the cafeteria and library. When all was said and done, Harris and Klebold killed 12 students and 1 faculty member and injured 24 others before taking their own lives.

Although there was a massive response by police officers, law enforcement agencies were criticized by the community for not neutralizing the threat quickly enough and bringing an end to the violent rampage in a timely manner. To be fair, the situation was very complicated. Bombs hindered entrance into the building, and there was no way for those responding to know if the perpetrators were blending in with the other students who were evacuating the premises. There was also insufficient knowledge about the layout of the building. While SWAT members were able to clear the building over a 90-minute period, this shooting prompted a national review of emergency procedures in these types of situations. The standard protocol since this event is to enter the building immediately and neutralize the threat as soon as possible.

Unfortunately, these principles were not followed on May 24, 2022, when a gunman entered the Robb Elementary School in Uvalde, Texas. Salvador Ramos was in the school for more than one hour before the U.S. Border Patrol Tactical Unit was sent to the scene to stop the perpetrator. Local and state officers had not acted quickly enough, so 19 people were killed and 17 were injured. Law enforcement officials were criticized again for not adhering to what is regarded to be a best practice in these disturbing incidents.

A few years later, a slew of mass shootings were witnessed at the Marjory Stoneman Douglas High School in Parkland, Florida; at the West Freeway Church of Christ in White Settlement, Texas; at a Walmart in El Paso, Texas; at the Molson Coors Beverage Company in Milwaukee, Wisconsin; at an Asian massage parlor in Atlanta, Georgia; and at the King Soopers grocery store in Boulder, Colorado. Taken together, these shootings resulted in the murder of at least 65 people and injured at least 40 more.

At the time of this writing, a mass shooting occurred in Lewiston, Maine. Robert Card, a skilled marksman and outdoorsman, started a violent rampage at 6:56 p.m. on October 25, 2023. He carried an AR-style rifle as he entered various businesses in this jurisdiction. He then shot and killed 18 people at a bar and bowling alley. At least 14 more were injured in the violent episode. Card was able to escape but not before video footage helped to identify him as the perpetrator. Residents of the area were asked to shelter in place while a massive manhunt was underway to find him. Fortunately, the threat ended when Card was found dead of a self-inflected gunshot wound to the head.

As tragic and disturbing as these prior incidents were, none compares to the worst mass shooting in U.S. history. After checking into the Mandalay Bay Hotel in Las Vegas and stockpiling numerous weapons in his room on the 32nd floor, a man named Stephen Paddock started to fire his weapons from the windows into a large crowd that was enjoying the Route 91 Harvest Festival on October 1, 2017. From his elevated perch, he shot over 1,000 rounds, which killed 60 people and injured 411 others.

15.1.1.1 Responding to Mass Shootings

If mass shootings occur, potential victims are told to "run, hide, fight." In other words, people are admonished to leave the area, barricade themselves in a closet or room, and do anything they can to stop the perpetrator. For instance, throwing books, fire extinguishers, or any other item in a school or office setting could distract the shooter enough to allow possible victims to escape, stop the attack, or even gain power over the offender.

Numerous lessons have been gleaned from active shooter incidents over the years (see **https://www.dhs.gov/xlibrary/assets/active_shooter_booklet.pdf**). For instance, if negotiation is not possible, the first law enforcement personnel on the scene are advised to neutralize the threat as soon as possible. When a standoff ensues, heavily equipped SWAT members should be dispatched. Tactical EMS personnel and other emergency medical providers should be deployed to care for the wounded. Finally, the public should be kept informed about what has happened, how to protect themselves, and what the government is doing.

In terms of the Las Vegas shooting, this heinous shooting prompted a massive response of hundreds of officers including 4 SWAT units and 50 strike teams. These law enforcement personnel were able to identify where the shooting was coming from. They ascended to the correct floor in the hotel and breached the room door. While this was occurring, Paddock killed himself. Meanwhile, 15 dispatch personnel, 15 engine companies, 3 private ambulance companies, and 28 communications specialists worked heroically to evacuate and treat the injured and share information with the public.

When the dust settled from this appalling event, the Federal Emergency Management Agency (FEMA) conducted a study of the incident and completed an after-action report in 2018. The report noted the need to improve the performance of response operations in regard to mutual aid, communication with medical providers, and coordination with other responding agencies. Other suggestions pertained to victim services, public information, and responder wellness (FEMA 2018). Further recommendations for improvement include implementing the unified command, curbing self-dispatch, keeping ingress/egress routes clear, setting up staging areas, controlling the location of the media, distributing disposable first-aid kits, and training citizens to respond (Police 1 2019). Of course, all of this is very difficult to accomplish when mass shootings are taking place.

15.1.1.2 Prevention of Mass Shootings

Although numerous suggestions have been given to better respond to mass shootings like the one that occurred in Las Vegas, there is also an urgent need to comprehend the causes of such actions and determine what can be done to prevent them in the first place. In the case of the Route 91 Festival shooting, the reasons that Stephen Paddock perpetrated these unspeakable acts of evil are unclear, but the FBI suspects Paddock may have been angry at how casinos were treating wealthy clients like himself (Kovaleski and Baker 2023).

In light of these revelations, there is a common belief that severe mental illness is a major risk factor in the probability of someone committing a mass shooting. This is not always the full story, however. "Other risk factors, such as a history of legal problems, challenges coping with severe and acute life-stressors, and the epidemic of the combination of nihilism, emptiness, anger, and a desire for notoriety among young men" are also likely explanations (Columbia University Department of Psychiatry 2022).

Additional variables must also be taken into account. Bosses, co-workers, parents, school administrators, and students and teachers have not always been sufficiently engaged with and observant of the behavior of others. Notifications have not always been relayed to proper authorities when justified. The lack of intelligence, failure to follow up on reported concerns, and inadequate security are often believed to be mistakes that allow mass shootings to occur (Koranyi 2012). And, of course, our violent culture and the ease of accessing weapons are commonly conveyed as notable factors in mass shootings. The argument here is that movies, television, and video games that all-too-frequently glorify violence played a role in these horrendous acts as may have been the case in the gun rampage that killed 50 people in two mosques in New Zealand on March 15, 2019 (Afriatni 2019). In addition, many victims, advocacy groups, and politicians have called for further gun control measures (e.g., ban personal weapons altogether, improve background checks, limit certain types of weapons, and reduce the capacity of magazines). These latter recommendations are almost always rejected by certain interest groups (e.g., the National Rifle Association) due to the **Second Amendment**, which is an addendum to the constitution that allows people in the United States to keep and bear arms. Nevertheless, certain measures—like the general prohibition of bump stocks—have been implemented to reduce the likelihood of mass shootings.

15.1.2 Rising Crime and Its Impact

Besides mass shootings, there are many other types of crime that occur, and they each have a bearing on the personal security of individuals and the entire nation. This includes general crimes, misdemeanors, and felonies. **General crimes** include minor infractions and violations (such as jaywalking and motor vehicle offenses). More serious **misdemeanors** include driving under the influence, theft and burglary, or assault. **Major felonies** include rape and murder. Several of these categories of crime have witnessed a dramatic surge in frequency since 2020, and certain jurisdictions have been plagued with these types of problems at overwhelming levels.

For example, statistics for certain cities and states in 2021 and 2022 reveal:

- San Francisco witnessed 7,000 burglaries, 6,050 motor vehicle thefts, and 26,000 other thefts (Property Club Team 2023).
- New York recorded 26,039 violent assaults (Fox 5 Staff 2023).
- California witnessed 18,116 rapes (Wisevoter 2023).
- Chicago had the highest number of homicides (697), followed by Philadelphia (516), New York (438), Houston (435), and Los Angeles (382) (Rosenberg, Dabrowski, and Klingner 2023).

The national numbers are also appalling—even if the statistics are complex and do not always present a clear or consistent picture of overall trends. Some types of crime have declined over the past 15 years, but the rates spiked again considerably from 2020 to 2022. As an example, over 7,000 incidents of hate crime were reported in 2021 (Federal Bureau of Investigation 2021). There were 690,158 violent crime incidents reported for the years 2020 to 2021 (Federal Bureau of Investigation 2021). And the number of murders in the nation rose by 29.4% in 2020 and another 4.3% in 2021 (Homeland Security Digital Library 2022).

The accuracy of the reporting has been called into question (Krishnakumar 2022). And some indications illustrate that the numbers are improving in specific categories. In other words, certain crimes have been falling from the recent peak in 2020, 2021, and 2022. Nevertheless, 70% of U.S. adults state that they feel anxious about their personal safety (American Psychiatric Association 2023). Many retailers (including Anthropologie, Gap, Nordstrom, Office Depot, Target, Walgreens, and Whole Foods,) have shut down operations in major cities like San Francisco because of repeated and unprosecuted thefts. Even Alvin Bragg—the Manhattan District Attorney—has publicly stated that he worries that his family could be victims of crime on subways in New York City. There are other impacts that must be considered as well. Crime is costing Americans between $4.71 and $5.76 trillion each year (Anderson 2021).

15.1.3 The Causes of Criminal Behavior

With this disturbing picture of crime in mind, more attention has been given to the causes of this deviant and illegal behavior. Several theories have been shared, and they are as numerous as the diverse types of crimes being committed. For instance, Marcus Aurelius, the Roman Emperor, once stated that "poverty is the mother of crime." Others argue that failing public schools and the lack of educational opportunities are producing high crime rates in inner cities (Smith 2022) or that homeless individuals are 514 times more likely to commit crimes (Noh 2022).

In The Real World

Policies on Crime

Individuals and research think tanks have examined the causes of crime, and in recent years they frequently point to the failures of public policy. As an example, California has reclassified a law which specifies that the stealing of merchandise under $950 is a misdemeanor, which most likely does not result in investigation and prosecution (Ohanian 2021). New York loosened its requirement for bail, and as a result perpetrators have repeatedly committed crimes and been arrested multiple times for the same offense (Lehman 2022). The media along with politicians from many jurisdictions and states demonized and defunded the police. This situation has demoralized peace officers, undercut their ability to enforce the law, and resulted in a mass exodus of personnel from this important profession. There is growing public sentiment that these and other policies are emboldening criminals and diminishing adherence to law in the United States. It will be interesting to see if the United States can do anything to reverse its problem with crime.

In other studies, the Department of Justice has observed that "the most reliable indicator of violent crime in a community is the proportion of fatherless families" (Johnson 2022). Different researchers suggest that those addicted to alcohol and drugs are more likely to participate in criminal activity and about 21% of those incarcerated committed crimes to obtain illegal drugs (Martens and Generes 2022; Sawyer 2017). Therefore, economic conditions, education, family circumstances, and programs to counter drug and alcohol consumption may be needed to reduce criminal activity.

15.1.4 Organized Crime (Drugs and Human Trafficking)

Any type of criminal behavior is problematic, but organized crime is especially troubling. **Organized crime** is the name given to the powerful groups that carefully plan and execute criminal activities usually with the purpose of economic gain. Illicit drugs and human trafficking are two of the most prevalent activities undertaken by cartels and other organized criminals. Each will be discussed in turn.

15.1.4.1 America's Addiction to Illicit Drugs

Illicit drugs include illegal narcotics and the abuse of legal prescription pharmaceuticals. The use of illegal drugs has been a major problem in the United States since the 1960s, and it is becoming even more alarming over time. The statistics provided by the National Center for Drug Abuse (2023) are staggering:

- 13.5% of Americans have used illegal drugs or misused prescriptions in the past month.
- 21.4% of Americans have used illegal drugs within the last year.
- 50% of Americans (or 138 million people) have used drugs illegally in their lifetime.
- Over 37 million Americans currently use illegal drugs or misuse prescriptions.

The most common recreational drug used is marijuana, and this will likely grow in popularity due to its legalization in about half of the states in our country. Other illegal drugs used and abused include cocaine, LSD, ecstasy, methamphetamine, and opioids such as morphine, heroin, oxycodone, and fentanyl.

The use of illicit drugs and misuse of prescriptions is sounding several alarms for government leaders, public health organizations, and the nation at large. While some politicians favor the legalization of drugs to better control these substances and to obtain additional tax revenue, others are concerned about the impact of drug use on people's education, their work ethic, and other aspects of national culture (e.g., instability of marriage and families). Medical and public health professionals likewise see drug use as one of the greatest failures of our time because over 100,000 people die from overdoses every year (National Institute on Drug Abuse 2023). People are also worried about illicit drug use because addictions are often related to psychotic episodes, schizophrenia, and criminal behavior while also putting a drain on the national economic engine or resulting in higher taxes to treat the addicted, address homelessness, care for those who develop mental illnesses, and so on.

15.1.4.2 Drug Production and Distribution

Some of the drugs are produced in the United States in an attempt to meet the voracious demand. However, "almost all of the illicit drugs causing the unprecedented rise in American deaths from drug overdoses are produced outside of the United States" (Bureau of International Narcotics and Law Enforcement Affairs 2023). For instance, marijuana, cocaine, heroin, and meth often come to the United States from Latin America. In the past, drug lords such as Pablo Escobar oversaw the production of cocaine in Peru, and it was then sent to the United States via Colombia. Mexico has also had heavy involvement in the drug trade, and several cartels have vied for control over this lucrative illicit activity (e.g., Sinaloa, Jalisco New Generation, Juarez, Gulf, Los Zetas, and Beltran-Leyva Organization). These organizations have fought each other to maximize profits. In fact, over 150,000 people have been killed in drug wars since 2006. The Zetas are one of the most violent groups in this country, and they operated under a man named Osiel Cardenas. This and other cartels continually bribe Mexican politicians, the national police, and the military to turn a blind eye to what is happening.

It is true that other drugs are often produced in different parts of the world (e.g., opium from Afghanistan or a variety of drugs from African and Asian nations). However, most of these drugs eventually end up in Mexico because of the ease of entering the United States from the porous border. For instance, the precursor chemicals for fentanyl are often shipped from China and dropped off in jugs near the shores of Mexico. The containers of liquid are then collected, taken to labs, processed into tablets (often with other ingredients to increase bulk or addictiveness), and packaged for smuggling and distribution.

Smugglers are typically—but not always—males in their 20s and 30s, and about 70% of the offenders are American citizens. These criminals are often eager to participate because they can make three to four times what they could earn in a legitimate employment opportunity. Drugs are packaged in such a way as to avoid detection and then hidden in tires, seats, or other locations on vehicles (e.g., side panels). Much of the drugs come through border crossings in South Texas, West Texas, New Mexico, and California. But it is important to point out that drugs can also be smuggled in makeshift submarines and private and commercial aircraft.

Drug cartels —organized crime syndicates that cultivate, manufacture, and distribute drugs—have also used tunnels to distribute drugs into the United States. In one instance, Joaquín "El Chapo" Guzmán hired 30 men to dig a tunnel, and then he killed these workers to ensure the location was kept a secret. This tunnel allowed the shipment of a truckload of drugs every day at a value of $40 million per month. The United States has discovered many similar tunnels operating under the border and has worked to capture affiliated drug lords and smugglers. This brings up what can be done to stop this illicit business activity.

15.1.4.3 Efforts to Prevent the Distribution and Use of Drugs

There are many things that can and must be done to prevent and limit the manufacturing, distribution, and use/abuse of drugs. Some of these measures must be taken domestically, while others have international implications.

One of the most important things that can be done is to educate people about the dangers of using illicit drugs and misusing legal drugs. Education programs must continually be a focus in schools and by governments to teach children, youth, and adults about the dangers of drug use. Doctors should also limit the prescription of certain opioids to reduce addiction that may come from these medications. Society must also view the addicted differently (without stigma) and find ways to treat their addictions and help return these individuals to being productive members of society. Local and state governments may also want to reconsider the legalization of marijuana in light of the additional social and public health problems it is creating.

In The Real World

Legalization of Drugs?

There is a growing call among people and politicians to legalize drugs. There are several arguments associated with this request. Some assert that legalizing drugs will lead to the downfall of drug cartels and reduce the violence associated with the drug trade. Others believe the distribution and use of drugs can be regulated, thereby promoting safety and even providing increased government revenue due to the taxation of drugs. With this in mind, several states have legalized marijuana, and some of these political units are providing free syringe services to users of other types of drugs. However, opponents suggest that this is leading to a major public health crisis and that the official or unofficial endorsement of drugs is resulting in serious social problems including unemployment, homelessness, and crime. In addition, there is increasing evidence that the potent cannabis that is being legally harvested and distributed is causing severe mental illness including schizophrenia. These opponents point out that California and New York appear to be causing more addictions and societal problems through their syringe service programs rather than reducing them.

Another important step is to continue working with other governments—whether that is with collected groups like **INTERPOL** (the International Criminal Police Organization) or individual countries such as Mexico. Because drugs are produced and distributed internationally, a global approach will be required to address this problem.

Since "drug trafficking organizations are among the most significant and well-resourced transnational criminal threats facing the United States" (The White House 2021a, b), other measures will be required. Senators Rick Scott and Roger Marshall asserted in March 2023 that Congress should designate drug cartels as Foreign Terrorist Organizations. This would ban them from entering the United States, allow the Secretary of Treasuring to freeze their assets, and make it illegal for anyone to support these types of organizations and their illicit activities. It could also result in aggressive intervention on the part of the United States military against drug cartels in Mexico and elsewhere.

President Biden has also issued two related executive orders to help counter drug cartel activity. The first is titled "Establishing the U.S. Council on Transnational Organized Crime." President Biden created a council of advisors that includes the Secretary of State, the Secretary of Treasuring, the Secretary of Defense, the Attorney General, the Secretary of DHS, and the Director of National Intelligence. The goal of this group is to oversee the gathering of intelligence to target, disrupt, and denigrate criminal networks. To stop the production of drugs at its source and limit distribution to and in the United States, efforts are being taken to go after producers and distributors by collaborating further with law enforcement agencies and even the private sector. The second executive order, titled "Imposing Sanctions on Foreign Persons Involved in the Global Illicit Drug Trade," increases punishments for those individuals and groups that participate in drug cartels, produce illegal substances, launder money, or distribute drugs.

Aside from these activities, other steps must be taken. The U.S. Customs and Border Protection and U.S. Coast Guard will need additional personnel and equipment to identify and confiscate drugs being smuggled into the country. This is important since the challenge is monumental. In 2022, 656,000 pounds of illegal drugs were confiscated, including 155,000 pounds of marijuana, 175,000 pounds of methamphetamine, and 14,700 pounds of fentanyl.

15.1.4.4 Human Trafficking

Human trafficking is another major threat that has recently garnered the attention of those involved in homeland security, and this type of crime has both international and domestic implications. **Human trafficking** is defined as "modern day slavery where people are exploited through force, fraud or coercion" (Department of Homeland Security 2023). The Blue Campaign (see **https://www.dhs.gov/blue-campaign/tools**) points out that human trafficking is related to human smuggling—or the illegal movement of people across national borders. However, human trafficking is an even more significant crime in that it is a major violation of human rights where people are required to engage in sexual acts, forced labor, or domestic servitude.

Those apprehended and subjected to human trafficking may be transported to foreign countries or to other jurisdictions to evade detection, and they are forced to sell sex on the Internet or street corners as well as at personal residences, truck stops, clubs, and massage parlors. Others are forced to work on farms, in mining operations, or in factories and sweatshops manufacturing products that are sold around the world. Some victims work in people's homes performing a variety of tasks including cooking, cleaning, babysitting, and doing the laundry among other services.

It is important to recognize that anyone could become a victim of human trafficking—regardless of their race, gender, nationality, ethnicity, and citizenship. However, women, teenage girls, homosexual boys, children, runaways, and homeless individuals are particularly vulnerable to human trafficking. Some individuals are kidnapped and forced into human trafficking through violence, threats, drug addiction the withholding of passports. Others are vulnerable to fraud and deception based on false promises of jobs, income, citizenship in another country, or love and marriage. Once involved, the victims become prisoners with little or no ability to escape or pursue their own interests.

15.1.4.5 Examples of Human Trafficking

Documentaries by Frontline and Nightline reveal some of the ways human traffickers operate and the difficulty of those escaping this form of modern-day slavery (PBS 2019; ABC NEWS 2015). While boys and men have been trafficked, girls and women are the most likely victims. Pimps find lonely girls on the Internet in chatrooms or young women at bus stops who have run away from home or foster care. The pimps develop a false sense of trust with the teenage girls and promise them attention, affection, and employment opportunities. The victims are then taken to other cities and states and dressed up and photographed for advertisements on Craigslist, Backpage, MeetMe, and so on.

The pimps may get the young women addicted to drugs so they cannot think clearly or leave the promise of the next fix. The handlers also threaten the victims with violence if they do not have sex with men multiple times a night. In many cases the sexual acts occur in sketchy motels, but in other situations the victims are transported to higher-end hotels during major sporting events such as the NCAA Final Four or the Superbowl. If the victims do not bring in a quota of money (e.g., $1,000 over a 24-hour period), they are beaten up. The pimps also play on the girls' fear of being caught by law enforcement personnel, which inhibits their ability to seek outside help.

The young women are thus isolated from those who might care for them, and they feel socially trapped. The pimps reiterate that "they own the girls' bodies," and attempt to weld enormous power over the victims. In addition, the sense of embarrassment, need for money or drugs, and worries about being arrested for prostitution keep individuals from reporting their exploiters who oversee this type of human trafficking.

15.1.4.6 Appalling Statistics

Regardless of the situation, the perpetrators are motivated by selfishness and greed, and human trafficking has therefore become a major problem and very profitable shadow economic activity. While exact numbers are hard to quantify due to the hidden nature of this crime, there are some disturbing statistics:

- A study by the International Labor Organization and the International Organization for Migration asserts that 27.6 million people were victims of forced labor and sexual exploitation globally at any given time in 2021 (Department of State 2023).
- According to a White House report, $150 billion in illicit profits are derived from human trafficking worldwide (The White House 2021b).
- The Department of Health and Human Services (2019) asserts that 50,000 people are trafficked into the United States each year.
- Other reports from the Department of Health and Human Services and the University of Pennsylvania suggest that between 100,000 and 325,000 are forced into sexual slavery in the United States every year (Deliver Fund 2020).
- The Director of the DHS Center for Countering Human Trafficking states that 80% of the victims are involved in forced labor and the remaining 20% are involved in sexual exploitation (Department of Homeland Security 2022).

15.1.4.7 What Is Being Done to Prevent Human Trafficking

Because of the disturbing numbers mentioned earlier, the attention being given to human trafficking is at an all-time high. Recent presidents and congressional officials are implementing executive orders and passing new laws against human trafficking. For instance, the Countering Human Trafficking Act of 2021 has been implemented to address this growing problem.

Government agencies are doing all they can to prevent this reprehensible crime and help the victims who have been trapped in the web of traffickers. Besides local law enforcement and state agencies who are trying to catch those involved in human trafficking, there are many other federal partners. The Department of Homeland Security has trained over 200,000 employees about human trafficking. U.S. Customers and Border Protection and the U.S. Immigration and Customs Enforcement are constantly looking for human traffickers and the victims who are being transported across the border. The Federal Bureau of Investigation also has a Crimes Against Children and Human Trafficking Unit. As a result of these agencies and their efforts, the Department of Homeland Security (2022) reports that 2,360 arrests were made in 2021. The Bureau of Justice (2022) indicates that the number of persons prosecuted for human trafficking rose by 84% from 2011 to 2020.

In The Real World

Operation Underground Railroad

Chances are that you have heard about Tim Ballard and his organization, Operation Underground Railroad (OUR), because of the movie *The Sound of Freedom*. After working for over a decade as a special agent in the Department of Homeland Security, Tim Ballard decided to leave this government entity. He was concerned that not enough was being done to stop human trafficking and sexual tourism in other nations since the federal government was constrained in what it could do outside of the United States. He obtained funding from wealthy donors and started a nonprofit organization named **Operation Underground Railroad** to stop human traffickers and rescue the victims of this disturbing type of organized crime. Ballard, in conjunction with his specially trained personnel and law enforcement partners in other nations, identifies missing children that have been reported to public officials, investigates possible traffickers, and works to shut down such activities. In one case, Ballard risked his life to enter territory controlled by drug lords in Colombia to rescue a girl who was taken into human trafficking. In another instance, his organization worked to stop a human trafficking ring in Haiti. Ballard even adopted some of the victims from this, the poorest country in the Americas. According to OUR website, Ballard and his organization have been able to free over 7,500 victims and help law enforcement officials arrest over 6,000 perpetrators in the United States and elsewhere around the world. The future of OUR is now less certain, however. Both Tim Ballard and OUR have recently been under investigation for alleged fraud and other questionable practices.

Faith-based and nonprofit organizations are also involved in addressing the problem of human trafficking, and they pay special attention to the victims of horrendous crimes. Some of these organizations include Free the Slaves, International Justice Mission, Logos Wilderness Therapy, the National Center for Missing and Exploited Children, the Nomi Network, Shared Hope, and the Polaris Project. The goal of these nonprofits is to support law enforcement intelligence-gathering efforts and provide financial assistance to victims along with psychological counseling. The organizations also help victims overcome drug addiction, advance their education and training, and pursue employment opportunities. This type of collective effort between the public and nonprofit organizations is required to prevent human trafficking and deal with the effects it has on its many victims.

Self-Check

1. The commission of crime generally appears to have decreased in recent years. True or false?
2. Drug cartels are organized crime syndicates that cultivate, manufacture, and distribute illegal drugs. True or false?
3. Which of the following are accurate statements?
   a. Anyone could become a victim of human trafficking.
   b. Human trafficking deals only with prostitution.

    c. The government is taking measures alone to stop human trafficking.

    d. Politicians have not yet taken any measures to reduce human trafficking.

4. What measures can be taken to reduce mass shootings, the smuggling of drugs, and human trafficking?

15.2 Social Disturbances

Another potential or actual problem facing those involved in homeland security relates to **social disturbances** such as protests and riots. One of the great benefits provided to U.S. citizens is the Constitution, which grants citizens' rights such as freedom of speech and freedom of assembly. These rights permit people to voice their concerns in demonstrations about policies, laws, regulations, government actions, and other social of political conditions. Freedom of speech and the right to assemble are two of the central hallmarks of effective democracies. However, peaceful protests can sometimes turn violent and pose a threat to other people, government institutions, and the broader society. For this reason, it is important to understand protests and riots and what—if anything—should be done about them from a homeland security perspective.

15.2.1 Protests

The United States has a long history of protests. In fact, our nation was founded on the expression of grievances against England due to "taxation without representation." The disgruntled feelings against King George III and colonial administrators ultimately led to the Revolutionary War and the Declaration of Independence. Since this time, Americans have protested abusive political decisions and social injustices (Dudenhoefer 2020). For instance, there have been protests about working conditions, slavery, women's rights, racial discrimination, gay rights, the Vietnam War, environmental disasters, nuclear weapons, anti-abortion/abortion rights, the war in Iraq, police brutality, and whole host of other issues and social, political, and economic conditions.

In the past few decades, the **Occupy Movement**—the gathering of protesters that decry social and economic inequality around the world—has produced notable protests in New York and in Seattle. After experiencing hardships after the economic recession in 2008 and 2009, young people began to congregate to protest the income gap and global capitalist activities. This was known as **Occupy Wall Street**.

For instance, on September 17, 2011, about 1,000 young adults gathered in Zuccotti Park in New York City to protest the "1%"—the millionaires and billionaires whose income doubled and tripled in prior decades. In contrast, those entering the workforce were concerned about the large student debts they possessed and the lack of employment opportunities in a very volatile economic market (Vara 2014). Consequently, the protesters set up tents and over the next three weeks they used social media and live-stream technology to spread their message in New York and to other parts of the nation. Republicans ignored or denounced the protesters while their efforts were praised by Democrats including Elizabeth Warren, Bernie Sanders, and Alexandria Ocasio-Cortez. Nevertheless, Mayor Bloomberg eventually cited health and fire concerns and shut down the movement in New York City. However, the protests were successful in raising wages in many service industries and other sectors of the economy (e.g., fast food, retail, and education) (Anderson 2021).

On June 10, 2020, another Occupy Movement began in Seattle, Washington. In some ways the young adults involved in this protest were motivated by concerns that were similar to those voiced by the Occupy Wallstreet movement. For instance, the protesters were frustrated with economic conditions and demanded rent control and free college tuition for Washington State residents. However, the protests were also very different in that they were directed at police brutality as witnessed in the death of George Floyd in Minneapolis, Minnesota, on May 25, 2020 (Clark 2020).

George Floyd was a 46-year-old African American man who was killed by Derek Chauvin, a police officer in Minneapolis. Floyd had tried to make a purchase at a convenience store with a counterfeit $20 bill and refused to give the cigarettes back when a store clerk asked him to do so. A call to 911 was made to report the incident, and the employee stated that Floyd was likely drunk or under the influence of drugs. Several officers were dispatched to the scene to find Floyd and question him. When officers approached his vehicle, Floyd ignored, delayed, and/or refused to follow all police instructions. He was therefore forcibly removed from his vehicle, wrestled to the ground, handcuffed, and restrained by officer Chauvin who dug his knee into Floyd's neck for over eight minutes. George Floyd initially expressed his inability to breathe due to Chauvin's actions, and bystanders complained to the four police involved that Chauvin should remove his knee as a result. Chauvin did not comply with this request. Floyd went unconscious, and emergency medical technicians tried to revive him with cardiopulmonary resuscitation while en route to the hospital. Floyd was pronounced dead at the Hennepin County Medical Center a short time later.

Much of the arrest and the overly aggressive restraint with Chauvin's knee was caught on cameras and distributed widely through social media in Minneapolis and around the nation. Impassioned protests broke out in Minnesota and over 2,000 cities across the United States. In fact, it is estimated that between 15 and 25 million people took to the streets to express anger at the police brutality against Floyd and the long-standing history of racial discrimination in the United States (Harvard 2020).

Occupy Seattle was one of the most notable protests that occurred after the death of George Floyd (Clark 2020). Those involved marched on the police precinct and entered city hall to demand a review of police tactics, stop excessive force, improve law enforcement training, and even dismantle and "defund the police." The protesters took over an area of Seattle and set up the People's Republic of Capitol Hill or the **CHAZ** —the Capital Hill Autonomous Zone. Over the next several weeks, the protesters controlled access to the area. Many acts of vandalism and violence occurred during this time, and businesses began to complain about government inaction regarding the situation. A few people were killed by protesters in the autonomous area, and this led to various skirmishes with the police. Because this situation disrupted business and posed a growing security threat, Mayor Jenny Durkin had enough and issued an executive order and demanded that police take back the area and that order be restored to the commercial district (KING 5 Staff 2020). In early July, the protesters dissipated and control over the autonomous zone was dismantled. Whether justified or not, laws enforcement's use of pepper spray, concussion grenades, tear gas, and rubber bullets has now resulted in a $10 million settlement with 50 Black Lives Matter protesters. Occupy Seattle illustrated how easy it is for protests to turn violent and deadly. This brings up the topic of riots.

In The Real World

Trucker Protest in Canada

After months of frustration due to the COVID-19 vaccine mandate imposed by Prime Minister Trudeau and the carbon tax on the transportation industry, numerous Canadian truck drivers started to move from British Columbia to the capital city of Ottawa to share their outrage at the new government policies. These and other drivers in Toronto, Quebec, and Calgary met in designated locations in January 2022 and blasted their horns for up to 16 hours a day. The drivers also blocked traffic in many areas. In one case, the Ambassador Bridge near Detroit—which carries nearly 25% of the trade between the United States and Canada—was shut down for six days. Conservative politicians expressed support for the truckers, and protesters set up a "Go Fund Me" website through which they raised over $7.8 million for the cause. Prime Minister Trudeau initially dismissed the truckers as a "small fringe minority," but he quickly acted to halt the transfer of funds to the protesters (thereby raising concerns about limits on the freedom of speech and the right to assemble). By employing the Emergencies Act (which has since been questioned in legal proceedings), the government also mobilized over 900 police officers to arrest the truckers and anyone who provided food, fuel, or other types of material support. The truckers soon disbanded, and it is unclear if this event had any impact on Canadian policies. However, the protests and demonstrations created a loss of $2.3 billion in trade. They also led to copycat convoys in Australia and New Zealand and to a lesser extent in the United States.

15.2.2 Riots

Riots are another type of social disturbance, and they are closely related to protests. However, riots are different from protests in that they include more antisocial behavior and law-breaking activity. This conduct includes rock throwing, tipping over vehicles, looting, starting fires, and attacking law enforcement personnel (Britannica 2023). Riots may therefore produce significant property damage and a large number of injuries and even death.

There are many causes of riots. For instance, drunk fans may engage in riotous behavior after the Superbowl, regardless of whether their team won or lost this ultimate competition in the National Football League. However, most riots are extensions of protests where grievances reach a tipping point. Often, this is due to political and economic conditions (e.g., repression and poverty) or racial tensions and cultural conflict (e.g., discrimination and disagreements about values and priorities).

Numerous riots have occurred in the United States throughout its history (Daniel 2013). For instance, in 1791, farmers in the newly established nation violently resisted the imposition of a tax on liquor in what is known as the Whiskey Rebellion. Those in opposition broke into the homes of government officials to tar and feather tax collectors. In 1894, the unemployed in Cleveland, Ohio, skirmished with the police while expressing extreme dissatisfaction with the collapse of the stock market and the failure of banks. The May Day riots as they were known included repeated fights with police who were armed with clubs, and the protesters destroyed property in various buildings and factories in the city. In 1905, the Chicago Teamsters' strike turned deadly; 21 people were killed and another 416 were injured when clothing workers protested the hiring of non-union personnel at Montgomery Ward & Co. In 1931, coal miners in Harlan, Kentucky, went on strike to decry horrible working conditions, limited pay, and insufficient housing. The Battle of Evarts broke out with a 15-minute exchange of gunfire that took place between protesters and the police. When all was said and done, three deputies and one miner were dead. There have been many other riots owing to military drafts, immigration policies, and other objections about domestic and international policies.

Another notable riot occurred in Seattle. During a World Trade Organization conference in November 1999, leaders of 135 nations gathered to discuss international economic policies. It is estimated that 50,000 protesters gathered to express their displeasure with worker exploitation, economic inequality, and environmental degradation (Brunner 2019). Most steelworkers, machinists, teachers, and other union members were peaceful, although the Direct Action Network blocked streets and intersections to prevent people from reaching the convention center. There were black-clad anarchists who broke windows in various buildings, which prompted the police to use pepper spray, tear gas, and stun grenades. The clash between the two groups escalated and the rioters threw bottles at the police. The opening ceremony of the meeting was cancelled, and the National Guard and State Troopers were called in to help deal with the violent protests over a four-day period. Over 500 people were arrested for their illegal activities. The riot caused an estimated $20 million in damages and required an additional $3 million in expenses by the local government. The nature of this event illustrated that law enforcement was not prepared (Associated Press 2000). They did not have enough personnel, food, water, or equipment and supplies.

In The Real World

Concerns for Homeland Security and National Security?

Homeland Security and national security are in some ways distinct disciplines and professions. For instance, homeland security is more focused on internal risks while national security gives greater attention to external threats. However, both are similar in that they are concerned about the physical safety and general well-being of the United States.

Some of the threats and hazards are obvious and result in the creation of new policies, the expansion of bureaucratic organizations, and further federal spending. However, are there other less visible issues that could also impact homeland security and national security? Consider if the following problems are serious and if they might have a negative consequence on the ability to secure our country domestically and protect ourselves from outside enemies:

Unresolved Problem	Probable Impact
Poverty	Lack of financial resources is often accompanied by poor educational systems, increased crime, political unrest, and poor health.
National debt	Increased debt will augment payments both principal and interest and will likely limit spending on future problems and priorities including homeland security and national security objectives.
Decline in educational performance	Dropouts and low test scores in reading, science, computers, and math will have a drag on the economy and hinder advancements in military technology.

(continued)

In The Real World *(Continued)*

Unresolved Problem	Probable Impact
Environmental degradation	Industrial disasters and accidents can create environmental refugees and lead to long-lasting health consequences.
Food insecurity	Insufficient food is a likely predictor of political unrest, civil war, and international conflict.
Water insecurity	Insufficient water is a likely predictor of political unrest, civil war, and international conflict.
Energy insecurity	Insufficient energy resources are a likely predictor of political unrest, civil war, and international conflict.
Lack of fitness among citizens	Obesity and other health concerns will reduce life expectancy and limit the ability of law enforcement and military personnel to perform their job functions.
Overreliance on foreign manufacturing	Failure to produce goods in the United States can result in a scarcity of essential items such as microchips and computers, which have military applications.
Falling behind on military advancements	Failure to fund and develop weapons systems could put the United States behind Russia and China (e.g., hypersonic weapons).
Politician lies, dishonesty, and corruption	Illicit political behavior is a likely predictor of political unrest, civil war, and international conflict.
The prevalence of drones	Increased personal drones can interfere with civilian flights and emergency operations, and these aircraft be used for criminal and terrorist purposes.
Unidentified anomalous phenomena	Unidentified flying objects could present major homeland and national security risks that are not completely understood.
Censorship of the press or social media	Shutting down, controlling, or limiting freedom of speech is a likely predictor of political unrest, civil war, and international conflict.
Debates about voting rights and processes	Continual concerns about voter irregularities or contested elections undermine democracy and could lead to more autocratic forms of government (which are often politically unstable).
Unabated crime	Unaddressed property crimes and physical assaults are likely to lead to vigilante behavior and discredit the legitimacy of government, law enforcement agencies, and the criminal justice system.

While social unrest pertaining to working and economic conditions has occurred repeatedly over time, the most frequent and problematic riots in the United States have centered on slavery and race relations (see Black Past 2023). There have been countless violent demonstrations relating to slave rebellions, abolitionist movements, counter-abolitionist protests, segregation policies, black voter rights, and the like. The history of the United States is unfortunately full of violent racial relations, including the riots that took place in several cities after the assassination of the well-known equal rights advocate Marin Luther King, Jr., in Memphis, Tennessee.

One of the most well-known riots took place in Watts, California (a neighborhood of Los Angeles). A young African American man named Marquette Frye was pulled over for drunk driving on August 11, 1965. After failing a sobriety test, police officers tried to apprehend the individual and he resisted arrest. Marquette was hit in the face with a baton, and the group of people who gathered at the scene began to protest police brutality. Widespread rioting, looting, assault, and arson occurred. Thirty-four people died in the riots, and the disturbance resulted in over $40 million in damages.

Fast forward two decades and similar riots occurred in California and elsewhere as a result of police brutality. For instance, prior to the 1984 Olympic games in Los Angeles, law enforcement was instructed to crack down on crime and gang violence in this part of Southern California. Over 50,000 people (mostly male minorities) were arrested as a part of Operation Hammer. This led to increased tensions between blacks and whites and blacks and Koreans during this period of time.

The mounting anger quickly exploded into violent riot behavior when four public safety officials were not punished for police brutality. A video tape was released showing several police officers beating a man named Rodney King, who was drunk and involved in a high-speed chase. Even though King was kicked and hit with batons at least 56 times, the officers involved were acquitted of any wrongdoing. The verdict was unbelievable for most people and consequently followed by various assaults in the streets of police officers and against three men (two white and one Hispanic) who were driving trucks. The riots spread rapidly, and over 3,600 fires were set, which destroyed over 1,100 buildings and caused over $1 billion in damages.

Local police officers and federal troops were sent in to establish order, but not before 63 people were killed and another 2,400 were injured. Public officials and the police department were criticized for the slow reaction and the inability to control the situation for the span of six days.

Over the next few decades, actual episodes of police brutality and other allegations continued, which fueled increased racial tensions. Michael Brown—a black teenager—was shot and killed by a police officer on August 9, 2014, in Ferguson, Missouri. Eric Garner, a black man, was put in a chokehold by a New York City officer and died as a result on July 17, 2014. Freddie Gray was fatally injured by Baltimore police officers while being transported in a van on April 12, 2015. On August 10, 2016, Anthony "Tony" Timpa was killed by a Dallas police officer while being restrained on the ground. And, on March 13, 2020, Breonna Taylor was shot and killed by police during a drug investigation that turned violent.

These and many other occurrences (e.g., the death of Trevon Martin by a security guard in Florida in February 2012) caused serious concerns about systemic racism and the abuse of power by those in positions of authority. As noted earlier, the rising tension came to a head on May 25, 2020, when George Floyd was killed by Derek Chauvin. While millions of protesters in Minneapolis and around the nation were peaceful, riots did ensue—and they were accompanied by assaults and other types of violence. Nineteen people were confirmed to have died during the riots, and over 2,000 people were injured. 62,000 National Guard personnel were mobilized, and 14,000 people were arrested by local police and other state law enforcement officials. The riots from May to June 2020 produced significant damages. The police department in Minnesota was vandalized, and buildings were gutted by arson (resulting in $500 million in losses in this city). Similar actions took place in cities around the nation, costing between $1 and $2 billion.

Not all of the recent riots have been a direct result of racial relations, although some assert that other violent political activity continues to demonstrate racial bias. On **January 6**, a mob of supporters of Donald J. Trump gathered at, attacked, and took over the U.S. Capitol building to protest Trump's defeat in the 2020 election and stop the certification of the electoral college votes. Trump and his ardent supporters felt that the victory of Joe Biden was illegitimate and that the election was beset with fraudulent voting, ballot stuffing, and other forms of election tampering. Trump told his supporters to "fight like hell," although he also asked them to avoid violent behavior (Naylor 2021). Some politicians and media personalities assert that mixed messages were therefore given to the protesters.

Regardless, about 2,000 to 2,500 individuals—including those from the Oath Keepers, Proud Boys, QAnon, and Three Percenters—showed up with signs and even a gallows on display. Many individuals fought with police, breached the perimeter by force, or were allowed to enter the Capitol. Law makers were evacuated in these emergency conditions, and property was stolen or damaged in the riots. A few people died in or after the event, although the direct causation to January 6 is being debated in some cases. One of the protestors was shot. One of the officers had a stroke the next day and died. Two officers passed away due to complications with heart disease. Four other law enforcement personnel committed suicide shortly after the riots.

As a result of this disturbing attack on our democratic institutions, Trump was impeached by the House of Representatives and remains the subject of many lengthy investigations that will last for several years. The lack of security at the Capitol, combined with the lessons of the 2020 riots, brought up many questions about what should be done to react more effectively to riot situations.

In The Real World

Attack on the U.S. Capitol

Not everyone who arrived at the U.S. Capitol on January 6 acted violently. Many individuals simply wanted to voice concerns about the future of the country or desired to protest what they regarded to be injustice regarding allegations about Trump's involvement with Russia or irregularities regarding voting and election results (e.g., see the documentary *2000 Mules* about alleged or actual ballot harvesting) (D-Souza 2022). However, much of the behavior could be classified as terrorism or a riot. For instance, two bombs were placed near the headquarters of both the Republican National Convention and the Democratic National Convention on the evening of January 5, but they were fortunately found the next day before being detonated. In addition, hundreds of people removed or jumped over fencing, broke windows, entered a restricted area, vandalized congressional offices, stole government property, and occupied offices and rooms in an illegal manner. Although there are various videos showing protesters being peacefully escorted into the Capitol by police officers without force, there were also situations where the actions were clearly violent. Some of those present carried chemical sprays, knives, batons, baseball bats, stun guns, and even loaded weapons with them as they made their way to the Capitol Building. They pushed police officers, sprayed them with chemical agents, hit them with pipes, and threw fire extinguishers at them. About this time, an unarmed Air Force Veteran was shot by a police officer while she was attempting to climb through a broken window. Members of the mob were heard asking where the vice president was and chanted "Hang Mike Pence." The situation would have likely been worse, but a Capitol Police officer named Eugene Goodman deliberately led the more violent element of the crowd away from the chamber, which allowed Mike Pence, Nancy Pelosi, and many other politicians to escape the building unharmed. By the time law enforcement had taken back control of the Capitol, an estimated $1.5 and $3 million in damages had been inflicted on this important building. More than 1,000 people were arrested for rioting and sedition, with about half of those receiving prison sentences thus far.

15.2.3 Dealing with Riots

Obviously, peaceful protests must be allowed as part of the First Amendment to the U.S. Constitution. And, where possible, political leaders should listen to concerns and find ways to negotiate solutions to grievances in society. Efforts should be made to dissipate tensions and find ways to peacefully remove protesters who create legitimate concerns relating to public transportation and public safety.

In other cases, political leaders and public safety officials must anticipate protests and riots and engage in advanced planning to curtail them or react more effectively (Institute for Constitutional Advocacy and Protection, no date). For instance, fences and barricades may need to be acquired and put in place along with armored vehicles and vehicles with water cannons. Additional and highly trained law enforcement personnel will be required with necessary intelligence, communications equipment, and other gear (e.g., helmets, riot shields, knee pads, elbow pads, holsters with positive locking mechanisms, zip ties, and non-lethal incapacitating agents such as tear gas, pepper spray as well as batons, tasers, plastic bullets, netguns, stink bombs, sticky foam, sound cannons, and active denials systems that direct electromagnetic radiation). Busses may also be required to carry law enforcement personnel and transport offenders, and other supplies and logistical support must be considered (e.g., food and water as well as firefighting and emergency medical care personnel). Jails or other locations must be identified and sufficient staffed to house the arrested until trials can occur and/or sentences are given. These are just some of the many steps that must be taken to deal with riots in an effective manner.

Self-Check

1. Freedom of speech and the right to assemble are protected under the amendments to the U.S. Constitution. True or false?
2. Many riots in U.S. history have focused on racial relations and discrimination. True or false?
3. Which of the following protests was more concerned with the excessive use of force by police officers?
   a. Occupy Wallstreet
   b. Occupy Seattle
   c. Trucker Protest
   d. The January 6 attack on the U.S. Capitol Building
4. What measures can be taken by law enforcement to prepare for possible riots?

15.3   Epidemics and Pandemics

The world and the United States have experienced a number of epidemics and pandemics over time, and these events have had a significant impact on the lives and well-being of people (Healthline 2023). An **epidemic** is a smaller-scale public health emergency where a disease is spreading actively in a specific region. In contrast, a **pandemic** is a larger event where a disease is spreading internationally and affecting many countries around the globe.

Epidemics and pandemics result from bacterial and viral infections. **Bacterial infections** are diseases that result from cells that can live outside the body. Bacterial infections have caused some of the deadliest public health disasters in history. Cholera is one example of this type of infection. **Viral infections** are diseases resulting from nonliving molecules that need a host to survive. Hemorrhagic fever is an example of a deadly virus.

15.3.1 Examples Throughout History

Diseases of various kinds have been recorded in Egypt, Asia, and elsewhere around the world throughout history. Some of the worst disease outbreaks in the Western world include smallpox, the Bubonic Plague, and the Spanish Flu. In terms of smallpox, from 1870 to 1874, Europeans experienced fever, chills, body aches, and especially rashes, blisters, and scabbing. Three out of ten people who contracted the disease died, and an estimated 400,000 perished.

In 1346, the Bubonic Plague, or Black Death as it was known, ravaged Europe for a decade. Those who contracted this disease experienced fever, fatigue, vomiting, aches and pains, and buboes (which were swollen lymph nodes that became tender in the groin or armpits). It is estimated that 50 million people in Europe died, which amounted to about 60% of the population on this continent.

From 1918 to 1920, the world experienced the Spanish Flu, also known as the Great Influenza Pandemic. This was caused by the H1N1 influenza A virus. It was accompanied by a very high fever, body aches, runny nose, and a dry cough. Estimates of deaths vary widely and range from 25 to 50 million.

In The Real World

Ebola in Africa and in Dallas, Texas

In 2014, a major outbreak of Ebola (hemorrhagic fever) occurred in central and western Africa. The disease infected over 28,000 people on this continent and was accompanied by fever, fatigue, diarrhea, dehydration, and bleeding from the ears, eyes, nose, and mouth. Because the disease in Africa had a very high mortality rate that killed over 11,000, public health officials became extremely concerned when a man named Thomas Duncan brought the disease to the United States in September 2014. Duncan was a Liberian citizen who may have contracted Ebola from a girlfriend or family member and then traveled to visit his family in Dallas, Texas. After arriving in the country, Duncan fell ill and went to the hospital for treatment. Medical personnel released him since his symptoms initially appeared to resemble the flu. Duncan's condition worsened, and he was brought back to the hospital and diagnosed with Ebola. Duncan later succumbed to Ebola, and some nurses who provided care to him also contracted the virus. The CDC blamed the hospital for not following proper protocol, but it appeared that the nurses complied with the existing standards for protective that clothing but they were proven inadequate. Messaging to the public was frequently inconsistent as well—as political leaders and medical personnel expressed no concerns for transmission at times while the media was concomitantly showing video footage of people in Tyvek suits extracting personal property in Duncan's family apartment. The response to this disease included the isolation of a personal pet and the disposal of medical waste, which cost hundreds of thousands of dollars. The Ebola incident seemed to reveal that the medical community and government officials were not as prepared as they should be for these types of medical emergencies (McEntire 2019).

The loss of life from the Bubonic Plague and the Spanish Flu is disturbing but perhaps expected due to the combination of unsanitary and crowded conditions, limited scientific knowledge, and inadequate medical treatment options that existed at the time. In spite of incredible progress in these areas, many disease outbreaks still occur in our modern era. The United States has witnessed several major public health emergencies in prior decades including polio and HIV/AIDS. Polio killed about 3% of children and about 20% of adults who were impacted by the disease. HIV/AIDS has resulted in the deaths of about 700,000 people. Other diseases, such as West Nile Virus, SARS, MERS, and Monkeypox have affected the nation as well. Some of these are less deadly but still contagious and have concerning effects. For instance, ZIKA can cause birth defects and lead to problems with a baby's brain development and vision.

COVID-19

The most concerning pandemic in recent years, of course, has been caused by the **coronavirus** —also known as SARS-CoV-2 or COVID-19 for short. The COVID-19 pandemic originated in China and soon made its way to other parts of the world. The virus had a major impact on contraction and death rates, and quickly became a politically charged public health emergency (McEntire 2022a).

The coronavirus is highly infectious and can be spread by very small airborne particles emitted from those who contracted the virus or by touching the eyes, nose, or mouth with fingers that have the virus on them. At the time of this writing, over 770,000,000 people have contracted the virus worldwide, with over 100,000,000 in the United States. Symptoms are similar to other types of viruses, although this one also included loss of taste and smell and serious respiratory problems. Over 3.3 million people have died from the virus around the world, including 1.1 million in the United States.

15.3.2 The Politics of COVID-19

As soon as COVID-19 made its appearance, it was accompanied by significant political disagreements. This included debates about the cause, the instituted travel ban, and policies on masks, vaccines, social isolation, and school and work closures.

As mentioned, COVID-19 first appeared in Wuhan, China. There was initial evidence that it came from the Wuhan Institute of Virology (e.g., U.S. funding of coronavirus studies in China, safety concerns in the lab, confirmed victims who worked in the lab, the loss of lab specimens and record keeping, and China's refusal to share information with the rest of the world) (McGhee 2021). This caused then President Donald J. Trump to label the disease as the "China virus." Anthony Fauci, who was the former Director of the National Institute of Allergy and Infectious Diseases, first entertained the lab leak but then dismissed it in favor of the wet market theory (i.e. natural origin and transmission through animals). There is still debate about the origins among various government agencies, although both the FBI, the CIA, and a few other departments now believe the virus emanated from the lab through an accident (Washington Desk 2023).

The travel ban instituted by then President Trump was also very political. Trump wanted to stop the spread of the virus and therefore prohibited travel from China to the United States. In response, Democrats criticized the president for being xenophobic. But it was the decisions and actions related to masks, vaccines, social isolation, and school and work closures that were especially controversial.

Although Trump implemented a travel ban, he and other Republican leaders did not seem to take the virus seriously. They argued against or did not fully support masks, vaccines, and school closures. Meanwhile, Democratic leaders asserted that it was the responsibility of every American to wear a mask to flatten the curve and limit the spread. Left-leaning politicians criticized what they regarded to be "anti-science" sentiments of Republicans, although evidence increasingly demonstrates that masks are not extremely effective against this type of virus.

The vaccine mandates were equally political. Virtually all democratic politicians and public health agencies required that all people obtain various doses of the coronavirus vaccine. While most people did get the first few doses of the vaccine, others expressed concerns about the trampling of freedoms or possible side effects including possibly negative impacts on the heart. There were also complaints that different treatments (e.g., monoclonal antibodies) were dismissed without warrant. Later on, the effectiveness of the vaccine was called into question and many citizens complained that politicians were benefiting financially from the pharmaceutical industry.

Another political controversy related to policies of social isolation. Public health organizations encouraged people to maintain a distance of at least 6 feet and avoid congregating in crowded areas. While many people complied, various governors, congressional leaders, and other politicians of both parties were seen at parties, eating at restaurants, getting their hair cut at salons, dancing at New Year's Eve celebrations, and so on. Meanwhile, New York implemented a policy that would retain the most vulnerable people in nursing homes throughout the state. People started to complain about the double standard that was being proposed by elected officials and the dangerous mistakes made by certain public bureaucrats.

A final topic of significant debate pertained to school and work closures. As part of the effort to socially isolate, democratic leaders and many public health agencies recommended that school and work activities be moved to online and virtual formats. Republicans argued that the virus was not as contagious as initially believed and worried about the impact of closures on student and economic performance. They argued that people suffered unnecessarily due to a perceived or actual overreaction.

Thus, almost all of the major decisions made by politicians and leaders became of subject of great debate and substantial policy disagreement (McEntire 2022b).

15.3.3 Responding to Epidemics and Pandemics

The experience of prior and recent epidemics and pandemics reveals several important lessons to react in the most effective manner possible (see Fauci and Folkers 2023). Recommendations include:

1. Adequate nutrition, sufficient exercise, proper sanitary conditions, and the promotion of overall physical well-being must become higher priorities for all people going forward.
2. Understanding of scientists, public health officials, and medical personnel, to the greatest extent possible, the exact nature of infections, their impacts, and possible treatments.
3. Judiciously weighing all possible positive and negative consequences of decisions regarding masks, vaccines, social isolation, and school and work closures.

4. Carefully crafting messages and distributing in a consistent manner to the public.

5. Augment the capacity of public health and medical facilities as well as doctors and nurses to meet the extraordinary demands that are placed upon them when epidemics and pandemics occur.

Making these changes will reduce the impact of contagious diseases and improve responses to major public health emergencies.

Self-Check

1. An epidemic is more widespread than a pandemic. True or false?

2. The treatment of Thomas Duncan's infection illustrates that public health officials and hospitals were prepared for Ebola. True or false?

3. Which of the following is/are true statements about COVID-19?

   a. There are no debates about the origin of the coronavirus.

   b. President Trump instituted a travel ban on flights coming into the United States from China.

   c. Republicans may have underreacted to COVID-19.

   d. Democrats may have overreacted to COVID-19.

   e. b, c, and d.

4. How can public health officials prepare for pandemics?

15.4   Global Warming and Climate Change

Another issue that is frequently mentioned by those working in homeland security and emergency management is climate change (Schneider 2011). It is believed that global warming and other climate variations are increasing the likelihood of various disasters along with the possibility of serious social disruption. These topics are politically charged, of course, but there is a need to consider policies on how to address such problems, nonetheless.

15.4.1   Definitions and Evidence

Global warming and climate change are frequently discussed in educational settings and in public policy these days. But what exactly do these concepts mean or imply? **Global warming** may be defined as the rise of temperatures in the earth's atmosphere, whereas **climate change** may include episodes of global warming and/or global cooling. Such changes—warming or cooling—are noteworthy since global temperature affects weather patterns around the world.

According to scientists, the earth's temperature "has risen by an average of 0.14° Fahrenheit (0.08°Celsius) per decade since 1880" (Lindsey and Dahlman 2023). Put differently, the average global temperature "has increased in total by a little more than 1 degree Celsius, or about 2 degrees Fahrenheit" (MacMillan and Turrentine 2021).

There is growing and irrefutable evidence that the earth has been warming over the past several decades. NASA (2023) asserts that temperature is rising, the ocean is getting warmer, ice sheets/glaciers/snow cover are retreating, and sea level is rising. But what is causing this to happen and is it permanent?

15.4.2 Climate Change Explained

Global warming may have multiple causes, but many scientists argue that it is generally driven by human activity including the use of fossil fuels (Bergquist et al. 2022). In some ways, different amounts of solar radiation enter the earth's atmosphere and may lead to variations in global temperature. Energy from the sun fluctuates on a daily basis and arrives at the earth as light and radio/x-rays and gamma rays. This causes the temperature of the earth to change in minute ways over time. In other ways, the actions of people over time have produced what is known as the "**Greenhouse Effect**" or the collection of heat in our atmosphere much like a greenhouse building retains heat for plants because the sun warms the air inside the enclosed structure.

While the sun is certainly a cause of climate change along with the greenhouse effect, many scientists argue that human actions are the primary culprit today. As an example, the Natural Resources Defense Council asserts:

> *Global warming occurs when carbon dioxide (CO2) and other air pollutants collect in the atmosphere and absorb sunlight and solar radiation that have bounced off the earth's surface. Normally this radiation would escape into space, but these pollutants, which can last for years to centuries in the atmosphere, trap the heat and cause the planet to get hotter. (MacMillan and Turrentine 2021)*

In other words, the belief is that the burning of fossil fuels in industrial manufacturing, home heating and cooling, and vehicle usage are creating harmful emissions and polluting gasses that trap the heat in the atmosphere and cause the temperature to rise.

15.4.3 Concerns About Global Warming

Rising temperatures are of the utmost concern because they can lead to a variety of hazards and disasters. For instance, global warming may result in additional and more intense heat waves, droughts, wildfires, atmospheric rivers, and tropical flooding as well as the occurrence of sea level rise and more frequent and extreme hurricanes and tornadoes (Bullock, Haddow, and Haddow 2008). California provides a unique case in point.

California experienced major droughts in the 1920s and 1930s, from 1976 to 1977, from 1987 to 1992, and from 2007 to 2009. These episodes may be related to El Niño or La Niña cycles in the Pacific Ocean, but the overall trend is disturbing nonetheless. In 2021, the state experienced the second driest year due to excessive heat and limited rains and snow. This has made California a tinderbox. The California Air Resources Board (2023) has noted that since the 1950s the "area burned by ... wildfires has been increasing" and that "8 of the 20 largest fires in state history have occurred since 2017."

In 2017, California witnessed over 9,500 forest fires including the Thomas Fire. This became the largest fire in California history but it was later surpassed by the Ranch Fire in 2018. Other major fires include the Kinkade fire in 2019, the Complex fire in 2020, the Dixie fire in 2021, and the Mosquito fire in 2022.

These disastrous events in California have burned millions of acres, destroyed thousands of homes, taken hundreds of lives, and resulted in unfathomable expenses and negative economic repercussions. This does not include the impacts of other hazards or the grave consequences that would be seen if sea level rise inundates coastal communities.

In The Real World

In 2023, California was battered by major storms, damaging winds, and a deluge of rain that affected communities across the state. At least 12 atmospheric fueled rivers—"plumes of tropical moisture that stretch thousands of miles across the Pacific Ocean"—"dropped trillions of gallons of water on" California (Prociv and Bush 2023). As a result, roads were damaged by avalanches and mudslides, homes were flooded, and many people lost their lives overflowing rivers with treacherous currents. These types of events caused more than $1 billion each year and resulted in 84% of the flood damage in Western states.

Some people argue that these types of events are the negative result of global warming. Others might have different explanations. For instance, some could argue that the Hunga Tonga-Hunga Ha'api underwater volcanic eruption on January 15, 2022, was to blame. This event produced the highest volcanic plume in satellite records and released more than 146 teragrams of water into the atmosphere (Patel 2022). The eruption may have increased water vapor in the atmosphere by 10% and resulted in higher global temperatures. Whatever the cause, those working in homeland security and emergency management must do all they can to prevent and prepare for these types of disasters.

15.4.4   The Politics of Global Warming and Climate Change

Although scientific evidence accurately denotes rising temperatures and the consequences of climate change are indeed worrisome, this does not mean that global warming is without controversy.

Critics of climate change have levied a number of arguments against the argument and evidence of fossil fuel–induced global warming (see Shellenberger 2019). These cannot be treated in this chapter in full detail due to space limitations, but there are several that can be mentioned here. Opponents of global warming argue that the earth's temperatures have always varied over time and that the current rise in temperatures is just one part of that recurring trend. These individuals assert that scientists must ask themselves how they are measuring global warming since the starting date can have an enormous impact on the amount of increase witnessed at any given point in time. In addition, critics reveal that volcanic eruptions can influence greenhouse gas emissions and they suggest that environmental protection and mismanagement of forests have led to fires that produce more emissions than all vehicles in the United States in an entire year. Furthermore, critics assert that the United States has made enormous progress in reducing greenhouse emissions and that "climate change warriors" are not consistent in their expectations for China, India, and other nations that also create significant pollution. Critics also call attention to the benefits

In The Real World

Extreme Weather and Infrastructure

Natural disasters do not result from global warming and climate change alone. There are many other types of hazards that may trigger disasters (e.g., earthquakes, industrial accidents, terrorist attacks). In addition, there is also the likelihood of catastrophic outcomes when freezing weather interacts with old, outdated, and fragile critical infrastructure. The impact of Winter Storm Uri in Texas provides a vivid example of the cascading effects and disruptive outcomes when extreme weather curtails the production and distribution of electricity (Spractes and McEntire 2023; Chand and McEntire 2023).

From February 13 through February 17, 2021, the state of Texas experienced a major storm system which blanketed the entire state with snow and ice. This created a major car pile-up in Fort Worth that involved more than 135 vehicles as well as countless fender benders in cities throughout the region. In addition, freezing temperatures were experience everywhere in the state and often at record-breaking levels. For instance, places as far south as Galveston, Houston and College Station saw the thermometer plunge to 20, 13, and 5 degrees Fahrenheit respectively. This situation along with the wind chill created a huge demand for power as people tried to stay warm at home and at work. At the same time, more than 1,000 power generating stations throughout Texas started to fail because they were not sufficiently winterized.

Seeing this major shortfall between supply and demand, the Electric Reliability Council of Texas (ERCOT) decided to implement rolling blackouts which were supposed to last between 20 minutes to one hour. It is estimated that as much as 69% of Texans did not have power when they needed it the most. Unfortunately, the blackouts lasted much longer than predicted (even multiple days in some locations) and this produced devastating consequences. The loss of power rendered many gas stations unable to operate, and this generated a shortage of fuel for emergency personnel and others. There were instances where some individuals froze to death due to the combination of frigid weather and lack of heating in their homes. In other cases, people tried to stay warm by sitting in a running car in their garage or by lighting candles and bar-b-que grills in their homes and apartments. This resulted in numerous fires that gutted various structures and at least 121 deaths from carbon monoxide poisoning.

In addition, because of the lack of heating, water lines started to freeze up in homes, businesses, and government facilities. Pipes burst in countless buildings while water mains also broke under streets in various locations in the state. Flood damage was commonplace and a boil-water notice was issued to more than 14 million people. Critical facilities like hospitals did not have water to care for patients or to keep server rooms at proper temperatures due to the loss of chillers. Nursing homes had to close down because of the lack of heat and water, but there were not enough shelters open due to the icy conditions of roads, the loss of electricity, and the presence of COVID-19. As can be seen, this cascading disaster produced severe consequences for nearly everyone in the state of Texas.

Sadly, at least some of the negative outcomes could have been prevented. ERCOT had experienced similar problems with electricity due to hurricanes or prior winter storms and excessive summer heat. This organization was advised by government officials to update their power systems but they did not follow the recommendations in some cases. Winter Storm Uri was a good reminder to prepare for the impacts of extreme weather and fortify critical infrastructure against the formidable power of mother nature.

of fossil fuels and underscore the fact that the number of fatalities owing to disasters has diminished over time. Finally, critics assert that environmentally friendly sources of energy are limited (e.g., the sun is not always shining, and the wind is not always blowing). This brings up possible solutions to global warming and climate change.

15.4.5 How to Address Climate Change

To limit greenhouse gas emissions, environmental advocates argue that fundamental changes are needed in regard to society (Moser and Dilling 2007) and to energy production and usage. The argument is that the United States and countries around the world must end or reduce their reliance on fossil fuels such as oil, gasoline, and natural gas. Instead of relying on these traditional sources of energy, the world must promote green, renewable energy from solar farms, wind turbines, and hydrologic sources.

Furthermore, the transition of the economy from relying on fossil fuels to renewable sources would require significant cultural changes regarding political values and economic development (Natural Resource Defense Council 2023). People would have to be more concerned about energy usage and pollution. This would require ongoing education about the negative impacts of fossil fuels and other measures to limit the out-of-control consumption mentality that permeates our capitalist society. In addition, developed nations would have to transfer funds to developing nations so they have adequate resources to alter their means of production. Everyone must work together to fundamentally transform how our world operates. In short, a global solution involving individuals, families, businesses, nonprofit organizations, and governments will be required.

Of course, it is true that not everyone may be willing to make such changes. For instance, critics of global warming environmental policies point out certain ironies and inconsistencies of their counterparts. These individuals and groups assert that battery-operated green vehicles are very expensive and are actually powered by fossil fuels (e.g., coal and natural gas provides electricity). In addition, people have pointed out that producing batteries for electric vehicles is extremely damaging to the environment (e.g., lithium extraction mining processes use lots of fossil fuels and leave a wasteland in their wake) (Lakshmi 2023). This is to say nothing about the impact of disposing of batteries that no longer function. Some companies like Hertz have recently sold their electric vehicles due to a variety of concerns about their operation and others assert that our infrastructure is not sufficient to address battery recharging. Consequently, it remains to be seen to what degree and how fast people are willing to address the problem of global warming and climate change.

Self-Check

1. Science reveals that average global temperatures have generally risen since 1880. True or false?

2. Everyone agrees with the cause or causes of global warming and what to do about it? True or false?

3. What is/are the possible or actual impacts of climate change?
   a. Sea level rise
   b. More droughts and forest fires
   c. Stronger storms and increased flooding
   d. All of the above

4. What steps could or should be taken to reduce global warming?

Summary

As has been illustrated in this chapter, those working in homeland security have many responsibilities beyond terrorism, disasters, and sector security. In fact, the United States is witnessing many problems that are creating apprehension about our safety and well-being. Mass shootings have been increasing in frequency over the past several decades. General crimes are rising, but organized crime involving drugs and human trafficking is spiraling out of control. The United States has also experienced substantial social disturbances in recent years. Americans' right to protest is guaranteed by the Constitution, but riots have resulted from grievances. These riots are troubling and must be dealt with due to the likelihood of injuries and property damage. A variety of epidemics and pandemics have also occurred in recent decades and the COVID-19 public health emergency resulted in hundreds of thousands of deaths and a shutdown of schools and the economy. In spite of passioned disagreement, global warming and climate change have increasingly become prevalent topics in homeland security policy due to the actual or potential impact they are having on a variety of hazards. For these reasons, homeland security must take several steps in conjunction with various personnel from law enforcement, public health, and environmental stakeholders. A team effort will be needed to address these unique problems in homeland security.

Assess Your Understanding

Understand: What Have You Learned?

Go to **www.wiley.com/go/mcentire/homelandsecurity3e** to assess your knowledge of how to evaluate other pressing problems.

Summary Questions

1. After the Columbine School shooting, the standard protocol was to neutralize the threat posed by mass shooters. True or false?
2. The worst mass shooting in history occurred at a King Soopers grocery store in Colorado. True or false?
3. "Run, hide, fight" is the public education slogan to help people know what to do in case of a mass shooting. True or false?
4. Crime statistics generally rose in dramatic fashion from 2020 to 2022. True or false?
5. The legalization of marijuana in some states may create increased usage throughout the nation. True or false?
6. Drug cartels have dug tunnels into the United States to keep the pipeline of illegal narcotics open. True or false?
7. Human traffickers only profit from forced prostitution. True or false?
8. INTERPOL is an organization devoted to minimizing protests and riots. True or false?
9. A protest is more violent than a riot. True or false?

10. The January 6 protest was especially concerning since the violent entry into the U.S. Capital was initiated to prevent certification of the national election. True or false?

11. A pandemic is a large public health emergency where a disease is spreading around the world. True or false?

12. Global warming is equivalent to climate change. True or false?

13. The "greenhouse" effect suggests that heat from the sun is being retained in the earth's atmosphere. True or false?

14. Proposals to minimize mass shootings include which of the following?

    a. More observant school administrators, teachers, and parents
    b. Increased gun control
    c. Further treatment for mental illnesses
    d. All of the above

15. Which of the following is considered a misdemeanor?

    a. Rape
    b. Murder
    c. Jay walking
    d. Theft and burglary

16. What percentage of Americans have experimented with drugs in their lifetime?

    a. 10
    b. 25
    c. 35
    d. 50
    e. 62

17. Organized crime syndicates from which two countries work together to manufacture fentanyl and distribute it in the United States?

    a. Afghanistan and Albania
    b. Mali and Afghanistan
    c. Afghanistan and Mexico
    d. Mexico and Canada
    e. China and Mexico

18. Which nonprofit organization was created to combat human trafficking?

    a. Stop Human Trafficking
    b. Act Now
    c. Operation Underground Railroad
    d. CARE
    e. A Higher Calling

19. Causes of recent protests and riots include:

    a. Economic inequality
    b. Racial tensions
    c. Policies relating to COVID-19
    d. All of the above
    e. b and c only

20. Which Occupy Movement was in reaction to the death of George Floyd?
   a. Occupy Seattle
   b. Occupy Wallstreet
   c. Occupy Minnesota
   d. Occupy the Nation
   e. Occupy Police Precinct 22
21. Which disease made its way to the United States from Africa in 2014?
   a. Ebola
   b. Coronavirus
   c. SARS
   d. West Nile
   e. MERS
22. Critics of climate change would make which of the following arguments?
   a. The climate has always changed over time.
   b. Things like volcanic eruptions and forest fires may increase the presence of CO_2 in the atmosphere.
   c. Some practices to curb global warming (e.g., the use of electric vehicles) may ironically pose hazards to the natural environment.
   d. Policies to reduce pollution may not be equally applied in countries around the world.
   e. All of the above.

Applying This Chapter

1. Explain the lessons derived from major mass shootings in the United States over the past three decades.
2. Describe how the drug trade and human trafficking underscore the importance of border control.
3. How are protests and riots similar? How are they different?
4. Illustrate why COVID-19 became a politicized public health emergency.

Be a Homeland Security Professional

Preventing and Reducing Crime

You have just been appointed as the police chief in Los Angeles, and you are proposing a budget to a city council that has been extremely critical of police brutality. What could you say in your presentation to illustrate the importance of investing money in your police department?

Learning from the Response to COVID-19

As the new Director of the National Institute of Allergy and Infectious Diseases, you have been tasked with writing a report to make sure we do not prevent the mistakes associated with COVID-19. What are three or four important points you would make?

Making the Case for Global Warming Policies

Channel 4 News has asked to interview you on the importance of addressing global warming because of your position in the Environmental Protection Agency. What would you say in the interview to illustrate why global warming is a significant threat, and what steps would you recommend to counter it?

Key Terms

Bacterial infections   A disease that results from cells that can live outside the body

CHAZ   The Capital Hill Autonomous Zone in Seattle, Washington

Climate change   The name given to episodes of global warming and/or global cooling

Coronavirus   The virus known as SARS-CoV-2 or COVID-19 that caused the most concerning pandemic in recent years

Drug cartels   Organized crime syndicates that cultivate, manufacture, and distribute drugs

Epidemic   A smaller-scale public health emergency where a disease is spreading actively in a specific region

General crimes   Minor infractions and violations (such as jaywalking and motor vehicle offenses)

George Floyd   A 46-year-old African American man who was killed by Derek Chauvin, a police officer in Minneapolis

Global warming   The rise of temperatures in the earth's atmosphere

Greenhouse effect   The collection of heat in our atmosphere much like a greenhouse building retains heat for plants because the sun warms the air inside the enclosed structure

Human trafficking   Modern day slavery where people are exploited through force, fraud or coercion

Illicit drugs   Illegal narcotics and the abuse of legal prescription pharmaceuticals

INTERPOL   The International Criminal Police Organization

January 6   The name given to the riot due to the mob that gathered at, attacked, and took over the U.S. Capitol building to protest Trump's defeat in the 2020 election and stop the certification of the electoral college votes

Major felonies   Major offenses such as rape and murder

Mass shooting   An attack that is committed by a perpetrator with a pistol, shotgun, or rifle that injures and kills multiple individuals

Misdemeanors   More serious offences such as driving under the influence, theft and burglary, or assault

Occupy Movement   The gathering of protesters that decry social and economic inequality around the world

Occupy Seattle   One of the most notable protests in Seattle, Washington, that occurred after the death of George Floyd

Occupy Wall Street   A protest in New York City by young people to bring attention to the income gap and global capitalist activities

Operation Underground Railroad A nonprofit organization that works tirelessly to stop human traffickers and rescue the victims of this disturbing type of organized crime

Organized crime The name given to the powerful groups that carefully plan and execute criminal activities usually with the purpose of economic gain

Pandemic A larger public health emergency where a disease is spreading internationally and affecting many countries around the globe

Riots A type of social disturbance that includes rock throwing, tipping over vehicles, looting, starting fires, and attacking law enforcement personnel

Second Amendment An addendum to the constitution that allows people in the United States to keep and bear arms

Social disturbances Mass gathering events including protests and riots

Viral infections A disease that results from nonliving molecules that need a host to survive

References

ABC News. (2015). "Hidden America: Chilling New Look at Sex Trafficking in the U.S." Nightline. https://www.youtube.com/watch?v=SgTmcq-bBk. (Accessed July 20, 2023).

Afriatni, A. (2019). "After the New Zealand Shootings, Indonesia's Islamic Council Mulls Videogame Fatwa." Benar News. https://www.benarnews.org/english/news/indonesian/game-fatwa-03262019170244.html. (Accessed July 18, 2019).

American Psychiatric Association. (2023). "Americans Express Worry Over Personal Safety in Annual Anxiety and Mental Health Poll." https://www.psychiatry.org/news-room/news-releases/annual-anxiety-and-mental-health-poll-2023. (Accessed July 17, 2023).

Anderson, D.A. (2021). "The Aggregate Cost of Crime in the United States." *The Journal of Law and Economics* 64 (4): 661–885.

Anderson, James A. (2021). "Some Say Occupy Wall Street Did Nothing. It Changed Us More Than We Think." *Time* (November 15). https://time.com/6117696/occupy-wall-street-10-years-later/. (Accessed September 13, 2023).

Associated Press. (2000). "Seattle Police Admit They Were Unprepared for WTO Protests." *Los Angeles Times* (April 5) https://www.latimes.com/archives/la-xpm-2000-apr-05-mn-16203-story.html. (Accessed July 27, 2023).

Bergquist, P., Marlson, J.R., Goldberg, M.H., Gustafson, A., Rosenthal, S.A., and Leiserowitz, S.A. (2022). "Information About the Human Causes of Global Warming Influences Causal Attribution, Concern, and Policy Support Related to Global Warming." *Thinking & Reasoning* 28 (3): 465–486.

Black Past. (2023). "Racial Violence in the United States Since 1526." https://www.blackpast.org/special-features/racial-violence-united-states-1660/. (Accessed August 20, 2023).

Britannica. (2023). "Riot." August 17. https://www.britannica.com/topic/riot. (Accessed August 22, 2023).

Brunner, Jim. (2019). "50,000 Protesters, Tear Gas—and Madeleine Albright Trapped in Her Hotel: How the 1999 WTO Protests Changed Seattle." *The Seattle Times* (November 30) https://www.seattletimes.com/seattle-news/wto-seattle-protests-20-years-later-do-they-matter/. (Accessed August 20, 2023).

Bullock, J.A., Haddow, G.D., and Haddow, K.S. (2008). *Global Warming, Natural Hazards, and Emergency Management*. CRC Press: Boca Raton, FL.

Bureau of International Narcotics and Law Enforcement Affairs. (2023). Addressing Illicit Drug Challenges. https://www.state.gov/bureaus-offices/

under-secretary-for-civilian-security-democracy-and-human-rights/bureau-of-international-narcotics-and-law-enforcement-affairs/. (Accessed July 18, 2023).

Bureau of Justice. (2022). "Human Trafficking Data Collection Activities, 2022." https://bjs.ojp.gov/library/publications/human-trafficking-data-collection-activities-2022. (Accessed July 22, 2023).

California Air Resources Board. (2023). "Wildfires & Climate Change." https://ww2.arb.ca.gov/wildfires-climate-change. (Accessed August 20, 2023).

Chand, M. and D.A. McEntire. (2023). The February 2021 Winter Storm and its Impact on Texas infrastructure: Lessons for communities, emergency managers, and first responders. Journal of the Risks, Hazards, and Crisis in Public Policy. 1-20.

Clark, Peter Allen. (2020). "Seattle Protestors Are Occupying 6 City Blocks as an 'Autonomous Zone.' Some Fear a Police Crackdown Is Imminent." *Time* (June 11). https://time.com/5851774/seattle-police-capitol-hill-autonomous-zone/. (Accessed September 10, 2023).

Columbia University Department of Psychiatry. (2022). "Is There a Link Between Mental Health and Mass Shootings?" https://www.columbiapsychiatry.org/news/mass-shootings-and-mental-illness. (Accessed July 18, 2023).

Daniel, Bukszpan (2013). "America's Most Destructive Riots of All Time." CNBC (September 13). https://www.cnbc.com/2011/02/01/Americas-Most-Destructive-Riots-of-All-Time.html. (Accessed August 3, 2023).

Deliver Fund. (2023). "Facts About Human Trafficking in the US." https://www.deliverfund.org/biog/facts-about-huyman-trafficking-in-united-states. (Accessed July 22, 2023).

Department of Health and Human Services. (2019). "Human Trafficking Within and Into the United States: A Review of the Literature." Office of the Assistant Secretary for Planning and Evaluation. https://aspe.hhs.gov/report/human-trafficking-and-within-united-states-review-literature#Trafficking. (Accessed July 22, 2023).

Department of Homeland Security. (2022). "Countering Human Trafficking: Year in Review." https://www.dhs.gov/sites/default/files/2022-02/CCHT%20Annual%20Report.pdf. (Accessed July 22, 2023).

Department of Homeland Security. (2023). "Tools That Teach: What Is Human Trafficking?" https://www.dhs.gov/blue-campain/tools. (Accessed July 20, 2023).

Department of State. (2023). "About Human Trafficking." https://www.state.gov/humantrafficking-about-human-trafficking/#:~:text=This%20report%20estimates%20that%2C%20at,forced%20labour%20imposed%20by%20state. (Accessed July 22, 2023).

D'Souza, Dinesh. (2022). "2000 Mules." https://node-3.2000mules.com. (Accessed August 20, 2023).

Dudenhoefer, Nicole. (2020). "7 Influential Protests in American History." UCF Today (July 2). https://www.ucf.news/7-influential-protests-in-american-history/. (Accessed August 20, 2023).

Fauci, Anthony S. and Gregory K. Folkers. (2023). "Pandemic Preparedness and Response: Lessons From COVID-19. *The Journal of Infectious Diseases* 228 (4): 422–425.

Federal Bureau of Investigation. (2021). "National Incident-Based Reporting (NIBRS) Details Reported in the United States." https://cde.ucr.cjis.gov/LATEST/webapp?#?pages/explorere/crime/crime-trend. (Accessed July 17, 2023).

Federal Emergency Management Agency. (2018). 1 October After-Action Report [Las Vegas Shooting]. Homeland Security Digital Library. https://www.hsdl.org/c/abstract/?doic=814668. (Accessed August 13, 2023).

Fox 5 Staff. (2023). "NYPD Releases 2022 Crime Numbers." https://www.fox5ny.com/new/2022-nyc-crime-numbers. (Accessed July 17, 2023).

Harvard. (2020). "Black Livers Matter May Be the Largest Movement in U.S. History." Kennedy School, Carr Center for Human Rights Policy." July 3. https://carrcenter.hks.harvard.edu/news/black-lives-matter-may-be-largest-movement-us-history. (Accessed September 10, 2023).

Healthline. (2023). "The Worst Outbreaks in U.S. History." May 10. https://www.healthline.com/health/worse-disease-outbreaks-history. (Accessed August 20, 2023).

Homeland Security Digital Library. (2022). "FBI Releases 2021 Crime Statistics." https://www.hsdl.org/c/fbi-releases-2021-crime-statistics. (Accessed July 17, 2023).

Institute for Constitutional Advocacy and Protection. (no date). "Protests & Public Safety: A Guide for Cities and Citizens." *Georgetown Law.* https://constitutionalprotestguide.org. (Accessed September 1, 2023).

Johnson, B. (2022). "Fatherlessness One of the Biggest Factors Driving Unemployment, Crime." *The Washington Stand.* https://washingtonstand.com/news/fatherlessness-one-of-the-biggest-factors-driving-unemployment-crime-researcher#:~:text=But%20nearly%20one%2Dthird%20of,education%20and%20lack%20of%20employment. (Accessed July 17, 2023).

KING 5 Staff. (2020). "Seattle Wants to Dismantle 'CHOP' Zone and Return to East Precinct." KING 5 News. June 22. https://www.king5.com/article/news/local/seattle-mayor-to-make-city-budget-changes-for-black-communities-in-response-to-chop-protests/281-de995e1b-27fc-4089-bb9e-e68b389901a6. (Accessed September 13, 2023).

Koranyi, B. (2012). "Norway Could Have Prevented Breivik Massacre, Says Commission." Reuters. https://www.reuters.com/article/us-breivik-commission/norway-could-have-prevented-breivik-massacre-says-commission-idUSBRE87C0PE20120813. (Accessed July 18, 2023).

Kovaleski, S.F. and M. Baker. (2023). "Gunman in 2017 Las Vegas Shooting Was Angry at Casinos, New F.B.I. Files Show." *The New York Times.* https://www.nytimes.com/2023/03/30/us/las-vegas-shooting-gunman.html. (Accessed July 18, 2023).

Krishnakumar, P. (2022). "The FBI Released Its Crime Report for 2021—But It Tells Us Less About the Overall State of Crime in the US Than Ever." https://www.cnn.com/2022/10/05/us/fbi-national-crime-report-2021-data/index.html. (Accessed July 17, 2023).

Laksmi, R.B. (2023). "The Environmental Impact of Battery Production for Electric Vehicles." Earth.org (January 11) https://earth.org/environmental-impact-of-battery-production/. (Accessed August 20, 2023).

Lehman, C.F. (2022). "Yes, New York's Bail Reform Has Increased Crime." City Journal. https://www.city-journal.org/article/yes-new-yorks-bail-reform-has-increased-crime. (Accessed July 17, 2023).

Lindsey, Rebecca and Luann Dahlman. (2023). "Climate Change: Global Temperature." NASA. https://www.climate.gov/news-features/understanding-climate/climate-change-global-temputerature#:~:text=Earth%27s%20temperature%20has%20risen%20by,0.18%20C)%20per%20decade. (Accessed August 20, 2023).

MacMillan, Amanda and Jeff Turrentine. (2021). "Global Warming 101." Natural Resources Defense Council. https://www.nrdc.org/stories/global-warming-101#warming. (Accessed September 13, 2023).

Martens, T. and W.M. Generes. (2022). "How Drugs & Alcohol Fuel Violent Behaviors." American Addiction Centers. https://americanaddictioncenters.org/rehab-guide/addiction-and-violence. (Accessed July 17, 2023).

McEntire, David A. (2019). "The Dallas Ebola Incident as an Indicator of the Bioterrorism Threat: An Assessment of Response with Implications for Security and Preparedness." *UVU Journal of National Security* 3 (2): 5–18.

McEntire, David A. (2022a). "When Emergencies and Disasters Collide: Lessons from the Responses to the Magna, Utah Earthquake During the Covid-19 Pandemic." *Journal of Emergency Management* 18 (7): 1–19.

McEntire, David A. (2022b). "An Assessment of Crisis Communication During the Covid-19 Pandemic." *Journal of Homeland Security and Emergency Management.* https://doi.org/10.1515/jhsem-2022-0013. (Accessed August 20, 2023).

McGhee, Kaylee. (2021). "More Evidence That the Wuhan Lab-Leak Theory Is the Correct One." *Washington Examiner* (May 7) https://www.washingtonexaminer.com/opinion/more-evidence-that-the-wuhan-lab-leak-theory-is-the-correct-one. (Accessed August 26, 2023).

Moser, Susanne C. and Lisa Dilling. (2007). "Toward the Social Tipping Point: Creating a Climate for Change." *Creating a Climate for Change: Communicating Climate Change and Facilitating Social Change.* 491–516.

NASA. (2023). "How Do We Know Climate Change Is Real?" https://climate.nasa.gov/evidence. (Accessed September 5, 2023).

National Center for Drug Abuse. (2023). Drug Abuse Statistics. https://drugabusestatistics.org. (Accessed July 18, 2023).

National Institute on Drug Abuse. (2023). Drug Overdose Rates. https://nida.nih.gov/research-topics/trends-statistics/overdose-death-rates. (Accessed July 18, 2023).

Natural Resource Defense Council. (2023). "How You Can Stop Global Warming." August 7. https://www.nrdc.org/stories/how-you-can-stop-global-warming. (Accessed August 20, 2023).

Naylor, Brian. (2021). "Read Trump's Jan.6 Speech, A Key Part of Impeachment Trial." NPR (February 10) https://www.npr.org/2021/02/10/966396848/read-trumps-jan-6-speech-a-key-part-of-impeachment-trial. (Accessed July 7, 2023).

Noh, A. (2022). "New Data Reveals Link Between Homelessness and Crime Wave in California." 600KOGO News Radio. https://kogo.iheart.com/featured/the-demaio-report/content/2022-03-29-new-data-reveals-link-between-homelessness-and-crime-wave-in-california/#:~:text=Crime%20data%20released%20by%20District,substance%20abuse%20treatment%20they%20need. (Accessed July 17, 2023).

Ohanian, L. (2021). "Why Shoplifting Is De Facto Legal in California." Hoover Institution. https://www.hoover.org/research/why-shoplifting-now-de-facto-legal-california#:~:text=Because%20state%20law%20holds%20that,want%20to%20take%20the%20risk. (Accessed July 17, 2023).

Patel, Kashna. (2022). "Tonga Volcano Blasted Unprecedented Amount of Water into the Atmosphere." *The Washington Post* (August 5) https://www.washingtonpost.com/climate-environment/2022/08/05/volcano-eruption-tonga-record-climate/. (Accessed August 22, 2023).

PBS. (2019). "Sex Trafficking in America." *Frontline*. https://www.youtube.com/watch?v=waRNXRaHH34. (Accessed August 20, 2023).

Police 1. (2019). "8 Recommendations From the FEMA Oct. 1 After-Action Report." https://www.police1.com/active-shooter/articles/8-recommendations-from-the-fema-oct-1-after-action-report-uiPfdApVGqVOpNJJ/. (Accessed July 18, 2023).

Prociv, Kathryn and Evan Bush. (2023). "Plagued by Drought, California Is Now Soaked After 12 Monstrous Storms." NBC News (March 23) https://www.yahoo.com/lifestyle/plagued-drought-california-now-soaked-225605641.html#:~:text=NBC%20News-,California%20is%20dangerously%20saturated%20after%2012%20monstrous,and%20more%20water%20is%20coming&text=Battered%20by%20a%20seemingly%20endless,atmospheric%20river%20storm%20since%20Christmas. (Accessed August 23, 2023).

Property Club Team. (2023). "Is San Francisco Safe? Crime Rates and Most Dangerous Neighborhoods." https://propertyclub.nyc/article/is-san-francisco-safe-crime-and-most-dangerous-neighborhoods. (Accessed July 17, 2023).

Rosenberg, M., Dabrowski, T. and Klingner, J. (2023). "Chicago, New Orleans Were the nation's Murder Capitals in 2022—A Wirepoints Survey of America's 75 Largest Cities." https://www.wirepoints.org/chicago-new-orleans-were-the-nations-murder-capitals-in-2022-a-wirepoints-survey-of-americas-75-largest-cities/. (Accessed July 17, 2023).

Sawyer, W. (2017). "BJS Report: Drug Abuse and Addiction at the Root of 21% of Crimes." Prison Policy Initiative. https://www.prisonpolicy.org/blog/2017/06/28/drugs/. (Accessed July 17, 2023).

Schneider, R.O. (2011). "Climate Change: An Emergency Management Perspective." *Disaster Prevention and Management* 20 (1): 53–62.

Shellenberger, Michael. (2019). "Why Apocalyptic Claims About Climate Change Are Wrong." *Forbes* (November 25) https://www.forbes.com/sites/michaelshellenberger/2019/11/25/why-everything-they-say-about-climate-change-is-wrong/?sh=5e9623ef12d6. (Accessed August 20, 2023).

Smith, J. (2022). "Many Argue Poor Education Is 'Root Cause' of Violent Crime." The National Desk. https://thenationaldesk.com/news/americas-news-now/many-argue-poor-education-is-root-cause-of-violent-crime-murders-chicago-baltimore-high-juvenile-crime-rates-muriel-bowser-statistics-carjackings-shootings-homicides-marty-walsh-fbi-data-homicide-rates-large-cities-high-crime. (Accessed July 17, 2023).

Spraktes, F. and D.A. McEntire. (2023). Responding to the Texas Freeze: A case study of the reaction to the cascading effects of a complex disasters. *Journal of Homeland Security and Emergency Management*. https://doi.org/10.1515/jhsem-2022-0025.

The White House. (2021a). "Fact Sheet: The Biden Administration Launches New Efforts to Counter Transnational Criminal Organizations and Illicit Drugs." https://www.whitehouse.gov/briefing-room/statements-releases/2021/12/15/fact-sheet-the-biden-administration-launches-new-efforts-to-counter-transnational-criminal-organizations-and-illicit-drugs/. (Accessed July 19, 2023).

The White House. (2021b). "The National Action Plan to Combat Human Trafficking." https://www.whitehouse.gov/wp-content/uploads/2021/12/National-Action-Plan-to-Combat-Human-Trafficking.pdf. (Accessed July 22, 2023).

Vara, Vauhini. (2014). "The Occupy Movement Takes on Student Debt." *New Yorker* (September 17) https://www.newyorker.com/business/currency/occupy-movement-takes-on-students-debt. (Accessed September 10, 2023).

Washington Desk. (2023). "Wray Publicly Comments on the FBI's Position on COVID's Origins, Adding Political Fire." NPR (February 28) https://www.tpr.org/2023-02-28/wray-publicly-comments-on-the-fbis-position-on-covids-origins-adding-political-fire. (Accessed August 10, 2023).

Wisevoter. (2023). "Rape Statistics by State." https://wisevoter.com/state-rankings/rape-statistics-by-state/. (Accessed July 17, 2023).

CHAPTER 16

Looking Toward the Future

Challenges and Opportunities

DO YOU ALREADY KNOW?

- The major lessons to be drawn from this book

- The need for accountability in homeland security

- What concepts can guide homeland security policy

- Where research in homeland security is incomplete

For additional questions to assess your current knowledge of the future of homeland security, go to **www.wiley.com/go/mcentire/homelandsecurity3e**

WHAT YOU WILL LEARN

16.1 **The need to review what is known about terrorism and other threats**

16.2 **Why accountability is imperative for homeland security**

16.3 **Alternative approaches to homeland security policy**

16.4 **What research will improve homeland security**

WHAT YOU WILL BE ABLE TO DO

- Recall the major recommendations for homeland security

- Promote responsible stewardship of government resources

- Assess how liability reduction and capacity building can improve homeland security

- Implement practical recommendations for homeland security

Introduction

As has been illustrated throughout this book, your role as a participant in homeland security is vital for the national interests of the United States and countries everywhere. Advancing the homeland security profession will occur when you and others remember and apply the major lessons in this book. Another important obligation you have is to promote accountability and develop coherent policies in homeland security to ensure that resources are spent carefully and that strategic priorities are clearly identified. It is also imperative that you understand what research is required to improve homeland security and how the principles from this scholarship can be applied to reduce the threat and consequences of terrorism. The central theme of this chapter is that success is more likely to be achieved when those working in homeland security are knowledgeable, when they use resources wisely, when they recognize opportunities to improve policy, and when they work diligently to accomplish strategic goals.

16.1 The Lessons of This Book

If you already work in homeland security or will be employed in this profession in the future, chances are that you perform or will have a very specialized function. That is to say, you probably will work in one agency and within a limited scope of responsibility. Nevertheless, it is vital that you gain a broad understanding of terrorism and what to do about it. The threat of terrorism is extremely complex, and countless measures must be taken by many departments, agencies, and individuals to prevent attacks or successfully respond if they occur. Many of these activities are highly interdependent and overlap with the various partners with whom you will collaborate. For this reason, it is wise to review the major lessons of this book in this concluding chapter.

Chapter 1 introduced you to the threat of terrorism along with the impact of 9/11 and the nature of homeland security and emergency management. You learned that terrorism is a serious problem and that it should not be discounted going forward. Events like 9/11 illustrate the significant impact of terrorism on society. Accordingly, the profession of homeland security is vital to reduce the loss of life and disruption terrorist attacks produce. It, along with the discipline of emergency management, will help to reduce the probability and consequences of terrorism.

Chapter 2 defined terrorism as a violent behavior (or the threat of violence) that instills fear in others and aims to force people to acquiesce to the demands of ideological extremists. Terrorism is characterized by public attacks that coerce others to support the goals of the perpetrators. It may take on various forms including mass terror, assassination, random terror, or terror against the government. Although terrorism may be regarded as one of many different types of threats or hazards, it has elements of both conflict and consensus disasters. That is to say, discord leads to terrorism, but terrorist attacks are often followed by increased altruistic behavior as people work together to solve mutual problems.

Chapter 3 explored the factors that lead people to engage in terrorism. It was revealed that some individuals and groups are violent due to their frustrations with prior social interactions, while others focus on problematic foreign policy or their impoverished condition. Terrorism may also result from disagreements about politics or the negative consequences of ineffective political systems. Culture and religion are major explanations of terrorism today. However, it is clear that all terrorists share strong ideological stances that lead them to act violently. These attitudes (e.g., being unwilling to compromise on beliefs) seem to be more engrained than before.

In The Real World

Justice, Equity, Diversity, and Inclusion in Homeland Security

Increasing the amount of attention given to justice, equity, diversity, and inclusion (JEDI) has been a major priority of the Biden Administration. In fact, President Biden issued an Executive Order on January 20, 2021, to advance racial equity and federal government support for underserved populations (White House 2021). The goal of these initiatives is to ensure that government programs are fair to all Americans and that there is representation in how government services are administered. These objectives are especially important for those working in homeland security and emergency management (Rivera and Knox 2022).

If federal, tribal, state, and local governments are biased in how they apply the law or provide services, it is likely that this will lead to grievances, protests, and even terrorism. For instance, unwarranted police brutality against black Americans led to riots in the summer of 2020. Also, if individuals and groups—regardless of race, ethnicity, gender, sexual orientation, age, and so on—cannot meet basic needs or cannot participate in the decisions that affect them, it is possible that these dismissed persons will commit crimes, become victims of disasters. An example is the theft and robbery that have increased dramatically in recent years concurrent with the rising cost of living, and minority communities that have been plagued with recurring physical assaults. In addition, individuals and families in poor neighborhoods are often vulnerable to hazards whether they be hurricanes, floods, industrial accidents, and so on. Those frustrated by political decisions may engage in violent behaviour.

For these reasons, there is an increased focus on the "Whole Community." This concept implies that the needs of all residents of the United States must be taken into account and that everyone must be included in finding solutions to the problems we face (FEMA 2011). By integrating diverse people in government positions and including them at the table when making decisions, we are more likely to find solutions to the complex problems facing us. Along these lines, there is also a recognition that "poverty is a major driver of people's vulnerability to natural disasters, which in turn increase poverty in a measurable and significant way" (Hallegatte et al. 2020, 223). As can be seen, it is increasingly clear that homeland security must include JEDI in all that it does.

Chapter 4 concentrated on understanding who terrorists are and how they operate. It was illustrated that many individuals, groups, and even nation-states participate in terrorism. These terrorists are categorized as being criminals, crusaders, or crazies. However, contrary to popular belief, there is no typical stereotype of a terrorist. Nevertheless, the people who advocate terrorism see the world through a unique lens and assume that they are justified in attacking others to correct the wrongs they perceive and address the issues that are important to them. Terrorists are also very calculating in their planning operations, and their attacks may be manifested through assassination, mass shootings, arson, bombings, and other tactics. No one should discount the threats and actions of terrorists.

Chapter 5 outlined the evolution of terrorism over time. Violent behavior with the goal of creating fear in others has always existed, and terrorist attacks can specifically be traced back to the Roman and Greek republics. Nonetheless, the concept of terrorism did not appear until the period of the Enlightenment. Europe's bout with terrorism is long, but the United States has also experienced terrorism (as a victim and some would even argue as a participant). Terrorism has changed dramatically over time and is far different today than in the past. For instance, terrorists are less hierarchical, more independent, increasingly creative, and more intent on violence than their peers from decades past.

Chapter 6 exposed the unique relationship between terrorism and the media. The ability to share stories around the world in minutes and seconds is one of the reasons why terrorists want to portray their attacks through the media. Indeed, terrorists thrive on the publicity given to them through the media. Television, radio, and news organizations also benefit from the increased attention terrorist attacks bring even though they may be the target of attacks themselves. For their part, the government utilizes the media to discredit those who engage in terrorism and works with reporters to find and apprehend the perpetrators of attacks. The fact that terrorists rely so heavily on the media is one of the reasons why self-censorship may be required at times. Without responsible reporting, additional attacks may occur, and the rights of victims may be disregarded. Of course, social media has changed the relationship between terrorists and their level of visibility, and it will be interesting to see how this alters terrorist activity in the future.

Chapter 7 identified the major dilemma facing those who create and implement homeland security policy. This section of the book reveals that terrorists do not operate according to the Geneva Convention rules of warfare. Instead, those engaging in terrorism attack innocent civilians and do so in a brutal manner. Some people therefore advocate the goal of security—even if it may jeopardize rights and freedoms. Others reject this viewpoint and alternatively argue the virtues of maintaining liberty despite the threat of terrorism. Various cases indicate the challenges of balancing each of these priorities. For this reason, those working in homeland security should be ever vigilant of ways to prevent attacks while also protecting rights and liberties.

Chapter 8 discussed the causes of terrorism in the hopes of finding ways to reduce such violent behavior. It first expressed the need to address "root causes" relating to poverty, political oppression, and the socialization of violence. Without addressing these fundamental issues, terrorism will remain a never-ending problem. In addition, this chapter explained that laws are required to prohibit terrorism and prosecute those who support or engage in such dastardly deeds. Organizations like the CIA and FBI play an important role in gathering and distributing information about terrorists through the intelligence cycle. If it is determined that attacks have been planned, relying on the special forces or terrorism task forces may be needed. Counterterrorism operations will be required in cases where terrorists are intent on inflicting death, damage, and disruption upon others.

Chapter 9 considered measures to secure the nation as well as our vital businesses and societal interests. While controversial in the minds of some people, the policies and actions that enhance border control will be needed to keep terrorists out of the United States. In addition, steps will be required to defend various segments of the economy. Terrorists have targeted and will continue to attack railways, air transportation, and seaports. Chemical facilities are also vulnerable to terrorist attacks because of the hazardous materials they contain. For these reasons, efforts must be taken to increase transportation security and regulate requirements for those handling dangerous chemicals.

Chapter 10 reviewed ways to protect people and property from terrorist attacks. It suggested that threat assessments are important to understand the true risk facing critical infrastructure, key assets, and soft targets. It was argued that more should be done to defend power and water systems, government institutions, sporting venues, and other locations where large numbers of people congregate. Buildings must also be constructed in such a way as to reduce the likelihood or impact of attacks. Besides this type of structural mitigation technique, it will be vital to implement nonstructural mitigation actions like proper zoning as well. What is more, the use of landscaping, access control, lighting, cameras, security guards, and other measures may deter terrorist attacks in the future. The goal is to reduce the desirability of certain locations as possible targets to be attacked.

Chapter 11 concentrated on various emergency management functions relating to preparedness. The concept of preparedness was defined as activities that increase readiness for possible terrorist attacks. It was noted that the federal government has created a National Response Framework and a National Disaster Recovery Framework to facilitate planning relating to the threat of terrorism. State governments have also instituted the Emergency Management Accreditation Program and the Emergency Management Assistance Compact to increase capabilities when attacks do occur. At the local level, the establishment of broadly represented advisory councils, the passing of well-designed ordinances, and the acquisition of grants will increase preparedness as will comprehensive efforts relating to planning, training, exercises, and community education. The goal is to augment the capacity to deal with terrorism beyond the simple development of an emergency operations plan.

Chapter 12 reflected on the importance of response operations when terrorist attacks occur. The chapter exposed common behavioral patterns exhibited during extreme events (e.g., convergence and emergence). It was also illustrated that the first priority is to thwart attacks and/or apprehend would-be terrorists with intelligence and the deployment of specialized counterterrorism teams. If this is not possible, the primary responsibility of any official or unofficial first responder is to minimize impact and ensure the safety of personnel so further deaths and injuries can be avoided. Searching for victims, providing medical care, triaging and decontaminating victims, and investigating the attack to find the perpetrators are all necessary aspects of terrorism response. The use of the incident command system and emergency operations centers may help you to more successfully complete and coordinate the myriad of functions that have to be performed immediately after terrorist organizations strike.

Chapter 13 discussed the short- and long-term recovery phases. To increase resiliency to a terrorist attack, communities, states, and nations must adequately assess damages, human losses, unmet needs, and other impacts. Once the nature of the situation is understood, an emergency or disaster may need to be declared by the mayor, county officials, the governor, and the president. At this point, it will be easier for the impacted jurisdiction to focus on mass fatality operations, debris removal, psychological counseling, and volunteer and donations management. Federal programs, including individual assistance and public assistance, can also bring much-needed resources to victims, their families, businesses, and local and state governments. However, affected jurisdictions must be aware of potential recovery pitfalls and do all they can to rebound in an expedited manner.

Chapter 14 assessed the probability that terrorists will launch more unique and devastating attacks in the future. The concept of weapons of mass destruction was defined. And the threat of radiological, nuclear, biological, and chemical weapons was discussed along with numerous recommended actions to counter such assaults. In these situations, it was noted that homeland security officials should call upon the assistance of specialized military and public health officials since sophisticated technical knowledge and capabilities will be required. Cyberterrorism was also described, and the measures being taken by the government and even citizens were identified. One of the major lessons to be derived from Chapter 14 is that terrorists are likely to exploit cybertechnology in the future. This must be countered by those involved in homeland security.

Chapter 15 discussed a variety of other issues that also pose a threat to individuals, organizations, and our entire society. For instance, criminal activity has become a major problem over the past decade, with a dramatic rise in general property crime, assaults, drug trafficking, and human trafficking. Our nation has witnessed several episodes of social protests over the past few years, including notable riots to alter social, political, and economic conditions. Public health emergencies like the COVID-19 pandemic have become important priorities for those involved in homeland security. And there are increasing concerns about environmental degradation and the impact of climate change. Thus, the responsibilities of those involved in homeland security have grown exponentially since it was initially founded.

While the concerns described in the prior chapters and the related actions that must be undertaken are necessary to improve the homeland security profession, they are insufficient in and of themselves. You will therefore also need to increase accountability in homeland security, improve policy guidance, foster research, and implement well-established principles to deal with terrorism. Each of these will be discussed in turn.

Self-Check

1. Homeland security is a relatively simple profession. True or false?
2. Homeland security includes a broad range of activities to prevent attacks or improve post-event responses. True or false?
3. What functions are included in proactive homeland security measures and when dealing with attacks?
   a. Intelligence gathering
   b. Border control
   c. Community education and preparedness
   d. Emergency medical care
   e. All of the above
4. Is homeland security easily accomplished? Why or why not?

16.2 Accountability in Homeland Security

Those working to counter terrorist activities and minimize their impacts may have some—but not complete—control over the actions of terrorists. However, homeland security officials certainly have power over their efforts to direct and administer prevention, protection, and preparedness initiatives. National leaders can also shape the direction of policy in the areas of response and recovery in the future.

Unfortunately, it appears that limited vision and poor management have at times been weaknesses among those responsible for or involved in this national priority (Perrow 2006; Wise 2006). At its inception, homeland security suffered significant problems. Leaders and employees of homeland security have at times committed programmatic mistakes and/or ignored innumerable challenges. For instance:

- The burial of FEMA within a massive department that focused almost exclusively on security or law enforcement concerns jeopardized natural hazard mitigation, terrorism preparedness, and response coordination (Perrow 2006). The overarching focus on terrorism in this organization and the movement of different programs away from FEMA initially hurt the nation's ability to deal with all types of crisis events.

- The structural changes resulting from the creation of Department of Homeland Security also resulted in the loss of ties between the FEMA director and the president. As a case in point, the lack of communication between Michael Brown and President Bush was readily apparent after Hurricane Katrina struck the United States in 2005. This weakness was the result of additional layers of bureaucracy that were added to emergency management since the establishment of The Department Homeland Security.

- The rejection of emergency operations plans that proved to be effective in prior disasters and terrorist attacks resulted in unclear expectations for subsequent response activities. For example, the strategies of the National Response Plan were too complex, convoluted, and unclear. Fortunately, the National Response Framework and Natural Disaster Recovery Frameworks have corrected many of these deficiencies, and the initial indications are positive.

- The introduction of the Homeland Security Advisory System (HSAS) damaged the credibility of the government to issue warnings (Aguirre 2004). The HSAS was not based on the scientific literature on how to most effectively notify people of impending harm or suggest what should be done as a result. The National Terrorism Advisory System was created to overcome these prior mistakes.

- The distribution of $4.3 billion for communication equipment did not initially lead to any real improvement in interoperability (Laskow 2010). Jurisdictions now have more communication equipment, but the lack of concrete national standards hampered the ability to coordinate with each other. There is a need to ensure interoperability whether that relates to technology or organizational culture and collaboration.

- A portion of the billions of dollars dedicated to homeland security appears to have been squandered on questionable projects or spent in unscrupulous ways. Some budgetary watchdogs decry the fact that homeland security funds have been spent on baseball caps and leather bomber jackets instead of on major prevention, protection, or preparedness initiatives as intended.

- The failure to protect U.S. borders against outside infiltration has been one of the most glaring weaknesses of homeland security (Flynn 2002; Governmental Accountability Office 2011). In spite of the growing threat of terrorism around the world, it seems as if little has been done to prevent people from simply walking across national boundaries into the United States. These facts provide one of the clearest threats and opportunities for improvement going forward.

- Presidential and political decisions have substantial—and even notable negative—consequences. Some people felt that President Bush was too aggressive in the war on terrorism and believed that he aggravated the pre-existing conflicts. In contrast, President Obama often denied or downplayed the fact that terrorists want to kill Americans. Later on, President Trump manifested a greater degree of intolerance of terrorism than his predecessor. In recent years, it is believed that President Biden has not fulfilled his responsibility to secure the border and others argue that he has appeased states like Iran that support terrorist organizations including Hamas and Hezbollah. These types of pendulum shifts have a significant impact on coherent and consistent policies to build the homeland security enterprise and facilitate its success.

It appears, therefore, that many of the troubles we are currently facing are a result of incorrect planning assumptions as well as ineffective government oversight and follow-through. This is not to deny the significant threat of terrorism or the major trials of creating the Department of Homeland Security, which has undoubtedly been the most sweeping reform of government policies and organizational structure in history. Problems are to be expected when cunning enemies are ever-present and when a reorganization of this magnitude takes place. The criticisms against homeland security are therefore not meant to diminish the important roles of the military, intelligence, and law enforcement communities either. Fighting against terrorism would be impossible without these important stakeholders. It is also necessary to note that many of the aforementioned problems were corrected under the Post-Katrina Emergency Management Reform Act. For example, the president now has closer ties to the FEMA Administrator, and some of the preparedness programs have been put back into this importantthe the lack of attention given to disaster organization. However, long-standing problems—particularly in relation to the lack of attention given to mitigation and recovery—remain in emergency management.

Thus, accountability was lacking when homeland security initially emerged. **Accountability** is the expectation of being responsible for decisions and activities that impact citizens in democratic nations. It includes being answerable for failed policies, misused resources, and incomplete goal attainment (Peters 2014). In other words, accountability includes a constant review of goals and programs to overcome mistakes and capitalize on successes. Accountability in homeland security should be

---

CAREER OPPORTUNITY | Public Administrators

To run and operate homeland security, emergency management, and public safety programs in an effective manner, a whole host of public administrators will be required in the future.

Public administrators are public servants who manage jurisdictions, departments, and agencies at all levels of government. They work for city, county, state, and federal governments in organizations ranging from the mayor's office and emergency management to police departments and the Department of Homeland Security.

Those entering this profession typically have degrees in public administration or public service. In their coursework, students learn about organizational behavior, project management, budgeting, and human resources among other topics. Those graduating in this discipline acquire useful knowledge and skills pertaining to organizational leadership and management.

Employment opportunities in public administration will likely expand in notable ways in the future due to regular growth as well as ongoing retirements of existing personnel. In fact, the Bureau of Labor Statistics anticipates a growth rate of 5% over the next decade (which is faster than average). The median pay of over $100,000 annually makes this an attractive job market for those desiring to work in government.

For further information, see **https://www.bls.gov/ooh/management/administrative-services-managers.htm**.

a major priority for the future. However, accountability will not resolve all of the problems evidenced in the past. Success will also depend upon solid policies going forward.

In The Real World

Management and Stewardship in Practice

On February 16, 2017, the Oversight and Management Efficiency Subcommittee held a hearing to explore ways to better manage the Department of Homeland Security (DHS) to prevent fraud, waste, abuse, and dysfunction. Representative Scott Perry (R-PA) gave an opening statement about the problems in DHS. He noted that this department should do a better job of getting its employees the resources they need and do so in a more streamlined process. However, Perry also complained that DHS has wasted millions of dollars on the Human Information Technology Program as well as the Federal Protective Vehicle Program (in which it purchased more vehicles than drivers). Representative Perry also mentioned the failure to follow up on misconduct by members of the Secret Service. He argued that each of these types of mismanagement put the nation at risk. Consequently, Americans are demanding more of their government. Perry recommended that DHS do more to follow up on high-risk reports the Government Accountability Office produces each year to identify areas of improvement. He also suggested that DHS adhere to the recommendations of the Office of the Inspector General.

Self-Check

1. Poor management has been one of the glaring weaknesses demonstrated during the early years of homeland security. True or false?
2. Some of the money devoted to homeland security has been misdirected or used for fraudulent purposes. True or false?
3. Correcting problems and ensuring policies are effective may be labeled as:
   a. Accountability
   b. Adaptability
   c. Flexibility
   d. Command and control
4. Why is accountability important for homeland security?

16.3 Clarification of Homeland Security Policy

Perhaps one of the reasons why homeland security programs have failed at times is because our nation does not have a fully developed and comprehensive policy for homeland security (May, Jochim, and Sapotichne 2011). Without a definitive statement on the goals of homeland security and methods of attaining them, our efforts to deal with terrorism could flounder aimlessly. For this reason, it is vitally important that you also consider policy issues in homeland security.

In The Real World

Approaches to Homeland Security

As our nation considers the future of homeland security, some scholars and practitioners argue that we need a broader perspective that will allow us to better understand and prevent the causes of crime, violence, and terrorism. The concept of "human security" has been introduced as one possible approach to help us achieve this lofty goal.

In its simplest form, **human security** may be defined as meeting the basic needs of people and ensuring the protection of their human rights. In other words, human security is about avoiding "freedom from want" or "freedom of fear" (Fukuda-Parr and Messineo 2012, 3). The general premise of human security is that criminal activity, terrorism, and other types of conflict will subside if people have enough resources to sustain life and are able to influence the political decisions around them.

The concept of human security is actually far more complex than this. Human security has a lengthy history emanating from a critical reflection on the Cold War and the success or failure of international institutions. Many people and nations have contributed to our comprehension of this notion. It therefore also relates to many other variables ranging from the quality of our educational systems to the protection of our natural resources.

One of the most important individuals related to the development of this concept was Mahbub ul Haq, a Pakistani economist who wrote an article titled "New Imperatives of Human Security" (1994). In this work, ul Haq admonished political leaders to consider four important questions: (1) security for whom? (2) security of which values? (3) security from what threats? and (4) security by what means? In his treatise, he argued that security must be focused on people rather than nations and that individuals must feel safe in their homes, at work, and in their communities. He invited others to consider an expanded vision of security that included overall physical and mental well-being. Mahbub ul Haq asserted that there are a variety of threats beyond traditional weapons such as poverty, drug addictions, disease, and terrorism. Therefore, to address these problems, ul Haq believed governments need to spend less money on military institutions and concentrate more on addressing economic development and sustainability around the world while also establishing a global civil society that would expand democracy and reduce authoritarianism.

Shortly after the distribution of Mahbub ul Haq's article, the United Nations published its Human Development Report (1994), which included a chapter titled "Redefining Security: The Human Dimension." In this document, the United Nations suggested that:

> The concept of security has for too long been interpreted narrowly: as security of territory from external aggression, or as a protection of national interests in foreign policy or as global security from the threat of a nuclear holocaust. It has been related more to nation-states than to people. ... Forgotten were the legitimate concerns of ordinary people." (UNDP 1994, 22)

For this reason, the report asserted that notions of security must be expanded. In the view of the United Nations:

> Human security is a child who did not die, a disease that did not spread, a job that was not cut, and ethnic tension that did not explode in violence, a dissident who was not silenced. Human security is not a concern with weapons—it is a concern with human life and dignity. ... It is concerned with how people live and breathe in society, how freely they exercise their many choices, how much access they have to market and social opportunities—and where they live in conflict or peace. (UNDP 1994, 229)

Human security consequently gives more attention than homeland security or national security to issues such as precarious employment situations, the lack of food security, access

(continued)

In The Real World (Continued)

to health care, the quality of the environment and natural resources (e.g., disasters, air and water pollution, deforestation, and desertification), the prevalence of violent crime and drug trafficking, the abuse of women and children, the breakdown of the family, ethnic and racial discrimination, government repression, and human rights violations.

As can be imagined, human security has received both praise and criticism from those who espouse it or oppose it. Some view human security as the new "central objective of national and international security policy" (Fukuda-Parr and Messineo 2012, 2). Others are more skeptical of human security and even actively oppose it.

On the one hand, scholars including Liotta and Owen (2006, 52) have stated that human security is essential in that "the conception and apparatus of security should ... be capable of protecting people from most, if not all, of the serious harms they face." These authors remind us that "until we can ensure that people are safe ... from preventable disease, starvation, civil conflict, and terrorism, then we have failed in the primary objective of security—to protect" (Liotta and Owen 2006, 52).

On the other hand, others suggest that the concept of human security is vague, based on untested methodologies, and perhaps unrealistic in practice (see Henk 2005). The argument is that human security's comprehensiveness makes it a meaningless concept or that there is no specified connection between the various parts of the theory and the overarching goal of a secure and peaceful world. Others believe the vision of human security is naïve since people and their institutions are inherently flawed with imperfections or that the less military spending and the guarantee of economic security, health, and environmental protection do not automatically and inevitably result in an international system that is free from crime, strife, conflict, terrorism, and war (Christie 2010).

In light of this ongoing and passioned debate, the concept of human security will likely engender both hope and caution among those involved in homeland security. Homeland security officials and personnel should be aware of this concept along with any associated strengths and weaknesses.

One of the challenges regarding homeland security policy is that government agencies can be pulled in different directions due to growing threats such as rising crime, drug cartels, human trafficking, social disturbances, riots, pandemics, and global warming/climate change. There is no easy way of balancing these priorities and ensuring that each needs is being met in an era of tight budgets. Of course, that is no excuse for not trying to do our best to address these challenges.

In other ways, and as this book reveals, the objectives of homeland security are surprisingly straightforward – at least in terms of the terrorist threat. The goals of those working in this field are twofold: (1) reduce the probability of terrorism and (2) minimize the consequences of attacks that do occur. Put differently, homeland security attempts to limit both the possibility and impact of terrorist attacks. Such intentions will require not only prevention and protection measures but preparedness, response, and recovery activities as well. Accordingly, emergency management must therefore be viewed as an equal partner in the homeland security process.

Although clarifying the purpose of homeland security is imperative, this does not necessarily outline the means for implementing the desired priorities. For this reason, those involved in homeland security might want to consider two proposed concepts to help guide your work in homeland security. These principles are liability reduction and capacity building (McEntire 2005).

The means for reducing the *probability* of terrorism are policies and actions that focus on liability reduction. **Liability reduction** is the name given to the strategy that attempts to address the factors that result in or permit terrorist attacks. This includes both proactive actions and defensive measures. Examples include:

- Understanding what motivates terrorists and working to alleviate root causes.
- Enhancing national security while protecting personal liberty.
- Improving intelligence gathering and analysis.
- Reducing the permeable nature of the borders.
- Guarding vulnerable infrastructure and key assets.
- Stopping the proliferation of weapons of mass destruction.

Alternatively, the way to effectively deal with the *consequences* of terrorism is to enhance response and recovery capabilities. This is known as capacity building. **Capacity building** is a strategy that attempts to enhance the ability of the nation, states, and communities to effectively deal with terrorist attacks. This includes:

- Establishing laws and ordinances in homeland security and emergency management.
- Meeting with an advisory council to plan how to best react to terrorist attacks.
- Training responders on important functions including warning, evacuation, sheltering, decontamination, search and rescue, and emergency medical care.
- Conducting exercises and educating the community about how to prepare for terrorist attacks and other disasters.
- Helping leaders understand their roles in disaster declarations, EOC activities, debris management, and individual and public assistance programs.

In The Real World

DHS Performance

On September 6, 2007, Paul A. Schneider, the Under Secretary for Management in the Department of Homeland Security (DHS), testified before the U.S. Senate Committee on Homeland Security and Governmental Affairs. After reviewing 24 performance expectations dealing with emergency preparedness and response, he revealed that DHS has produced mixed results since it was established. The department succeeded in five areas including areas such as grant funding, exercise programs, and the development of a national incident management system. However, DHS was unsuccessful in 18 other areas. Risk assessments, planning, training programs, and interoperable communications ranked among the list of efforts regarded to be unsatisfactory. Other problems included the lack of an inventory of federal capabilities, unclear national goals, and a failure to provide assistance to individuals and communities during emergency events. The Government Accountability Office and other government oversight organizations were also highly critical of the activities of the DHS during its initial years of operation. Fortunately, recommendations are being made to correct prior problems. They include improved strategic planning, sharing information with key stakeholders, partnering with other agencies, and integrating DHS's management functions. It is anticipated that progress will be made as time proceeds.

While liability reduction and capacity building have thus far been treated as isolated strategies, the reality is that these processes are inherently intertwined. Minimizing liabilities necessitates the development of additional capacities. For instance, surprise attacks can only be averted by augmenting human intelligence. Strengthening capabilities can likewise limit liabilities. As an example, the provision of additional training for first responders could promote increased safety at the scene of an attack.

Other complicated relationships between liability reduction and capacity building are also possible. Counterterrorism activities could possibly foment additional terrorist attacks if they result in the death of innocent people, while well-justified and carefully executed counterterrorism operations could increase our ability to protect life and freedoms. In another example, media reports may intensify terrorist behavior unless reporters are aware of the potential negative impact their portrayals may have on such conduct. Failing to understand how terrorists operate will subsequently lead to future attacks, although improved recognition of the dynamic nature of terrorism can augment readiness for unprecedented violence. Liability reduction and capacity building could thus be seen as mutually reinforcing policies, leading to a more effective homeland security apparatus.

Self-Check

1. Homeland security has not lacked a coherent policy guide. True or false?
2. An example of liability reduction is efforts to minimize the root causes of terrorism. True or false?
3. Funding the Department of Homeland Security, hiring and training employees, and writing response and recovery plans are examples of:
   a. Risk reduction
   b. Capacity building
   c. The paper plan syndrome
   d. The National Response Framework
4. Explain how the concepts of liability reduction and capacity building could improve homeland security in the future.

16.4 Research Needs and Recommendations for the Future

If homeland security is to be successful in the future, your focus on liability reduction and capacity building must rely on the unique insights of knowledgable scholars and the recommendations of dedicated practitioners. For their part, researchers must increase knowledge about terrorism and share it with government officials, business leaders, and everyday citizens. On the other hand, professionals must concentrate efforts on ways to improve the performance of critical responsibilities in homeland security (Wise 2006). While countless suggestions for the future could be mentioned, this book will conclude with five recommendations for both scholars and professionals.

16.4.1 Direction for Researchers

To improve homeland security, scholars must advance knowledge about terrorism and homeland security in several areas.

For instance:

1. Homeland security will never maximize success unless we fully understand the causes of terrorism. It will be imperative that knowledge is generated about why people are willing to kill themselves and others for ideological goals (Pedahzur 2005). Without understanding this fundamental issue and how to counter it, terrorist attacks will continue to grow in frequency and intensity.

2. Security and liberty are important values for societies, but insufficient information exists about how to protect each of these vital goals. Scholars should explore how it will be possible to protect each objective simultaneously and promote ways to ensure neither perspective supersedes the other.

3. A major threat today relates to the anticipated use of radiological, nuclear, chemical, and biological weapons. Numerous research grants were given to scholars from the 1950s to the 1980s to understand human behavior in natural and technological disasters. Less is known about human behavior in response to weapons of mass destruction, and this shortfall needs to be corrected.

4. Terrorism has changed dramatically over time, but there are limited studies about modern-day threats. In particular, there is a dearth of awareness about cyberterrorism and how it relates to homeland security and emergency management (McEntire 2021). Research must uncover both the threat of cyberterrorism and provide solutions that pertain to government, the private sector, and everyday citizens.

5. The massive reorganization of government after 9/11 revealed several shortcomings relating to public administration. Additional scholarship must determine how best to improve government operations in relation to complex problems such as terrorism and the solutions that may be provided through homeland security and emergency management.

In The Real World

Working Across the Academic/Practitioner Divide

A Training and Education Synergy Focus Group was established with the support of the FEMA Higher Education Program. It provides a great example of how scholars and practitioners may work together (see Figure 16-1). Although this group focused on emergency management, the model is one that could be applied to homeland security. The group asserts that it is the responsibility of everyone in both higher education and practice to contribute to emergency management training and education synergy. Individuals do not have to do the same thing. The key is that everyone does at least one thing and sustains it over time.

Ideas for practitioners include:

- Meeting and developing a relationship with people in emergency management higher education programs.
- Offering to host student interns to support their professional development and enhance their classroom education.

(continued)

In The Real World (Continued)

- Mentoring college students who are interested in your emergency management career path.
- Inviting students to attend emergency management training.
- Encouraging student professional development by introducing students to your emergency management professional network.
- Promoting the value of an emergency management education as a complement to relevant training and experience.
- Asking students and faculty to participate in exercises; planning efforts; hazard, risk, and vulnerability assessments; and after-action reviews.
- Requesting higher education partners to make presentations at practitioner conferences.
- Suggesting that emergency management scholars design and deliver training.
- Allowing emergency management researchers to conduct research on your jurisdiction's emergency management activities.
- Subscribing to homeland security and emergency management academic journals for the benefit of all staff.
- Taking an emergency management higher education course. As personal and professional circumstances permit.

Ideas for education partners and scholars include:

- Meeting emergency managers from various sectors in your local area, state, and region.
- Forming an advisory board composed of practitioners from different emergency management practice settings and specialties.
- Promoting internships that allow students to develop skills and additional knowledge related to the sector in which they desire a career.
- Joining local, state, and/or regional emergency management associations and volunteer for committees.
- Attending practitioner conferences and pursue offers to present at those conferences.
- Collaborating with practitioners to identify research projects that would be useful to them.
- Sharing research findings in practitioner-valued outlets.
- Offering continuing education opportunities that would help practitioners earn and maintain emergency management certifications.
- Making students aware of various career paths in emergency management and professional development needs related to those paths.
- Inviting practitioners to be guest speakers in academic courses.
- Requesting that practitioners present at conferences or other academic meetings or, better yet, co-present with them.
- Identifying service learning opportunities that benefit both students and practitioners.
- Making internships in emergency management a degree requirement.

By working together, practitioners and scholars can improve homeland security and emergency management policies and functions.

FIGURE 16-1 Professors and practitioners must work with students to help advance knowledge about terrorism and homeland security. *Source: US Department of Homeland Security.*

16.4.2 Guidance for Practitioners

Professionals in homeland security may also wish to focus on several measures to improve their efforts to prevent or react to terrorist attacks, for instance:

1. Terrorism is ultimately a reflection of a breakdown of respect and an unwillingness to permit alternative lines of thinking. Therefore, ways must be found to promote mutual understanding and stop radicalization (see Figure 16-2). The socialization of young children into a life of terrorism is one of the greatest challenges to be overcome by current and future generations (Lombardi 2015).

2. Acts of violent extremism can only be halted if there is sufficient information about diabolical plans in advance of their implementation. For this reason,

FIGURE 16-2 We must reverse the socialization of children into violence. *Source: © ZouZou/ Shutterstock.*

intelligence gathering and information sharing must remain top priorities for organizations like the FBI and CIA.

3. The threat of terrorism is now self-evident and cannot be downplayed or ignored. It will be imperative that we build a modern intelligence, military, and police force to counter violent individuals and groups. The intelligence, armed services and law enforcement agencies must have the necessary equipment and training to react quickly and decisively to the danger posed at home or abroad. However, "the trick, if one has the political acumen to learn it, is to avoid fueling it while claiming to fight it" (Chalian and Blin 2007, p. 11).

4. Terrorists have repeatedly expressed their intentions to attack the U.S. homeland. Accordingly, the porous border must be tightly regulated, and infrastructure must be protected through more stringent antiterrorism measures. The consequences of failing to meet these obligations are dangerous and unacceptable.

5. In addition, it will unfortunately be impossible to prevent terrorism if people are committed to undertaking such atrocities. As a consequence, there will be a need to improve emergency preparedness at the federal, state, and local levels. Emergency managers should therefore be included as crucial partners in the development of homeland security policies and procedures. In addition, more attention should be given to education and training. Fortunately, FEMA recently established the National Training and Education System. The goal is to improve knowledge and capabilities, build and sustain a community of practice, and establish a defined career path for those working in homeland security.

In The Real World

Artificial Intelligence for Terrorism and Homeland Security

Americans are increasingly exploring the use and benefits of artificial intelligence. **Artificial intelligence (AI)** is the use of computer programs and large datasets to solve everyday problems. AI can be used for many tasks ranging from data collection and data analysis to technical writing and customer-centered business activities.

Unfortunately, "U.S. adversaries and criminal gangs are moving forward with plans to exploit the technology at Americans' expense" (Seldon 2023). FBI Director Christopher Wray spoke about this concern to attendees of a cybersecurity conference in Washington, D.C., on September 18, 2023. He mentioned that AI is bound to be abused by nefarious actors. He warned that people with mal intent could use AI in diverse ways to advance their agenda. For instance, AI may allow terrorists to generate malicious cyber codes or produce a deepfake (i.e., a digital video that may mimic another person to spread false information). In addition, it is possible that terrorists could use AI to distribute propaganda, increase radicalization and recruitment, improve covert funding, develop more sophisticated cyber weapons, and plan and launch attacks.

AI can be used as a solution to these threats, however. The federal government is also relying on AI to counter fentanyl trafficking and human trafficking and to protect critical infrastructure. The DHS knows that it must stay on top of advancements in this area to counter terrorist activities. This government organization also desires to implement measures with AI that ensure privacy and protect civil rights. Those working on homeland security must understand the pros and cons of AI as it relates to the threat of terrorism.

As can be seen, future success in homeland security will largely be dependent upon the ability of scholars and professionals to formulate a logical mission for homeland security, identify major priorities to reduce terrorism, and craft and implement appropriate policies to deal with ideologically motivated acts of violence. Ongoing support and monitoring of necessary programs as well as the intentional adaptation of strategy will be required if we are to thwart terrorism and cope with its adverse effects.

In The Real World

The Shortage of Professionals in Emergency Management and Homeland Security

In 2012, the author of this book wrote an editorial piece in the *Journal of Emergency Management* entitled "The Current Crisis and Impending Disaster" (McEntire 2012). In that article, McEntire argued that terrorist attacks, natural disasters, and industrial hazards are not the only concerns we face today. He asserted that there is an unrecognized and rising vulnerability that must also be addressed. This risk relates to the staffing of emergency management and homeland security positions across the nation.

In the editorial, McEntire pointed out that emergency management has historically been understaffed and that the workload has become more onerous since 9/11 (see Figure 16-3). He expressed worries that "the quantity of local emergency managers is not keeping up with the burgeoning demands placed upon them" (McEntire 2012, 317). He then went on to say that "the current situation is unsustainable" and that emergency managers will eventually burn out, leave the field, or selectively neglect some of their responsibilities that are essential for the benefit of society.

Unfortunately, it appears that these claims and predictions may have come to fruition. A recent study on mental health stressors facing emergency managers illustrates that many factors—including the $24 \times 7 \times 365$ nature of disasters, limited pay, and the political aspects of their job—are causing emergency managers to leave the profession (Patel et al. 2023). This departure of personnel is occurring among new employees as well as some of the most seasoned and educated emergency managers in the nation. The authors of this study therefore acknowledge that "it is challenging to have a resilient organization capable of solving current and future EM issues with a high turnover rate" (Patel et al. 2023, 383). The shortage of personnel is not limited to emergency management, however. Law enforcement organizations and homeland security agencies are also preoccupied with recruitment and retention due to the defund the police movement, ongoing retirements, and a host of other variables. Including limited salaries and benefits.

In short, it appears that employment in emergency management and homeland security will require a lot of work with limited resources—at least in the near term. There is good news, though, in that there should be ample employment opportunities for those wishing to enter these professions. The workforce entering these positions must consistently explain its multifaceted responsibilities, vigorously defend funding, and advocate for increased political support in an era of mounting terrorist attacks and other types of disasters. It will be up to the next generation to help us resolve these staffing shortages to ensure a brighter future.

FIGURE 16-3 Professionals are needed in homeland security to advance efforts to prevent attacks or respond effectively when they occur.

In short, the threat of future attacks is real and menacing. It is up to you and others involved in homeland security to meet this challenge and do so in an efficacious manner. In the words of a former director of the Defense Intelligence Agency, "The question is not whether such an attack will occur ..., but when and where. It is up to you ... to be prepared" (Chalian and Blin 2007, p. 2).

Self-Check

1. Scholars play no role in homeland security policy or its practical application. True or false?
2. Stopping the socialization of children into terrorist behavior is one of the most important solutions in homeland security. True or false?
3. There is currently insufficient information about how to deal with which type of terrorism?
   a. Conventional terrorism
   b. Terrorism involving explosives
   c. Cyberterrorism
   d. Shootings
   e. Hijackings
4. Describe why it is important for scholars and practitioners to work together to resolve the challenges facing homeland security.

Summary

This book has illustrated countless issues that must be considered and addressed if efforts to prevent or deal with terrorism are to be effective. Therefore, you should review the major lessons of homeland security often, so you do not forget all the things that need to be accomplished to minimize the probability and impact of terrorism. In addition, it is advisable that you promote continued accountability in homeland security and the development of clear policies so that resources can be maximized and strategy can be simplified. The goals of preventing attacks, limiting their negative consequences, and reacting to them effectively can only be achieved if you work astutely and diligently to counter the deadly intents of terrorists. For this reason, the concepts of liability reduction and capacity building may help to direct your actions in this important field and profession. Furthermore, cutting-edge research will help you to know what else needs to be done to counter the threat of terrorism. However, the application of existing principles will also determine if you will be successful in protecting the homeland against terrorist attacks. Consequently, your work in homeland security will be both challenging and rewarding. It is up to you to prevent terrorist attacks, prepare for adverse consequences, and react effectively.

Assess Your Understanding

Understand: What Have You Learned

Go to **www.wiley.com/go/mcentire/homelandsecurity3e** to assess your knowledge of the future of homeland security.

Summary Questions

1. Terrorist attacks have occurred more frequently in recent years and pose a serious risk to people. True or false?
2. Homeland security includes only a few functions that must be performed to prevent attacks or respond effectively. True or false?
3. The DHS has created and experienced a few problems since its inception as a government entity. True or false?
4. Accountability is the expectation of being responsible for decisions and activities that impact citizens in democratic nations. True or false?
5. The United States has a comprehensive and fully developed policy to guide homeland security. True or false?
6. Human security concentrates more on the root causes of terrorism than homeland security. True or false?
7. Liability reduction implies a strategy to minimize the probability of terrorist attacks. True or false?

8. Liability reduction and capacity building are mutually exclusive activities. True or false?

9. What systems can help to increase coordination after terrorist attacks?
   a. Incident command
   b. Emergency operations centers
   c. The homeland security consortium
   d. Answers a and b

10. Many of the shortcomings in homeland security result from a failure of:
    a. Terrorists to advocate nonviolent solutions
    b. Inaction
    c. Accountability
    d. Effort

11. Examples of historic failures in homeland security include:
    a. The introduction of the Homeland Security Advisory System
    b. The distribution of millions of dollars for interoperable communications
    c. An inability to control the borders of the United States
    d. All of the above

12. The central goals of homeland security include:
    a. Reducing the probability of attacks and minimizing consequences
    b. Providing intelligence and responding to attacks
    c. Engaging in counterterrorism and stopping cyberattacks
    d. Strengthening SWAT teams and facilitating recovery

13. Which of the following is representative of capacity building?
    a. Launching a preemptive strike against known terrorists
    b. Sealing off the borders so terrorists cannot enter the United States
    c. Planning, training, and exercises
    d. Promoting tolerance among different religions

14. There is currently insufficient knowledge about which threat(s)?
    a. Cyberterrorism
    b. Bombings
    c. Arson
    d. None of the above

Applying This Chapter

1. What are the major lessons of this book?
2. Why is accountability important in homeland security?
3. What can be done to reduce fraud, waste, abuse, and mismanagement in homeland security programs?
4. What is liability reduction and capacity building, and how are these concepts related?
5. What are some recommendations for future research and practical application?

Be a Homeland Security Professional

Learn About Homeland Security Resources

Go to the library or look online to become familiar with homeland security and emergency management journals. What ones exist? How are they similar or different from others?

Accountability in Homeland Security

Review some of the government reports on the status of homeland security in the United States. Explain whether or not you think the DHS has been effective or ineffective in its responsibilities.

Future Research Needs

You have been asked to recommend future studies to assist the DHS. What types of research projects would you recommend?

Key Terms

Accountability   The expectation of being responsible for decisions and activities

Artificial intelligence (AI)   The use of computer programs and large datasets to solve everyday problems

Capacity building   A strategy that attempts to enhance the ability of the nation, states, and communities to effectively deal with terrorist attacks

Human security   Meeting the basic needs of people and ensuring the protection of their human rights

Liability reduction   A strategy that attempts to address the factors that result in or permit terrorist attacks

References

Aguirre, B.E. (2004). Homeland security warnings: lessons learned and unlearned. *International Journal of Mass Emergencies and Disasters* 22 (2): 103–115.

Chalian, G. and A. Blin. (2007). Introduction. In: *The History of Terrorism: From Antiquity to Al Qaeda* (ed. G. Chalian and A. Blin), 1–11. Berkeley, CA: University of California Press.

Christie, R. (2010). Critical voices and human security: to endure, to engage or to critique? *Security Dialogue* 41 (2): 169–190.

FEMA. (2011). *A Whole Community Approach to Emergency Management: Principles, Themes, and Pathways for Action*. FDOC 104-008-1. FEMA: Washington, D.C.

Flynn, S.E. (2002). America the vulnerable. *Foreign Affairs* 81 (1): 60–74.

Fukuda-Parr, S., and C. Messineo. (2012). Human security: a critical review of the literature. CRPD Working Paper No. 11. Center for Research and Peace and Development. Leuven, Belgium.

Governmental Accountability Office. (2011). *Border Security: Additional Steps Needed to Ensure That Officers Are Fully Trained*. Washington, DC: USGAO.

Hallegatte, S., Vogt-Schilb, A., Rozenberg, J., Bangalore, M., and C. Beaudet. (2020). From poverty to disaster and back: a review from the literature. *Economics of Disasters and Climate Change* 4: 223–247.

Henk, D. (2005). Human security: relevance and implications. *The US Army War College Quarterly* February 20, 2017: *Parameters* 35(2). https://press.armywarcollege.edu/parameters/vol35/Iss2/8.

Laskow, S. (2010). Homeland security's billion-dollar bet on better communications. *The Center for Public Integrity* (February 17). www.publicintegrity.org. (Accessed February 20, 2017).

Liotta, P.H. and T. Owen. (2006). Why human security? *The Whitehead Journal of Diplomacy and International Relations* 2: 37–54.

Lombardi, M. (2015). *Countering Radicalization and Violent Extremism Among Youth to Prevent Terrorism*. Amsterdam: IOS Press.

May, P.J., Jochim, A.E., and Sapotichne, J. (2011). Constructing homeland security: an anemic policy regime. *Policy Studies Journal* 39 (2): 285–307.

McEntire, D.A. (2005). Why vulnerability matters: exploring the merit of an inclusive disaster reduction concept. *Disaster Prevention and Management* 14 (2): 206–222.

McEntire, D.A. (2012). The current crisis and impending disaster. *Journal of Emergency Management* 10 (5): 317–318.

McEntire, D.A. (2021). Cyberattacks and their implications for emergency management. *Journal of Emergency Management* 1: 7–8.

Patel, S.S., Guevara, K. Hollar, T.L., DeVito, R.A., and T.B. Erickson. (2023). Surveying mental health stressors of emergency management professionals: factors in recruiting and retaining emergency managers in an era of disasters and pandemics. *Journal of Emergency Management* 21 (5): 375–384.

Pedahzur, A. (2005). *Suicide Terrorism*. Malden, MA: Polity Press.

Perrow, C. (2006). The disaster after 9/11: the department of homeland security and the intelligence reorganization. *Homeland Security Affairs* 1 (11).

Peters, B.G. (2014). Accountability in public administration. In: *The Oxford Handbook of Public Accountability* (ed. M. Bovens, R.E. Goodin, and T. Schillemans). Oxford: Oxford University Press.

Rivera, J.D. and C.C. Knox. (2022). Defining social equity in emergency management: a critical first step in the nexus. *Public Administration Review* 83 (5): 1170–1185.

Seldon, J. (2023). FBI echoes warning on danger of artificial intelligence. VOA (September 18). https://www.voanews.com/a/fbi-echoes-warning-on-danger-of-artificial-intelligence-/7273751.html#:~:text=FBI%20Director%20Christopher%20Wray%20issued,Wray%20said%2C%20without%20sharing%20specifics. (Accessed October 26, 2024).

Ul Haq, M. (1994). New imperatives of human security. *RGICS Paper No. 17*. New Delhi, India: Rajiv Gandhi Institute for Contemporary Studies, Gandhi Foundation.

UNDP (United Nations Development Program). (1994). Human Development Report 1994: New Dimensions of Human Security. United Nations Development Program, New York.

White House. (2021). Executive Order on Advancing Racial Equity and Support for Underserved Communities Through the Federal Government August 20, 2023. https://www.whitehouse.gov/briefing-rooom/presidential-actions/2021/01/20/executive-order-advancing-equity-and-support-for-underserved-communities-through-the-federal-government/.

Wise, C.R. (2006). Organizing for homeland security after Katrina: is adaptive management what's missing? *Public Administration Review* May/June: 302–318.

Glossary

A

9/11 the most consequential terrorist attack involving hijacked planes against the United States.

Absolute poverty a situation where people lack so many resources that they cannot even meet basic necessities such as food, clothing, and shelter.

Abu Musab al-Zarqawi a Sunni terrorist responsible for many atrocities in Iraq, including the beheading of an American businessman named Nicolas Berg.

Abu Sayyaf an Islamic separatist group in the Philippines that desires an independent state in Mindanao.

Accountability the expectation of being responsible for decisions and activities that impact citizens in democratic nations.

Affect dimension specific feelings or emotions that are generated in conjunction with an ideology.

Agro-terrorism terrorism against farming industries and products.

Al Jazeera a TV station based in Qatar that is popular in the Middle East and used to disseminate terrorist information.

Al-Qaeda a well-known extreme Islamic fundamentalist organization whose name refers to "the base"—the location from which its supporters attacked the Soviet Union to free Afghanistan.

Anarchists those opposing specific governments or all governments.

Anders Behring Breivik a terrorist who espoused far-right ideology and conducted one the worst attacks in Norway.

Animal Liberation Front a terrorist organization that opposes cruelty to animals.

Annexes a portion of the emergency operations plan that discusses specific hazards or functions that will need to be addressed if an event takes place.

Appendices additional information at the end of the emergency operations plan that includes resource and contact lists, maps, standard operating procedures, and checklists.

Arab Spring a series of uprisings and armed rebellions that spread across the Middle East in 2010 and 2011.

Arousal hypothesis media reports on terrorism can increase people's interest in acting aggressively.

Area command an ICS organization that supervises several incident command posts.

Armed Forces of National Liberation a Puerto Rican terrorist organization seeking the liberation of Puerto Rico from the United States.

Artificial intelligence (AI) the use of computer programs and large datasets to solve everyday problems.

Asymmetrical warfare terrorist attacks on the part of the militarily weak against those who are powerful.

B

Bacteria a single-cell organism that causes disease in plants, animals, and humans.

Bacterial infections a disease that results from cells that can live outside the body.

Basic plan an overview of the entire emergency operations plan.

Biological weapons living organisms or agency produced by living organisms that may be used in terrorist attacks.

Biological Weapons Convention (BWC) an international treaty designed to prevent the proliferation of biological agents around the world.

Bioterrorism terrorism that employs biological weapons.

Black Panthers an organization composed of African Americans to counter the actions of the KKK and other white supremacists.

Black September an operational unit of the Al-Fatah terrorism organization that launched a terrorist attack against Israeli athletes during the Munich Olympics.

Blood agents chemical weapons that prevent the flow of oxygen in the blood.

Bojinka plot a planned attack on airliners over the Pacific Ocean.

Bollards metal or concrete posts installed into the ground or cement, used to keep vehicles from entering restricted areas.

Border the territorial boundary of any nation along with its various points of entry.

Built-in escalation hypothesis more deadly and visible attacks are required to get equal media coverage in the future.

C

Capacity building a strategy that attempts to enhance the ability of the nation, states, and communities to effectively deal with terrorist attacks.

Category A agents biological weapons that pose a serious risk to people because they are easily transmitted to others and result in high mortality rates.

Category B agents biological weapons that have a moderate chance of contagion and generally result in lower mortality rates than category A agents.

Category C agents biological weapons that could be used for mass dissemination and high morbidity if engineered for that purpose.

CBRNE an acronym for chemical, biological, radiological, nuclear, or explosive devices.

Cell a terrorist branch or unit operating in locations away from the organization's headquarters.

Censorship the withholding, banning, or altering of information the media shares with the public.

Central Intelligence Agency (CIA) the federal government agency that is responsible for intelligence collection outside of the United States.

CHAZ the Capitol Hill Autonomous Zone in Seattle, Washington.

Chemical weapons lethal human-made poisons that can be disseminated as gasses, liquids, or aerosols.

Chemical Facility Anti-Terrorism Standards (CFATS) a program that identifies risk at chemical facilities (as well as at power plants, refineries, and universities) and regulates stands to ensure sufficient security measures are in place.

Choking agents chemical weapons that cause respiratory distress.

Civil defense the government's initiative to prepare communities and citizens to react effectively to nuclear war against the Soviet Union.

Civil Support Teams specialized military units that assist local and state governments that have been affected by weapons of mass destruction.

Classified intelligence information given only to a very specific and limited number of people to protect sources of acquisition and deny adversaries information that would lead them to alter their communications or operations.

Climate change the name given to episodes of global warming and/or global cooling.

Code/Ordinance an authoritative order or law issued by a government.

Cognitive dimension the knowledge and beliefs of the ideology.

Coherence the internal logic of an ideology.

Cold zone the uncontaminated area where responders and victims may enter and leave.

Communism an ideology that sympathizes with the poor and downtrodden and attempts to do away with private property and capitalism.

Community Emergency Response Team (CERT) a group of citizens who receive basic training on response operations.

Comprehensive Homeland Security Act a law passed in 2003 containing new regulations for critical infrastructure security, railroad security, and more stringent measures related to border control and weapons of mass destruction.

Computer Assisted Passenger Prescreening System II (CAPPS II) a former computer program utilized by the TSA to screen passengers against lists of known terrorists and others with criminal records.

Conflict disaster a socially disruptive event that involves a riot, violence, or some type of warfare.

Consensus disaster a socially disruptive event like an earthquake or tornado that brings the community together.

Consequence management an emergency management function that stresses planning, emergency medical response and public health, disaster relief, and restoration of communities.

Container Security Initiative (CSI) one of the first measures taken by the government to protect maritime trade and ports against terrorism.

Continuity of operations the maintenance of government functions after terrorist attacks through the identification of leader succession, alternate work sites, and resumption of operational practices.

Convergence the flow of people and resources to the scene of an emergency or disaster.

Coordination cooperative efforts to pursue common goals in the wake of terrorist attacks.

Coronavirus the most concerning pandemic in recent years, known as SARS-CoV-2 or COVID-19 for short.

Corporatist model a model that stresses the integration of various components of society into the government (e.g., close ties to business, churches, and clubs).

Counterterrorism the active pursuit of known terrorists that includes preemptive military strikes or the involvement of law enforcement officials.

Crazy a terrorist who is regarded to be psychologically disturbed (e.g., Ted Kaczynski).

Criminal a person who seeks personal gain by breaking the law.

Criminal Justice a discipline and profession interested in intelligence gathering, terrorist investigation, prosecution, border control and other security measures.

Crisis counseling the treatment of psychological problems that may arise from the stress produced by terrorism.

Crisis management a law enforcement function that concentrates on identifying, anticipating, preventing, and prosecuting those involved in terrorism.

Critical incident stress (CIS) the inability of emergency service personnel to cope with the trauma that is experienced while on the job.

Critical infrastructure interdependent networks composed of industrial, utility, transportation, and other distribution systems.

Crusader a terrorist that promotes high moral goals (e.g., Islamic fundamentalism).

Crusades wars endorsed by the pope to recapture the Holy Land of Jerusalem from Muslims.

Culture the lifestyle of groups based on their shared history, language, religion, and moral system.

Customs and Border Protection (CPB) the largest organization within the Department of Homeland Security that enforces immigration laws at and between the ports of entry.

Customs-Trade Partnership Against Terrorism (C-TPAT) an agreement between the public and private sectors to protect international commerce from terrorist attacks.

Cyberterrorism terrorist activity that utilizes or attacks computer networks to instill fear and force some type of change.

Cybersecurity & Infrastructure Agency (CISA) a unit within the Department of Homeland Security that is responsible for promoting cybersecurity in the United States.

D

Damage assessment a detailed survey of physical destruction, economic losses deaths, social disruption, and recovery needs resulting from a terrorist attack.

Daniel Pearl a reporter with the *Wall Street Journal* who was captured and killed by the National Movement for the Restoration of Pakistani Sovereignty.

Debriefing a recurring and more in-depth discussion designed to redirect harmful thinking and develop improved coping mechanisms.

Debris management the removal, storage, disposal, or recycling of rubble produced from terrorist attacks.

Decontamination the removal of hazardous materials from victims through clothing removal and the washing of bodies.

Defusing a short, unstructured meeting to allow a person to discuss an experience as soon as it takes place.

Denial of service a type of attack where a server is overwhelmed with illegitimate requests to the point where legitimate users cannot access the service.

Department of Defense the government agency responsible for the military and its operations.

Department of Homeland Security the government organization charged with preventing terrorist attacks or react effectively to their consequences.

Department of State the government agency in charge of diplomatic relationships among nations.

Dirty bombs explosive laden with dangerous chemicals or radioactive material.

Disaster declaration an acknowledgment of the severity of the event and that response and recovery assistance are required.

Disaster Mortuary Operational Response Team (DMORT) a group of private citizens from around the nation who may be activated by the federal government to assist with mass fatality incidents.

Disaster Recovery Center (DRC) a temporary facility near the attack location where victims can seek information about federal assistance programs.

Disinhibition hypothesis violence portrayed by the media may weaken the inhibition of others to participate in terrorism.

Domestic terrorism terrorism that occurs within a single country.

Donations management the collection, sorting, and distribution of goods and money for the benefit of victims of terrorist attacks.

Drug cartels   organized crime syndicates that cultivate, manufacture, and distribute drugs.

Dynastic assassination   the murder of the head official in government.

E

Earth Liberation Front   a terrorist organization that opposes environmental degradation.

Economic class model   a model that suggests a division of society based on the amount of wealth one possesses (e.g., bourgeoisie, middle class, proletariat)

Emergence   the appearance of altruistic behavior that is unfamiliar to the participants.

Emergency alert system   an announcement that interrupts TV and radio programs and relays information about what is taking place and what people should do for protection.

Emergency assistance   financial help given to local governments to take care of immediate needs such as debris removal or safety precautions.

Emergency Management Accreditation Program   a standard-based assessment and certification initiative for local and state emergency management agencies.

Emergency Management Assistance Compact (EMAC)   an agreement among states to render assistance to one another in times of disaster.

Emergency management   a discipline and profession that addresses to prevent or react successfully to various types of disasters.

Emergency manager   a local government official in charge of disaster mitigation, preparedness, response, and recovery.

Emergency operations center (EOC)   a location from which disaster response and recovery activities can be overseen and managed.

Emergency operations plan (EOP)   a document that describes what may be anticipated in terms of homeland security and emergency management and how best to react.

Enlightenment   a period in history when a new way of looking at the world emerged.

Epidemic   a smaller-scale public health emergency where a disease is spreading actively in a specific region.

Evacuation   the movement of people away from hazardous areas or situations.

Exercises   drills and mock events that test the knowledge and skills of those in charge of reacting to attacks.

Extensiveness   how many people share a particular ideology.

F

Fascism   an ideology that promotes the uniting of citizens in support of the state.

Fatwa   a religious edict.

Federal Air Marshal Service   an organization of a few thousand marshals who act under the Transportation Security Administration to prevent hijacking of aircraft.

Federal Bureau of Investigation   the federal organization that enforces U.S. law.

Federal Emergency Management Agency   the national entity in charge of disaster management.

Finance/Administration   the final section under ICS. This section tracks the expenses associated with response operations and logistics.

First responders the first official governmental responders in the field including police, firefighters, and emergency medical technicians.

Focused terror terrorism directed toward a specific group of people deemed as the enemy.

Freedom of religion people cannot be denied their right to worship according to the dictates of their own conscience.

Freedom of speech people are allowed to express their opinions, even when they criticize the government.

Full-scale exercises major scenarios that test many functions or the entire response system.

Functional exercises practice scenarios that explore one or a few of the annexes in the plan.

G

General crimes minor infractions and violations of the law such as jaywalking and motor vehicle violations.

George Floyd a 46-year-old African American man who was killed by Derek Chauvin, a police officer in Minneapolis.

Geneva Conventions a set of internationally accepted laws pertaining to the conduct of war.

Global warming the rise of temperature in the earth's atmosphere.

Grants funds given to local governments to support or enhance homeland security and emergency management programs.

Greenhouse effect the collection of heat in our atmosphere much like a green house building retains heat for plants because the sun warms the air inside the enclosed structure.

Group competition model a model that asserts that interest groups interact with or counteract one another in their attempt to sway government policy.

GSG9 a German counterterrorism organization, whose name means "Border Guards, Group 9."

Guerilla spanish term for little war, which is an armed protest of occupying forces.

Gunpowder plot an attempted terrorist attack in 1605 against King James I and other leaders of Parliament to reinstate Catholic Involvement in England.

H

Hazard the physical or other agent(s) that may trigger or initiate disaster events and processes.

Holocaust the extermination of approximately six million Jews by the Nazi regime during World War II.

Homeland security a concerted national effort to prevent terrorist attacks within the United States, reduce America's vulnerability to terrorism, and recover from and minimize the damage of attacks that do occur.

Homeland Security Act a law passed in 2002 that mandated the creation of the Department of Homeland Security.

Homeland Security Advisory System the nation's method for warning the population of potential and actual terrorist attacks after 9/11.

Homeland Security Exercise and Evaluation Program a federal program that provides guiding principles for exercises.

Hot zone the area contaminated by the terrorist attack.

Human security meeting the basic needs of people and ensuring the protection of their human rights.

Human trafficking modern day slavery where people are exploited through force, fraud or coercion.

HUMINT intelligence collected by people from people (and can be done overtly or covertly).

I

Ideology a set of beliefs related to values, attitudes, ways of thinking, and goals.

IED improvised explosive devices.

Illicit drugs illegal narcotics and the abuse of legal prescription pharmaceuticals.

IMINT geospatial imagery collected by satellites and aircraft.

Immigration and Customs Enforcement (ICE) an organization within the Department of Homeland Security that enforces immigration and customs laws in the internal portions of the United States (i.e., away from the border).

Impact assessment a broad evaluation of the number of deaths and degree of social disruption caused by terrorists.

Incident command the on-scene leader or leaders in the incident command post.

Incident Command System (ICS) a set of personnel and procedures that helps facilitate coordination among first responders.

Individual assistance relief programs for citizens and businesses impacted by terrorist attacks.

In-kind donations physical donations including food, water, clothing, supplies, and equipment.

Input functions activities that influence priorities in the political system.

Integrated Public Alert & Warning System (IPAWS) a warning system that shares information with the public through cell phones.

Intensiveness the strength of attachment to an ideology.

Intelligence the function of collecting, assessing, and distributing information about an enemy, criminal, or terrorist.

Intelligence adjustment adaptation of the intelligence cycle that is required when collection is incomplete, analysis seeks to "connect the dots," production generates new questions, and dissemination results in the anticipation of future concerns.

Intelligence analysis efforts to make sense of the voluminous data that is gathered from the field.

Intelligence collection activities to gather information about terrorist organizations and their operations and potential attacks.

Intelligence cycle a four-step process of gathering, understanding, and synthesizing data and then sharing it with those who will use it.

Intelligence dissemination sharing information with end users (e.g., policy makers, FBI special agents, and homeland security personnel).

Intelligence production the creation of written reports, briefings, images, or maps to influence operational decisions.

International Atomic Energy Agency (IAEA) the international organization responsible for the enforcement of the Non-Proliferation Treaty.

International relations a discipline and profession that deals with the conflicts among nation-states and nonstate actors (e.g., why terrorism occurs and what governments are doing about it).

International terrorism terrorism that spans two or more nations.

INTERPOL an international police organization that is involved in intelligence.

Iran a state in the Middle East that denounces the United States and Israel, promotes anti-western propaganda, and has a long-standing history of participation in terrorism.

Irritants agents that led to allergic reactions.

Islamic fundamentalists individuals or groups of Muslims who violently oppose Israel and the United States.

Islamic State of Iraq and Syria a group that seeks to establish an Islamic government in the Middle East.

J

January 6th the name given to the riot due to the mob that gathered at, attacked, and took over the U.S. Capitol building to protest Donald Trump's defeat in the 2020 election and stop the certification of the electoral college votes.

Japanese Red Army a terrorist organization that protested the presence of the United States in Japan after World War II, disapproved of the Vietnam War, and rejected capitalism.

Jihad an internal struggle to pursue righteousness or a war of self-defense, but has been used by terrorists to denote an offensive attack.

Joint field office (JFO) a temporary federal office that is created with local, state, and federal representatives who will manage the paperwork regarding public assistance.

K

Key assets a variety of unique facilities, sites, and structures that require protection.

Kickoff meeting a gathering of local, state, and federal officials for the purpose of explaining public assistance programs.

Ku Klux Klan a white supremacist group that has been involved in terrorism in the United States since the Civil War.

L

Liability reduction a strategy that attempts to address the factors that result in or permit terrorist attacks.

Liaison officer the person who serves as the link between the incident commander(s) and other organizations.

Libya a country in Northern Africa that supported terrorism heavily in the 1980s.

Local Emergency Planning Committees (LEPCs) preparedness councils promoted in the 1980s to help communities prepare for hazardous materials releases.

Logistics a section that supports operations. It acquires people, equipment, and other resources need by those responding to the attack.

Lone-wolf terrorists individual terrorist who act alone.

M

Madrasahs schools that offer basic education in the Middle East and are exploited by terrorists.

Major felonies major crimes such as rape and murder.

Maintenance the feedback function of the political system.

MASINT measurement and signature intelligence that looks for the characteristics of certain types of actions (e.g., the presence of nuclear material when one is trying to develop a nuclear weapon).

Mass fatality incident   an attack that creates so many deaths that the processing of remains is beyond the capability of local government.

Mass shooting   an attack by a perpetrator with a pistol, shotgun, or rifle that injures and kills multiple individuals.

Mass terror   terrorism by the government in power against its own citizens.

Metadata   information gathered from communication through electronic devices.

Minuteman Project   activities to promote border security carried out by a group of volunteers who founded the Minuteman Civil Defense Corps in Arizona.

Misdemeanors   more serious crimes such as driving under the influence, theft and burglary, or assault.

Mitigation   activity that attempts to avoid disasters or minimize negative consequences.

Molly Maguires   a group of Irish citizens who joined together to dispute the treatment of coal mine workers in the United States.

Money laundering   the process of hiding where money is coming from and what it is being used for.

Multiagency coordination centers (MACCs)   an ICS organization level that supervises command across several jurisdictions.

Muslims   those following the Prophet Muhammad and adhering to the religion of Islam.

Mutual aid   a collaborative agreement between jurisdictions when external help is warranted.

N

National Cybersecurity and Communications Integration Center (NCCIC)   a 24 × 7 center composed of officials from the federal government, the intelligence communities, and law enforcement officials to monitor the Internet, respond to threats, and manage incidents.

National Emergency Management Association   a professional association of state emergency management agencies.

National Incident Management System   a comprehensive national approach for incident management in the United States.

National Processing Service Center   a FEMA office set up to help victims apply for federal assistance programs.

National Response Framework (NRF)   the successor to the National Response Plan that describes the principles, roles, and structures of response operations.

National Response Plan (NRP)   a document that describes the procedures for responding to all types of hazards with a multidisciplinary perspective.

National Strategy to Secure Cyberspace   a component under the Information Analysis and Infrastructure Protection Unit and takes into account home users/ small businesses, larger enterprises, critical infrastructure, and national and international Internets.

Nationalist movements   efforts on the part of a group or nation to obtain political independence and autonomy.

NBC   nuclear, biological, and chemical weapons.

Nerve agents   chemical weapons that prevent the transmission of electrical signals in the nervous system.

Non-Proliferation Treaty (NPT)   an international regime designed to prevent nuclear states from giving nuclear weapons or materials to those who do not possess them.

Nonstructural mitigation methods beyond construction that may limit the possibility or consequences of terrorist attacks.

Nuclear weapons weapons that produce massive explosions due to the release of vast amounts of energy through fission or fusion.

O

Occupy movement a movement of protesters that decry social and economic inequality around the world.

Occupy Seattle one of the most notable protests in Seattle, Washington, that occurred after the death of George Floyd.

Occupy Wallstreet a protest in New York City by young people to bring attention to the income gap and global capitalist activities.

Operation Mongoose an attempt by the United States to kill Cuban leader Fidel Castro with a poisoned cigar.

Operation Underground Railroad a nonprofit organization that works tirelessly to stop human traffickers and rescue the victims of this disturbing type of organized crime.

Operations the name given to the section under ICS that is in charge of implementing the strategy created by those in planning.

OPINT open-source intelligence acquired through publicly available materials including academic research, newspaper articles, library books, and so on.

Organized crime the name given to the powerful groups that carefully plan and execute criminal activities usually with the purpose of economic gain.

Output functions activities emanating from the political system.

P

Pandemic a larger public health emergency where a disease is spreading internationally and affecting many countries around the globe.

Paper plan syndrome an attitude that assumes that having a plan ensures you are prepared to deal with terrorism and other types of disasters.

Permanent assistance financial aid for the repairing of publicly owned critical infrastructure and key assets.

Pervasiveness how long an ideology has been in existence.

Planning the section under ICS in charge of collecting information about the terrorist attack, including operational priorities.

Point of distribution locations where medicines may be given to victims.

Political elite model a model in which the leaders are ruling over the masses.

Political system a government that operates in a self-contained environment (e.g., a national territory).

Politics the authoritative allocation of values and resources in society.

Post-Katrina Emergency Management Reform Act a law that specifies ways to avert the slow and disjointed federal response that was witnessed after the catastrophe in New Orleans, Louisiana.

Post-traumatic stress disorder (PTSD) the clinical diagnosis for individuals who become depressed due to a traumatic event in their lives.

Preliminary damage assessment (PDA) a more detailed assessment of impacts that typically takes place within days or weeks of the event; it determines the possibility and extent of outside federal assistance.

Preparedness concerted efforts to improve response and recovery capabilities.

Preparedness council   a group of individuals who come together in an advisory role to share recommendations for policy and assist emergency managers with program administration.

Prevention   counterterrorism operations such as intelligence gathering and preventive strike activities.

Profiling   the practice of law enforcement officials (including security personnel) using race, ethnicity, religion, or national origin as the decisive factors in targeting an individual for suspicion of wrongdoing.

Programs dimension   the plans and actions to support ideological goals.

Proliferation   the acquisition, sharing, and spread of nuclear weapons and materials to those who do not currently possess them.

Protection   antiterrorism operations such as border control and infrastructure protection.

Public administration   a discipline and profession that directs attention to the formation of policy and the best organization to deal with difficult societal problems.

Public assistance   relief programs that make aid available to government entities that have been affected by terrorism.

Public health   a discipline and profession that concentrates on understanding diseases and how to treat them (e.g., identifying how to react from a medical standpoint to the use of nuclear, biological, chemical, and radiological weapons).

Public information officer   the person who gathers information for the incident commander(s) and shares information with the media or a city employee who specializes in working with the media.

R

Radiological weapons   weapons that spread dangerous radiological material but not result in a nuclear explosion. Radiological weapons are also known as radiological dispersion devices (RDDs) or dirty bombs.

Random terror   an attack on larger numbers of people wherever they gather.

Rapid assessment   a quick survey of impacts designed to gain an appreciation of the scope of the attack.

Recovery   long-term activities to rebound after disasters or terrorist attacks.

Red scare   senator McCarthy's fear of communist infiltration into the United States.

Reign of terror   a period during the French Revolution where an estimated 20,000 persons were killed by France's Committee of Public Safety.

Relative poverty   situation in which some people are less wealthy than their fellow citizens.

Religion   the beliefs and practices espoused by those sharing a common spiritual faith.

Response   the immediate reaction to an emergency situation, like a terrorist attack.

Reverse 911 systems   computerized messages sent over phone lines rapidly to anyone in a designated area.

Right to assemble   people are permitted to join in politically motivated gatherings.

Right to bear arms   guns can be purchased and owned without government interference.

Right-wing terrorist organizations   conservative groups that promote gun rights, oppose abortions, question the value of government and the United Nations, and discriminate against blacks, Jews, and homosexuals.

Riot a type of social disturbance that includes antisocial behavior and law-breaking activity such as rock throwing, tipping over vehicles, looting, starting fires, and attacking law enforcement personnel.

Risk a measure of probability and consequences.

S

Safety officer the person who evaluates the dangers at the scene and makes sure everyone is operating according to safety policies.

Sayeret Matkal a strike team devoted to finding terrorists before they attack Israeli interests.

Scope the subjects covered by an ideology.

SEAR special Events Assessment Rating is used by the FBI to identify the risk of terrorist attacks against major public gatherings.

Search and rescue (SAR) response activities undertaken to find disaster victims and remove them from danger or confinement.

Second Amendment an amendment to the constitution that allows people in the United States to keep and bear arms.

Secondary devices the detonation of other bombs to add to the disruption and fear of the initial attack.

Secure flight an advanced passenger screening program that is administered by the Transportation Security Administration.

Self-censorship media control over their reporting of news to the public.

Self-referred patients who arrive at the hospital whether they require immediate care or not.

Set-back requirements laws that describe the proximity of buildings to roads and parking lots.

Sheltering the location of individuals in places of safety or refuge.

SIGINT interception and interpretation of electronic communications such as phone conversations and e-mails.

Situational awareness continual monitoring of safety concerns at the scene of an attack.

Size-up the process of evaluating the nature of the attack site.

Social base dimension the individuals and groups that espouse an ideology.

Social disturbances mass gatherings including protests and riots.

Social learning theory observing terrorist attacks in the news may generate similar types of behavior among others

Soft targets potential sites of terrorist attacks because they are open and accessible to the public.

Special Air Service (SAS) a British counterterrorism organization.

Strategic national stockpile (SNS) a cache of medicines in secret locations that can be quickly sent to affected locations around that nation.

Strategic Rail Corridor Network a network of railroads that transport Department of Defense munitions and other materials, including hazardous items.

Structural mitigation special construction practices and materials to limit the impact of terrorist attacks.

Structure the organizational relationships within the political system in reference to the ability to acquire and wield power.

Suitcase bomb portable nuclear weapons that can be carried or rolled to the target location.

Surface Transportation and Public Transportation Information Sharing and Analysis Center (ST-PT ISAC) a system that interfaces with government leaders, intelligence agencies, law enforcement personnel, and computer emergency response teams to spread top-secret information among railroad operators.

SWAT   special Weapons and Tactics used to train police forces.

System requirements   functions that must be performed to maintain the operation of a political system.

T

Tabletop exercises   informal discussions about hypothetical scenarios that occur in an office setting.

Tactical emergency medical services   the name given to a team of paramedics that are armed and trained in weapons use.

Tactical terror   the use of attacks against the government for revolutionary or other purposes.

Taliban   the name of the government that provided safe haven for Al-Qaeda.

Technical assessment   a survey of damages that points out methods and costs for rebuilding.

Terrorism   the use or threat of violence to support ideological purposes.

Terrorist Threat Integration Center (TTIC)   a government organization that attempts to improve coordination between the FBI and CIA.

Theocracy   a government run by clerics in the name of God.

Theodore "Ted" Kaczynski   a terrorist known as the "Unabomber" who opposed technology and wrote a manifesto that decried advances in this area.

Threat and Hazard Identification and Risk Assessment (THIRA)   a comprehensive study that outlines what could happen along with possible consequences.

Threat assessment   a careful study of the targets that might be appealing to terrorists.

Toxin   a poison that is produced by plants or animals.

Training   information sharing in classroom or field settings to help familiarize people with protocol.

Transportation Security Act   a law designed to protect transportation systems in the United States.

Transportation Security Administration   a federal agency under the Department of Homeland Security created to protect our transportation systems from terrorist attacks.

Triage   the assessment, sorting, and treatment of injured in such a way as to maximize limited resources.

Tribal government   one of 574 federally recognized tribes that have relative autonomy in the United States.

Trojan horses   the use of what appears to be legitimate software to conceal malware of various types.

U

United Nations   an international organization that establishes security rules regarding access and security at ports.

United Nations International Maritime Organization   a UN organization that establishes security rules regarding access and security at ports.

United States Coast Guard (USCG)   a military branch within the Department of Homeland Security that is in charge of maritime law, environmental protection of waterways, search-and-rescue operations at sea, and interdiction of illegal aliens and contraband.

United States Visitor and Immigrant Status Indicator Technology (US-VISIT) a computer database used to screen passengers who wish to travel to the United States.

Unmet Needs Committee a group of concerned citizens and community leaders who work together to collect donations and address the long-term needs of victims.

USA PATRIOT Act a homeland security law that stands for "Uniting and Strengthening America by Providing Appropriate Tools Required to Intercept and Obstruct Terrorism" meant to prevent terrorist attacks and enhance law enforcement's ability to investigate and punish offenders.

V

Valuation dimension the norms and judgements of an ideology.

Vesicants blister agents that produce chemical burns.

Viral infections a disease owing to nonliving molecules that need a host to survive.

Virus (biological) a microscopic genetic particle that infects the cells of living organisms but cannot multiply outside a host cell.

Virus (computer) malware that may overload a single computer through replication or may spread to other computers and cause damage through other means.

Volunteer management the harnessing of volunteers to take advantage of their potential contributions while averting potential negative consequences.

Volunteer registration center the location where citizens fill out forms noting their skills and other information that can help you when making assignments.

Vulnerability a high degree of disaster proneness and/or limited disaster management capabilities.

W

Wahhabism a very stringent and legalistic religious movement that attempts to ensure the purity of the Muslim faith with no deviations whatsoever.

Warm zone the location where victims are washed for decontamination.

Warnings notifications sent out to the public so they can take protective measures.

Weather radios electronic devices that receive information from the National Weather Service to warn people of severe weather.

WMD acronym for weapons of mass destruction.

Weapons of Mass Destruction weaponry that will create major injuries, carnage, destruction, and disruption when utilized.

Worm malware that may overload a single computer through replication or may spread to other computers and cause disruption through other means.

Writ of habeas corpus a law protecting citizens from unlawful imprisonment.

Z

Zoning regulations that delineate where buildings can be located.

Index